Mathematical Engineering

Series Editors

Bernhard Weigand, Institute of Aerospace Thermodynamics, University of Stuttgart, Stuttgart, Germany

Jan-Philip Schmidt, Universität of Heidelberg, Heidelberg, Germany

Advisory Editors

Günter Brenn, Institut für Strömungslehre und Wärmeübertragung, TU Graz, Graz, Austria

David Katoshevski, Ben-Gurion University of the Negev, Beer-Sheva, Israel

Jean Levine, CAS-Mathematiques et Systemes, MINES-ParsTech, Fontainebleau, France

Jörg Schröder, Institute of Mechanics, University of Duisburg-Essen, Essen, Germany

Gabriel Wittum, Goethe-University Frankfurt am Main, Frankfurt am Main, Germany

Bassam Younis, Civil and Environmental Engineering, University of California, Davis, Davis, CA, USA

Today, the development of high-tech systems is unthinkable without mathematical modeling and analysis of system behavior. As such, many fields in the modern engineering sciences (e.g. control engineering, communications engineering, mechanical engineering, and robotics) call for sophisticated mathematical methods in order to solve the tasks at hand.

The series Mathematical Engineering presents new or heretofore little-known methods to support engineers in finding suitable answers to their questions, presenting those methods in such manner as to make them ideally comprehensible and applicable in practice.

Therefore, the primary focus is—without neglecting mathematical accuracy—on comprehensibility and real-world applicability.

To submit a proposal or request further information, please use the PDF Proposal Form or contact directly: Dr. Thomas Ditzinger (thomas.ditzinger@springer.com)

Indexed by SCOPUS, zbMATH, SCImago.

Tugal Zhanlav · Ochbadrakh Chuluunbaatar

New Developments of Newton-Type Iterations for Solving Nonlinear Problems

 Springer

Tugal Zhanlav
Institute of Mathematics and Digital
Technology
Mongolian Academy of Sciences
Ulaanbaatar, Mongolia

Ochbadrakh Chuluunbaatar
Institute of Mathematics and Digital
Technology
Mongolian Academy of Sciences
Ulaanbaatar, Mongolia

Meshcheryakov Laboratory of Information
Technologies
Joint Institute for Nuclear Research
Dubna, Russia

ISSN 2192-4732   ISSN 2192-4740   (electronic)
Mathematical Engineering
ISBN 978-3-031-63363-8   ISBN 978-3-031-63361-4   (eBook)
https://doi.org/10.1007/978-3-031-63361-4

Preface

In Newton-type methods, iteration parameters play a key role. Suitable choices of these parameters not only speed up convergence, but also extend the convergence domain. Moreover, special choices allow us to control the convergence behaviour of iterations. Over the last two decades, many papers on this topic have been published, and it is highly desirable to systematically collect these results. This motivates us to write this monograph based on our research results obtained in collaboration with the co-authors. Since the topic is of importance, we believe that this book will be useful to readers, graduate students and researchers interested in the field of applied mathematics, numerical analysis and applied sciences. The book is divided into three parts containing 8 chapters.

Part I is devoted to the construction of Newton-type iterations for solving nonlinear equations. In Chap. 1, we consider the τ-region of convergence and the convergence of the continuous analogue of Newton's method (CANM for short), the relationship between the inexact Newton method and CANM, the choice of iteration parameters, and the acceleration of Newton-type iterations. In Chap. 2, we treat the construction of two-sided approximations. Chapter 3 is devoted to the new development and extension of Newton-type iterations. The materials cover necessary and sufficient conditions for the convergence of optimal two- and three-point iterations with and without derivatives, generating function methods for constructing new iterations and extensions of some classes of optimal eighth-order methods. Chapter 4 is dedicated to the construction of derivative-free iterations.

Part II includes Chaps. 5 and 6, and is devoted to the construction of higher-order iterations for solving nonlinear systems of equations. As in Part I, necessary and sufficient convergence conditions for iterations to be p-order of convergence are written in terms of parameter matrices. These conditions allow us not only to check the convergence order of the considered iterations, but also to derive new iterations. In Chap. 6, we propose derivative-free higher-order iterations.

Part III is devoted to the applications of the proposed iterative methods. In Chap. 7, we apply the proposed iterative methods for solving some problems in linear algebra. Chapter 8 considers the applications of new iterations to the numerical solution of some nonlinear problems of theoretical physics. Some obtained theoretical results are compared with experimental physics data.

It should be noted that the performance and convergence order of all the presented iterative methods in this book have been verified by numerical experiments and different numerical examples many times, and for lack of space, we do not include numerical results. For interested readers, we refer to the corresponding original articles in References. The major results included in this monograph were obtained in our joint works.

Ulaanbaatar, Mongolia   Tugal Zhanlav
April 2024   Ochbadrakh Chuluunbaatar

Acknowledgements We particularly thank Prof. Changbum Chun (Republic of Korea), Doctor (Sc.D.) R. Mijiddorj, Doctor (Ph.D.) Kh. Otgondorj, Doctor (Ph.D.) V. Ulziibayar, Doctor (Ph.D.) D. Khongorzul, Doctor (Ph.D.) B. Saheya, and G. Ankhbayar for fruitful collaboration. The research was partially supported by the Mongolian Foundation for Science and Technology, the Institute of Mathematics of the National University of Mongolia and the Institute of Mathematics and Digital Technology, the Mongolian Academy of Sciences.

Contents

List of Figures

List of Tables

Part I
Newton-Type Iterations for Nonlinear Equations

Chapter 1
Newton-Type Iterations, Convergence and Accelerations

Abstract The damped Newton method or continuous analogue of Newton's method (CANM) for solving nonlinear equations is considered. The influence of the parameter τ_k on the convergence of the method is studied, and the τ-region of convergence is given. The semilocal convergence theorem for CANM is proved using a technique consisting of a new system of recurrence relations. A close relationship between the inexact Newton's method and CANM is revealed. Based on this relationship, it is concluded that the choice of τ_k from (0,2) can guarantee the convergence of the method. We developed an accelerating procedure for Newton-type iterations, which leads to speeding up the convergence by certain units.

1.1 A Local and Semilocal Convergence of Continuous Analogue of Newton's Method

One of the modifications of Newton method is the well-known continuous analogue of Newton's method (CANM) or damped Newton's method

$$F'(x_k)v_k = -F(x_k), \quad x_{k+1} = x_k + \tau_k v_k, \quad k = 0, 1, \ldots \tag{1.1}$$

to solve nonlinear equations $F(x) = 0$ in Banach spaces. In this case $F : \Omega \subseteq X \to Y$ is an operator defined in an open convex subset Ω of the Banach space X into the Banach space Y. If $\tau_k \equiv 1$ the CANM leads to the Newton method. In our paper [1] we have shown that the iteration (1.1) is convergent if $\tau_k \in (0, 2)$. Now, we give a τ-region of convergence of CANM and propose an optimal choice for the parameter τ_k in (1.1). We also prove the semilocal convergence theorem for the iteration (1.1) by using a technique consisting of a new system of recurrence relations [2].

T. Zhanlav and O. Chuluunbaatar, *New Developments of Newton-Type Iterations for Solving Nonlinear Problems*, Mathematical Engineering, https://doi.org/10.1007/978-3-031-63361-4_1

1.1.1 τ-Region of Convergence of CANM

We study the influence of the parameter τ on the convergence of method (1.1). In that follows we assume that

(c_1) $\|F'(y) - F'(x)\| \leq L \|y - x\|, \quad x, y \in \Omega,$

(c_2) $\|(F'(x_0))^{-1}\| \leq \beta,$

(c_3) $\|(F'(x_0))^{-1} F(x_0)\| \leq \eta, \quad a_0 = L\beta\eta,$

(c_4) $L \|(F'(x_k))^{-1}\| \|(F'(x_k))^{-1} F(x_k)\| \leq a_k, \quad k = 0, 1, \ldots,$

and consider the open level set $L_\alpha = \{x \in \Omega; \; \|F(x)\| < \eta/\beta = \alpha\}$.

Theorem 1.1 *Let us assume that the conditions $(c_1) - (c_4)$ are satisfied and $\tau_k \in I_k$, where*

$$I_k = \begin{cases} \left(0, \frac{-1+\sqrt{1+8a_k}}{2a_k}\right) \subseteq (0, 2), & for \; 0 < a_k \leq 1, \\ \left(0, \frac{1}{a_k}\right), & for \; 1 < a_k < M < \infty. \end{cases} \tag{1.2}$$

Then the sequence $\{x_k\}$ obtained by (1.1) is well defined and remains in L_α and converges to some x^* with $F(x^*) = 0$.

Proof We notice first of all that

$$\tau_k a_k < 1, \quad k = 0, 1, \ldots,$$

as $\tau_k \in I_k$. By induction we prove that there exists $\Gamma_{k+1} = (F'(x_{k+1}))^{-1}$, such that

$$\|\Gamma_{k+1}\| \leq \frac{\|\Gamma_k\|}{1 - a_k \tau_k}. \tag{1.3}$$

Now from the conditions (c_1) and (c_4) we have

$$\|I - \Gamma_k F'(x_{k+1})\| \leq \|\Gamma_k\| \|F'(x_k) - F'(x_{k+1})\| \leq L \|\Gamma_k\| \|x_{k+1} - x_k\|$$
$$= \tau_k L \|\Gamma_k\| \|\Gamma_k F(x_k)\| \leq \tau_k a_k < 1, \quad k = 0, 1, \ldots.$$

So, by the Banach perturbation lemma there should exist Γ_{k+1} satisfying (1.3). Thus, the sequence $\{x_k\}$ obtained from the damped Newton iteration is well defined. The Taylor expansion of $F(x)$ about x_k gives

$$F(x_{k+1}) = F(x_k) + F'(\xi_k)(x_{k+1} - x_k)$$
$$= (1 - \tau_k)F(x_k) + (F'(\xi_k) - F'(x_k))(x_{k+1} - x_k), \tag{1.4}$$

where $\xi_k = \theta x_k + (1 - \theta)x_{k+1}, \theta \in (0, 1)$. Using the conditions (c_1) and (c_4) in (1.4) we have

$$\|F(x_{k+1})\| \le \psi(\tau_k)\|F(x_k)\|, \tag{1.5}$$

$$\psi(\tau_k) = |1 - \tau_k| + a_k\tau_k^2. \tag{1.6}$$

It is easy to show that

$$\psi(\tau_k) < 1 \tag{1.7}$$

under condition $\tau_k \in I_k$. Therefore, from (1.5) we deduce the following

$$\|F(x_k)\| \le \psi(\tau_{k-1})\psi(\tau_{k-2}) \cdots \psi(\tau_0)\|F(x_0)\|. \tag{1.8}$$

According to (1.7) we have

$$\|F(x_k)\| < \|F(x_0)\| < \frac{\eta}{\beta}, \quad k = 1, 2, \ldots,$$

i.e., $x_k \in L_\alpha$ for all $k = 1, 2, \ldots$. From (1.8), we obtain

$$\|F(x_k)\| \to 0 \quad \text{at} \quad k \to \infty.$$

Since L_α is bounded, there exists an accumulation point x^* of the sequence $\{x_k\}$ with $F(x^*) = 0$. $\qquad\square$

We call the interval I_k by the τ-region of convergence, as well as call the value τ_k, such that

$$\psi(\tau_k) = \min_{\tau_k \in (0,2)} \psi(\tau),$$

the optimal one and denote it by τ_{opt}. The direct calculation gives us

$$\tau_{opt} = \begin{cases} 1, & \text{for } 0 < a_k \le \frac{1}{2}, \\ \frac{1}{2a_k}, & \text{for } \frac{1}{2} < a_k < 1, \\ \frac{1}{a_k} - \varepsilon, & \text{for } a_k \ge 1, \end{cases}$$

$$\psi(\tau_{opt}) = \begin{cases} a_k, & \text{for } 0 < a_k \le \frac{1}{2}, \\ 1 - \frac{1}{4a_k} < \frac{3}{4}, & \text{for } \frac{1}{2} < a_k < 1, \\ 1 - \varepsilon + \varepsilon^2 a_k, & \text{for } a_k \ge 1. \end{cases}$$

Here ε is a small number. Some particular choices of the parameter, in contrast to the optimal one, may be useful. We consider

$$\tau_k = \frac{-1 + \sqrt{1 + 8a_k}}{4a_k}, \tag{1.9}$$

which is a middle point of the admissible interval I_k. Since $\tau_k < 1$, then from (1.6) we obtain

$$\psi(\tau_k) = \frac{3}{2}(1 - \tau_k). \tag{1.10}$$

In (1.10), we can see that $\psi(\tau_k) < 1$ if $\tau_k > 1/3$. On the other hand, the τ_k is decreasing and $\psi(\tau_k)$ is increasing with respect to a_k. Therefore from (1.9) and (1.10) we get

$$\frac{1}{3} < \tau_k < 1, \quad 0 < \psi(\tau_k) < 1, \quad \tau_k a_k < 1$$

under condition $0 < a_k < 3$. Thus, the choice (1.9) allows us to weaken the condition imposed on a_k. From (1.9), (1.10) it follows that

$$\tau_k \to 1 \quad \text{and} \quad \psi(\tau_k) \to 0 \quad \text{at} \quad a_k \to 0.$$

Moreover, if $a_{k+1} < a_k$ for $k = 0, 1, \ldots$, then we have

$$\tau_0 < \tau_1 < \cdots < \tau_k < \ldots < 1, \quad 0 < \psi(\tau_k) < \cdots < \psi(\tau_1) < \psi(\tau_0).$$

It should be pointed out that, according to affine invariance property [3] we can also take

$$\overline{\tau}_k = \frac{-1 + \sqrt{1 + 2ba_k}}{ba_k} \quad \text{for} \quad \forall b > 0, \quad k = 0, 1, \ldots, \tag{1.11}$$

which has the same properties as τ_k given by (1.9), i.e., $\overline{\tau}_k : (0, \infty) \to (0, 1)$ and decreasing with respect to a_k.

Theorem 1.2 *Suppose that the conditions* $(c_1) - (c_4)$ *are satisfied and* τ_k *is given by (1.11) with* $0 < a_k < (b + 2)/2$. *Then the sequence* $\{x_k\}$ *obtained by (1.1) is well defined and remains in* L_α *and converges to some* x^* *with* $F(x^*) = 0$.

Proof By virtue of (1.6) and (1.11) we have

$$\psi(\overline{\tau}_k) = \frac{b + 2}{b}(1 - \overline{\tau}_k).$$

Since $\overline{\tau}_k(a_k)$ is a decreasing function with respect to a_k and $\psi(\overline{\tau}_k)$ is decreasing too with respect to $\overline{\tau}_k$, and $0 < a_k < (b + 2)/2$, then we get

$$\frac{2}{b + 2} < \overline{\tau}_k < 1,$$

and

$$0 < \psi(\overline{\tau}_k) < 1, \quad \overline{\tau}_k a_k < 1.$$

The proof follows immediately from (1.5) and (1.8). $\qquad\qquad\qquad\qquad\qquad\square$

The above derived admissible interval (1.2) and the theoretical optimal value as well as the particular choices (1.9), (1.11) cannot be implemented directly, since the arising quantities a_k are computationally unavailable due to the arising Lipschitz constant, $\|\Gamma_k\|$ and $\|\Gamma_k F(x_k)\|$. However, the obtained theoretical results can be useful for the construction of computational strategies. Namely, from the assumption (c_4) it is evident that $\|F(x_k)\| \to 0$ if $a_k \to 0$, and $a_k \to \infty$ if $\|F(x_k)\| \to \infty$. Therefore, replacing a_k in (1.11) by $y_k = \|F(x_k)\|$, we obtain

$$\tau_k^* = \frac{2}{1 + \sqrt{1 + 2by_k}} \in (0, 1), \quad b > 0. \tag{1.12}$$

The function $\tau_k^* = \tau_k^*(y_k)$ decreases with respect to y_k. From (1.12) it is also clear that $\tau_k^* \to 1$ iff $y_k \to 0$, and $\tau_k^* \to 0$ iff $y_k \to \infty$. We have the following theorem.

Theorem 1.3 *Suppose that the conditions (c_1)–(c_4) are satisfied and τ_k is given by (1.12). Suppose also that $0 < a_k < \sqrt{1 + 2by_k}/2$. Then the sequence $\{x_k\}$ obtained by (1.1) is well defined and remains in L_α and converges to some x^* with $F(x^*) = 0$.*

Proof By virtue of (1.6) and (1.12) we have

$$\psi(\tau_k^*) = \frac{by_k + 2a_k}{1 + by_k + \sqrt{1 + 2by_k}}. \tag{1.13}$$

From (1.12), (1.13) it is easy to show that the next two inequalities hold simultaneously

$$\tau_k^* a_k < 1 \quad \text{and} \quad \psi(\tau_k^*) < 1$$

under the condition

$$0 < a_k < \frac{1 + \sqrt{1 + 2by_k}}{2} \equiv M_k$$

or equivalently

$$y_k > \frac{2a_k(a_k - 1)}{b}.$$

Since $0 < \tau_k^* < 1$ and $\tau_k^* a_k < 1$ then $\tau_k^* \in I_k$. As a consequence, the assertion of the theorem follows directly from Theorem 1.1. $\qquad\square$

Corollary 1.1 *Theorem 1.3 is valid for any choice $\tau_k \in (0, \tau_k^*]$.*

Indeed, $\tau_k a_k \le \tau_k^* a_k < 1$ and for such τ_k we have $\psi(\tau_k) = 1 - \tau_k + a_k\tau_k^2 < 1$, that assures a reduction of the residual norm, i.e., $\|F(x_{k+1})\| < \|F(x_k)\|$. However the choice τ_k^* is preferable in the sense that $\tau_k^* \to 1$ as $y_k \to 0$ and the method (1.1) asymptotically leads to quadratically convergent Newton method. Nevertheless, there arises a question of the choice of the parameter b in (1.12). In our opinion, it must be done in such a way that the damped Newton's method converges rapidly.

1.1.2   Semilocal Convergence of CANM

We define the sequence

$$a_{k+1} = f(a_k, \tau_k)^2 g(a_k, \tau_k) a_k, \quad a_0 = L\beta\eta, \quad k = 0, 1, \ldots,$$

where

$$f(x, \tau) = \frac{1}{1 - \tau x}, \quad g(x, \tau) = |1 - \tau| + \tau^2 x. \tag{1.14}$$

We need the following auxiliary lemmas, whose proofs are trivial from [2].

Lemma 1.1 *Let f and g be two real functions given by (1.14). Then*

(i) *For a fixed $\tau \in (0, 1]$, $f(x, \tau)$ and $g(x, \tau)$ increase as a function of $x \in (0, 1/\tau)$ and $f(x, \tau) > 1$,*

(ii) *$f(\gamma x, \tau) < f(x, \tau)$ and $g(\gamma x, \tau) \leq \gamma^p g(x, \tau)$ for $\gamma \in (0, 1]$,*

where

$$p = \begin{cases} 0, & if \ \tau \neq 1, \\ 1, & if \ \tau = 1. \end{cases}$$

Lemma 1.2 *Let $0 < a_0 < (3 - \sqrt{5})/2$ and $\tau_k \in (0, 1)$ for all $k = 0, 1, \ldots$. Then the sequence $\{a_k\}$ is decreasing, i.e.*

$$0 < \cdots < a_k < a_{k-1} < \cdots < a_1 < a_0 < \frac{3 - \sqrt{5}}{2}.$$

Lemma 1.3 *Let us suppose that the conditions of Lemma 1.2 are satisfied and define $\gamma = a_1/a_0$. Then*

(i) *$\gamma = f(a_0, \tau_0)^2 g(a_0, \tau_0) \in (0, 1)$,*

(ii) *$a_k \leq \gamma^{(1+p)^{k-1}} a_{k-1} \leq \cdots \leq \gamma^{\frac{(1+p)^k - 1}{p}} a_0$,*

(iii) *$f(a_k, \tau_k) g(a_k, \tau_k) \leq \gamma^{(1+p)^k} \Delta$ with $\Delta = \frac{1}{f(a_0, \tau_0)} < 1$.*

Notice that

$$L \|\Gamma_0\| \|\Gamma_0 F(x_0)\| \leq a_0,$$

$$\|x_1 - x_0\| \leq \|\Gamma_0 F(x_0)\| \leq \eta < R\eta \quad \text{with} \quad R = \frac{1}{1 - \Delta\gamma} > 1,$$

i.e., $x_1 \in B(x_0, R\eta) = \{x \in X; \|x - x_0\| < R\eta\}$. In these conditions we prove, for $k \geq 1$, the following statements:

(I_k) $\|\Gamma_k\| \leq f(a_{k-1}, \tau_{k-1}) \|\Gamma_{k-1}\|,$

(II_k) $]\|\Gamma_k F(x_k)\| \leq f(a_{k-1}, \tau_{k-1}) g(a_{k-1}, \tau_{k-1}) \|\Gamma_{k-1} F(x_{k-1})\|,$

(III_k) $L \|\Gamma_k\| \|\Gamma_k F(x_k)\| \leq a_k,$

(IV_k) $x_{k+1} \in B(x_0, R\eta)$.

Assuming $\tau_0 a_0 < 1$ and $x_1 \in \Omega$, we have

$$\|I - \Gamma_0 F'(x_1)\| \le \|\Gamma_0\| \, \|F'(x_0) - F'(x_1)\| \le L \, \|\Gamma_0\| \, \|x_1 - x_0\|$$
$$= \tau_0 L \, \|\Gamma_0\| \, \|\Gamma_0 F(x_0)\| \le \tau_0 a_0 < 1.$$

Then by the Banach lemma, Γ_1 exists and

$$\|\Gamma_1\| \le \frac{\|\Gamma_0\|}{1 - \|\Gamma_0\| \, \|F'(x_0) - F'(x_1)\|} \le f(a_0, \tau_0) \, \|\Gamma_0\|,$$

and (I_1) is true. On the other hand, according to (1.1), we have

$$F(x_{k+1}) = (1 - \tau_k) F(x_k) + (F'(\xi_k) - F'(x_k))(x_{k+1} - x_k),$$

where $\xi_k = \theta x_k + (1 - \theta)x_{k+1}, \theta \in (0, 1)$, and

$$F'(x_{k+1}) = F'(x_k)(I - P_k), \quad P_k = \Gamma_k(F'(x_k) - F'(x_{k+1})).$$

If $x_k, x_{k+1} \in \Omega$ following

$$\|P_k\| \le L \, \|\Gamma_k\| \, \|x_{k+1} - x_k\| \le \tau_k L \, \|\Gamma_k\| \, \|\Gamma_k F(x_k)\| \le \tau_k a_k < 1.$$

Then $\Gamma_{k+1} = (I - P_k)^{-1} \Gamma_k$, and for $k = 0$, if $x_0, x_1 \in \Omega$, we have

$$\Gamma_1 F(x_1) = (I - P_0)^{-1} \Gamma_0 \{(1 - \tau_0) F(x_0) + (F'(\xi_0) - F'(x_0))(x_1 - x_0)\}$$
$$= (I - P_0)^{-1} \{(1 - \tau_0) - \tau_0 \Gamma_0 (F'(\xi_0) - F'(x_0))\} \Gamma_0 F(x_0).$$

Thus, we get

$$\|\Gamma_1 F(x_1)\| \le \frac{1}{1 - \tau_0 a_0} (|1 - \tau_0| + \tau_0^2 a_0) \, \|\Gamma_0 F(x_0)\|$$
$$= f(a_0, \tau_0) g(a_0, \tau_0) \, \|\Gamma_0 F(x_0)\|,$$

i.e., (II_1) is true. To prove (III_1) and (IV_1), we use

$$L \, \|\Gamma_1\| \, \|\Gamma_1 F(x_1)\| \le f(a_0, \tau_0)^2 g(a_0, \tau_0) L \, \|\Gamma_0\| \, \|\Gamma_0 F(x_0)\|$$
$$\le f(a_0, \tau_0)^2 g(a_0, \tau_0) a_0 = a_1$$

and

$$\|x_2 - x_0\| \le \|x_2 - x_1\| + \|x_1 - x_0\| \le \tau_1\|\Gamma_1 F(x_1)\| + \tau_0\|\Gamma_0 F(x_0)\|$$
$$\le (\tau_1 f(a_0, \tau_0) g(a_0, \tau_0) + \tau_0) \|\Gamma_0 F(x_0)\|$$
$$\le \eta\left(1 + \frac{\gamma}{f(a_0, \tau_0)}\right) = \eta(1 + \Delta\gamma) < \frac{\eta}{1 - \Delta\gamma} = R\eta,$$

i.e., $x_2 \in B(\alpha_0, R\eta)$ and this proof holds by induction for all $k \in \mathbb{N}$. Now following an inductive procedure and assuming $x_{k+1} \in \Omega$ the items (I_k)–(IV_k) are proved.

To establish the convergence of $\{x_k\}$, we only have to prove that it is a Cauchy sequence and that $x_{k+1} \in \Omega$. We note that

$$\|\Gamma_k F(x_k)\| \le \|\Gamma_0 F(x_0)\| G_k, \quad G_k = \prod_{i=0}^{k-1} f(a_i, \tau_i) g(a_i, \tau_i).$$

As a consequence of Lemma 1.3 it follows that

$$G_k \le \Delta^k \prod_{i=0}^{k-1} \gamma^{(1+p)^i} = \Delta^k \gamma^{\frac{(1+p)^k - 1}{p}}.$$

So, from $\Delta < 1$, we deduce that $\lim_{k\to\infty} G_k = 0$. We can now formulate the following result on convergence for the iteration (1.1).

Theorem 1.4 *In the conditions indicated for the operator F, let us assume that $\Gamma_0 = (F'(x_0))^{-1} \in L(Y, X)$ exists at some $x_0 \in \Omega$ and $(c_1) - (c_4)$ are satisfied. Suppose that τ_k is given by*

$$\tau_k = \begin{cases} \tau_k^* = \dfrac{-1 + \sqrt{1 + 2by_k}}{by_k}, & for\ 1 - \tau_k^* > \varepsilon, \\ 1, & for\ 1 - \tau_k^* \le \varepsilon. \end{cases} \tag{1.15}$$

Then if $\overline{B(x_0, R\eta)} = \{x \in X;\ \|x - x_0\| \le R\eta\} \in \Omega$, the sequence $\{x_k\}$ defined by (1.1) and starting at x_0 has, at least, R-order $(p + 1)$ and converges to a solution x^* of the equation $F(x) = 0$. In this case, the solution x^* and the iterations x_k belong to $\overline{B(x_0, R\eta)}$ and x^* is the only solution to $F(x) = 0$ in $B(x_0, 2/(L\beta) - R\eta) \cap \Omega$. Furthermore, we can give the following error estimates:

$$\|x^* - x_{k+m}\| \le \frac{\Delta^{k+m}}{1 - \Delta\gamma} \gamma^{2^k - 1 + m} \|\Gamma_0 F(x_0)\|, \quad \gamma = \frac{|1 - \tau_0| + a_0\tau_0^2}{(1 - \tau_0 a_0)^2}. \tag{1.16}$$

Proof We have proved above that the assumption (c_4) is satisfied. Then by virtue of Theorem 1.3 the damped Newton's iteration converges. Namely the residual norm $\|F(x_k)\|$ decreases as k increases. Then τ_k^* tends to units as k increases. Hence the condition $1 - \tau_k^* \leq \varepsilon$ will holds starting at some number $k = m$. Then by virtue of (1.15) we have $\tau_k \equiv 1$ for all $k \geq m$. On the other hand, $x_k \in B(x_0, R\eta)$ for all $k \in \mathbb{N}$, then $x_k \in \Omega$, $k \in \mathbb{N}$. Now we prove that $\{x_k\}$ is a Cauchy sequence. To do this, we consider $k, m \geq 1$:

$$\|x_{m+k+e} - x_{m+k}\| \leq \sum_{l=0}^{e-1} \|x_{m+k+l+1} - x_{m+k+l}\| \leq \sum_{l=0}^{e-1} \|\Gamma_{m+k+l} F(x_{m+k+l})\|$$

$$\leq \|\Gamma_{m+k} F(x_{m+k})\| \sum_{l=0}^{e-1} \Delta^l \gamma^{2^{m+k}(2^l - 1)} \leq \|\Gamma_{m+k} F(x_{m+k})\| \sum_{l=0}^{e-1} (\Delta\gamma)^{e-1}$$

$$= \frac{1 - (\Delta\gamma)^e}{1 - \Delta\gamma} \|\Gamma_{m+k} F(x_{m+k})\|, \tag{1.17}$$

in which we have used the well-known inequality $(1 + x)^k > 1 + kx$. According to Lemma 1.3, we have

$$\|\Gamma_{m+k} F(x_{m+k})\| \leq \Delta^k \gamma^{2^m (2^k - 1)} \|\Gamma_m F(x_m)\| < \Delta^k \gamma^{2^k - 1} \|\Gamma_m F(x_m)\| \tag{1.18}$$

and

$$\|\Gamma_m F(x_m)\| \leq (\Delta\gamma)^m \|\Gamma_0 F(x_0)\|. \tag{1.19}$$

Substituting (1.18), (1.19) into (1.17), we obtain

$$\|x_{m+k+e} - x_{m+k}\| \leq \frac{1 - (\Delta\gamma)^e}{1 - \delta\gamma} \Delta^{m+k} \gamma^{2^k - 1 + m} \|\Gamma_0 F(x_0)\|, \tag{1.20}$$

then $\{x_k\}$ is a Canchy sequence. Now by letting $e \to \infty$ in (1.20), we obtain (1.16).

To prove that $F(x^*) = 0$, notice that $\|\Gamma_k F(x_k)\| \to 0$ by letting $k \to \infty$. As $\|F(x_k)\| \leq \|F'(x_k)\| \|\Gamma_k F(x_k)\|$ and $F'(x_k)$ is a bounded sequence, we deduce $\|F(x_k)\| \to 0$ and then $F(x^*) = 0$ by the continuity of F. Now, to show the uniqueness, suppose that $y^* \in B(x_0, 2/(L\beta) - R\eta) \cap \Omega$ is another solution to $F(x) = 0$. Then

$$0 = F(y^*) - F(x^*) = \int_0^1 F'(x^* + t(y^* - x^*))dt(y^* - x^*). \tag{1.21}$$

Also we use the estimation

$$\left\| I - \Gamma_0 \int_0^1 F'(x^* + t(y^* - x^*))dt \right\| \leq \|\Gamma_0\| \int_0^1 \|F'(x^* + t(y^* - x^*)) - F'(x_0)\| dt$$

$$\leq L\beta \int_0^1 \|x^* - t(y^* - x^*) - x_0\| dt \leq L\beta \int_0^1 ((1 - t)\|x^* - x_0\| + t\|y^* - x_0\|)dt$$

$$< \frac{L\beta}{2} \left(R\eta + \frac{2}{L\beta} - R\eta \right) = 1.$$

We see that the operator $\int_0^1 F'(x^* + t(y^* - x^*))dt$ has an inverse and consequently from (1.21), we get $y^* = x^*$. $\qquad\square$

Remark 1.1 In [4] the following choice was proposed

$$\tau_k = \frac{\|F(x_{k-1})\|}{\|F(x_k)\|}\tau_{k-1}, \quad k = 1, 2, \ldots, \quad \tau_0 \approx 0.1, \tag{1.22}$$

and was the error bounds of kind of (1.16) derived under the conditions

$$\|F'(x)^{-1}\| \le \beta, \quad \|F''(x)\| \le M.$$

Here we derive the error bounds (1.16) without such conditions. There is a closed relationship between (1.12) and (1.22). Namely, (1.12) gives

$$\frac{\tau_k^*}{\tau_{k-1}^*} = \frac{\|F(x_{k-1})\|}{\|F(x_k)\|}A_k, \quad A_k = \frac{-1 + \sqrt{1 + 2b\|F(x_k)\|}}{-1 + \sqrt{1 + 2b\|F(x_{k-1})\|}}.$$

It is easy to show that

$$A_k \to 1 \quad \text{when} \quad \max(\|F(x_{k-1})\|, \|F(x_k)\|) \to 0,$$

i.e., in the limit these two choices coincide. Since $\tau_k \equiv 1$ for all $k \ge m$ our iteration (1) can be considered as a Newton iteration starting at $\tilde{a}_0 = a_m$, where

$$a_m = L\,\|\Gamma_m\|\,\|\Gamma_m F(x_m)\| \le a_0\gamma^m < \frac{3 - \sqrt{5}}{2}, \tag{1.23}$$

which obviously, holds for large m, and for any a_0. On the other hand, the Kantorovich semilocal convergence theorem was proved under the condition $a_0 < 1/2$, i.e., the local Newton methods, require sufficiently good initial guesses. Unlike, our iteration (1.1), as Newton's method is able to compensate for bad initial guesses by virtue of damping strategies. The inequality (1.23) has shown that the damped Newton's method at first m-stage allows one to extend the convergence domain of the initial guesses.

1.2   Relationship Between Inexact Newton's Method and CANM

Consider a nonlinear system

$$F(x) = 0, \tag{1.24}$$

where $F : D \subseteq R^n \rightarrow R^n$ is a continuously differentiable nonlinear mapping. Among all kinds of methods for solving the nonlinear equations (1.24), the Newton method is perhaps the most elementary, popular and important [4–9]. One of the advantages of the method is its local quadratic convergence. However, its computational cost is expensive, particularly when the size of the problem is very large, because the Newton equations

$$F(x_k) + F'(x_k)s_k = 0 \tag{1.25}$$

should be solved at each iteration step. To reduce the computational cost of the Newton method, Dembo et al. [10] proposed an inexact Newton (IN) method,

$$F'(x_k)\bar{s}_k = -F(x_k) + r_k,$$
$$x_{k+1} = x_k + \bar{s}_k, \quad k = 0, 1, \ldots, \quad x_0 \in D. \tag{1.26}$$

The terms $r_k \in R^n$ represent the residuals of the approximate solutions $\bar{s}_k$ of the Newton equation (1.25), i.e., inexactly solve the Newton equation (1.25) and obtain a step $\bar{s}_k$ such that

$$\|r_k\| = \|F(x_k) + F'(x_k)\bar{s}_k\| \leq \eta_k \|F(x_k)\|, \tag{1.27}$$

where $\eta_k \in [0, 1)$ is the forcing term. In each iteration step of the IN method, a real number η_k should be chosen first, and then an IN step $\bar{s}_k$ is obtained by solving the Newton equations approximately with an efficient iteration solver for systems of linear equations. The forcing terms play an important role both in reducing the residuals of Newton equations and in increasing accuracy of the method. In particular, if $\eta_k = 0$ for all k, then the IN method reduces to the Newton method. The IN method, like the Newton method, is locally convergent.

Theorem 1.5 ([10]) *Assume that the IN iterates converge to x^*. Then the convergence is superlinear if and only if*

$$\|r_k\| = o(\|F(x_k)\|) \quad as \quad k \rightarrow \infty.$$

Theorem 1.6 ([10]) *Given $\eta_k \leq \bar{\eta} < t < 1, k = 0, 1, \ldots$ there exists $\varepsilon > 0$ such that for any initial approximation x_0 with $\|x_0 - x^*\| \leq \varepsilon$, the sequence of the IN iteration x_k satisfying (1.27) converges to x^*. Moreover, the convergence is linear in the sense that*

$$\|x_{k+1} - x^*\| \leq t \|x_k - x^*\|, \quad k = 0, 1, \ldots.$$

Theorem 1.7 ([10]) *Assume that $F : R^n \rightarrow R^n$ is continuously differentiable, $x^* \in R^n$ such that $F(x^*) = 0$ and $F'(x^*)$ is nonsingular. If the sequence x_k generated by IN iterates converges to x^*, then*
(1) x_k converges to x^* superlinearly when $\eta_k \rightarrow 0$;
(2) x_k converges to x^* quadratically if $\eta_k = O(\|F(x_k)\|)$ and $F'(x)$ is Lipschitz continuous at x^*. In [11] the condition

$$\frac{\|r_k\|}{\|F(x_k)\| + \|\bar{s}_k\|} = O(\|F(x_k)\|) \quad as \quad k \to \infty$$

is proposed which characterizes the quadratic convergence of the IN iterations.

In [11] it is shown that the IN, inexact perturbed and quasi-Newton methods are equivalent models. At present there are some strategies for choosing the forcing terms [12–14].

One of the modifications of Newton method is the well known CANM or damped Newton method (1.1). The suitable choice of the parameter allows us to speed-up the convergence and to enlarge the convergence domain. There exists a closed relationship between the CANM and the IN iterates. Indeed, the CANM (1.1) can be considered as the IN iteration (1.26) with the residual

$$r_k = F'(x_k)\bar{s}_k + F(x_k) = (1 - \tau_k)F(x_k). \tag{1.28}$$

From here we get

$$\|r_k\| = \eta_k \|F(x_k)\|$$

with

$$\eta_k = |1 - \tau_k|.$$

We have a local and semi-local convergence results.

Theorem 1.8 *There exits $\varepsilon > 0$ such that for any initial approximation x_0 with $\|x_0 - x^*\| \le \varepsilon$, the sequence generated by (1.1) with parameter $\tau_k \in (0, 2)$ converges to x^*.*

Proof As mentioned above, the CANM is equivalent to IN iterates (1.28), for which the residual satisfies inequality (1.27) with forcing term $\eta_k \in [0, 1)$ under condition $\tau_k \in (0, 2)$. Then by Theorem 1.6 the sequence x_k generated by (1.1) converges to x^*. $\qquad\qquad\qquad\qquad\qquad\qquad\qquad\qquad\qquad\qquad\qquad\qquad\qquad\qquad\qquad\quad\Box$

Let D_0 be a convex set with $\overline{D}_0 \subseteq D$, and $(F'(x))^{-1}$ exists for all $x \in D_0$.

Theorem 1.9 *We assume that*

(i) $\|F''(x)\| \le M, \quad x \in D_0,$
(ii) $\|F'(x_0)^{-1}\| \le \beta,$
(iii) $\|F'(x_0)^{-1}F(x_0)\| \le \eta, \quad a_0 = M\beta\eta,$
(iv) $M\|F'(x_k)^{-1}\|\|F'(x_k)^{-1}F(x_k)\| \le a_k < 2, \quad k = 0, 1, \ldots,$

and

$$0 < \tau_k < \frac{-1 + \sqrt{1 + 4a_k}}{a_k}. \tag{1.29}$$

Then the sequence $\{x_k\}$ defined by (1.1) and starting at $x_0 \in D_0$ converges to a solution x^* of (1.24).

Proof The Taylor expansion gives

$$F(x_{k+1}) = (1 - \tau_k)F(x_k) + \frac{F''(\xi_k)}{2}\tau_k^2 (F'(x_k)^{-1}F(x_k))^2, \qquad (1.30)$$

where $\xi_k = \theta x_k + (1 - \theta)x_{k+1}, \theta \in (0, 1)$. If we use the assumption (iv), then from (1.30) it follows that

$$\|F(x_{k+1})\| < \left(|1 - \tau_k| + \frac{a_k}{2}\tau_k^2\right)\|F(x_k)\| < \|F(x_k)\|$$

under condition (1.29), i.e., $\{\|F(x_k)\|\}$ is a decreasing sequence provided the iteration parameter τ_k is chosen in the interval $\left(0, (-1 + \sqrt{1 + 4a_k})/a_k\right)$. $\square$

Remark 1.2 It is easy to show that

$$1 < \frac{-1 + \sqrt{1 + 4a_k}}{a_k} < 2, \quad \text{when} \quad 0 < a_k < 2.$$

This means that $\tau_k \in (0, 2)$.

Remark 1.3 Assume that the sequence x_k generated by CANM converges to x^*. Then

(1) $\{x_k\}$ converges to x^* superlinearly when $\tau_k \to 1$,
(2) $\{x_k\}$ converges to x^* quadratically if

$$|1 - \tau_k| = O(\|F(x_k)\|) \quad \text{or} \quad |1 - \tau_k| = O(\|F(x_{k-1})\|),$$

because of Theorem 1.7.

1.2.1   Some Choices of Iteration Parameters

From (1.30) it follows that

$$\frac{\|F(x_k)\| - |1 - \tau_{k-1}|\|F(x_{k-1})\|}{\|F(x_{k-1})\|} = O(\|F(x_{k-1})\|)$$

and

$$\frac{\|F(x_k)\| - |1 - \tau_{k-1}|\|F(x_{k-1})\|}{\|F(x_k)\|} = O(\|F(x_{k-1})\|).$$

Then by Remark 1.3, one can choose τ_k, such that

$$|1 - \tau_k| = \frac{\left| \|F(x_k)\| - |1 - \tau_{k-1}| \|F(x_{k-1})\| \right|}{\|F(x_{k-1})\|} \tag{1.31a}$$

or

$$|1 - \tau_k| = \frac{\left| \|F(x_k)\| - |1 - \tau_{k-1}| \|F(x_{k-1})\| \right|}{\|F(x_k)\|}, \tag{1.31b}$$

which allows us the quadratic convergence of CANM. The formulas (1.31) can be rewritten in term of the forcing term η_k as

$$\eta_k = \frac{|\alpha_k \eta_{k-1} - 1|}{\alpha_k} \tag{1.32a}$$

and

$$\eta_k = |\alpha_k \eta_{k-1} - 1|, \tag{1.32b}$$

respectively, where

$$\alpha_k = \frac{\|F(x_{k-1})\|}{\|F(x_k)\|}.$$

Assume that

$$\|F(x_k)\| \le \eta_{k-1} \|F(x_{k-1})\|, \quad 0 \le \eta_{k-1} < 1. \tag{1.33}$$

Then $\alpha_k > 1$ and $\alpha_k \eta_{k-1} \ge 1$. As a consequence, the minimum of the possible choices (1.32) is

$$\eta_k = \frac{\alpha_k \eta_{k-1} - 1}{\alpha_k}.$$

If the inequality (1.33) is not true, then (1.32) give us $\eta_k = 1 - \eta_{k-1}\alpha_k$. Thus we have

$$\eta_k = \begin{cases} 1 - \eta_{k-1}\alpha_k, & \text{when } \eta_{k-1}\alpha_k < 1, \\ -\frac{1-\eta_{k-1}\alpha_k}{\alpha_k}, & \text{when } \eta_{k-1}\alpha_k \ge 1. \end{cases} \tag{1.34}$$

The second choice in (1.34) allows us to decrease η_k, i.e., $0 < \eta_k < \eta_{k-1}$, while the first choice in (1.34) implies that $0 < \eta_k < 1$. In both cases we have $0 < \eta_k < 1$. According to (1.27), it is possible that (1.33) is true and thereby the second choice in (1.34) allows us to decrease η_k, i.e., $\eta_k \to 0$ as $k \to \infty$. In terms of τ_k we have the following choice

$$|1 - \tau_k| = \eta_k. \tag{1.35}$$

From this it follows that, if $0 < \eta_k < 1$ and $\eta_k \to 0$ as $k \to \infty$, then we have $0 < \tau_k < 2$ and $\tau_k \to 1$ as $k \to \infty$. Thus, the choice of the iteration parameter given by (1.34), (1.35) can be used in CANM.

In [9] (see (1.12)) another choice for CANM was proposed:

$$\tau_k = \frac{2}{1 + \sqrt{1 + 2b\|F(x_k)\|}}.$$

(1.36)

According to (1.35), the formula (1.36) in term of η_k reads as

$$\eta_k = \frac{\sqrt{1 + 2b\|F(x_k)\|} - 1}{\sqrt{1 + 2b\|F(x_k)\|} + 1}$$

(1.37)

which can be used in the IN as a forcing term. From (1.34) we see that the inexact Newton method with forcing term given by (1.34) is locally Q-superlinearly convergent, while IN with forcing term given by (1.37) converges quadratically because of $\eta_k = O(\|F(x_k)\|)$ (see Theorem 1.7).

1.3 Accelerating Convergence of Newton-Type Iterations

As is known, the monotone approximations interesting not only from theoretical, but also from practical view points. In particular, two-sided approximations can be efficiently used as a posteriori estimations for the desired solution. It means that one can control the error at each iteration step. In the last decade, many authors have developed new monotone iterative methods [15–17]. The main advantage of the monotone iterations is that it does not require good initial approximations contrary to what occurs in the other iteration methods, such as secant-like methods, Newton's methods and so on [18]. On the other hand, accelerating the convergence of iterative methods is also of interest both from theoretical and computational view points [19–28]. For example, in [18] was constructed a family of the predictor-corrector iterative method from the simplified secant method and a family of secant-like methods; the authors analyzed the initial conditions on the starting point in order to improve the semilocal convergence of the method.

In this section, we propose a new accelerating procedure for Newton-type methods. By virtue of this procedure, we obtain a higher order, in particular optimal order methods. The usage of the optimal choice of parameter allows us to improve the convergence speed. This may be also helpful in order to extend the domain of convergence.

1.3.1 Statement of Problems

Let $a, b \in \mathbb{R}$, $a < b$, $f : [a, b] \to \mathbb{R}$ and consider the following nonlinear equation

$$f(x) = 0.$$

(1.38)

Assume that $f(x) \in \mathbb{C}^4[a, b]$, $f'(x) \neq 0$, $x \in (a, b)$ and Eq. (1.38) has a unique root $x^* \in (a, b)$. In [16, 17] (see also (2.1)) the following iterations were proposed:

$$x_{2k+1} = x_{2k} - \tau_k \frac{f(x_{2k})}{f'(x_{2k})}, \tag{1.39a}$$

$$x_{2k+2} = x_{2k+1} - \frac{f(x_{2k+1})}{f'(x_{2k+1})}, \quad k = 0, 1, \ldots, \tag{1.39b}$$

and was shown that the iterations (1.39a) and (1.39b) monotone convergent under conditions

$$0 < \tau_k \leq 1, \quad a_k = \frac{M_2|f(x_k)|}{(f'(x_k))^2} < \frac{1}{2}, \quad M_2 = \sup_{x \in U_r(x^*)} |f''(x)|, \tag{1.40}$$

and under the

Assumption 1.1 $f'(x) \neq 0$, $f''(x)$ preserve sign in the small neighborhood $U_r(x^*) = \{x : |f(x)| < r\}$.

However the iterations (1.39a) and (1.39b) are not equipped with a suitable choice of parameter τ_k. It was shown that the iterations (1.39a) and (1.39b) have a two-sided approximation behavior under conditions

$$\tau_k \in I_{2k} = \left[\frac{1 - \sqrt{1 - 2a_{2k}}}{a_{2k}}, \frac{-1 + \sqrt{1 + 4a_{2k}}}{a_{2k}} \right) \subseteq (1, 2), \quad a_{2k} < \frac{4}{9}.$$

It is also proved that the convergence rate of these iterations is at least 2, and the order of convergence is increased up to 4, when $\tau_k \to 1$. From this it is clear that the accelerating of the convergence of iterations (1.39a) and (1.39b) is important, especially at the early stage of iterations.

1.3.2 Monotone and Two-Sided Convergence of Iterations and Their Acceleration

If $\tau_k \equiv 1$, then the iterations (1.39a) and (1.39b) become as Newton's one

$$x_{k+1} = x_k - \frac{f(x_k)}{f'(x_k)}, \quad k = 0, 1 \ldots . \tag{1.41}$$

According to [17] the iteration (1.41) is a monotone convergent under condition (1.40) and Assumption 1.1. Let $\theta_k = f(x_{k+1})/f(x_k)$. Then the Taylor expansion of $f(x_{k+1})$ gives

$$0 < \theta_k \leq \frac{a_k}{2} < \frac{1}{4}. \tag{1.42}$$

Now we proceed to accelerate the convergence of monotone iteration (1.41). To this end, we use two known approximations x_k, x_{k+m} satisfying either $x_k < x_{k+m} < x^*$ or $x^* < x_{k+m} < x_k$ and consider

$$y_k = x_k + t(x_{k+m} - x_k), \quad t > 1. \tag{1.43}$$

From (1.43) it is clear that y_k belongs to interval connecting x_k and x_{k+m} under condition $0 \le t \le 1$. Hence, the extrapolating approach corresponds to the case $t > 1$. Our aim is to find the optimal value of $t = t_{\text{opt}}$ in (1.43) such that the new approximation y_k given by (1.43) will be situated more close to x^* as compared with x_k and x_{k+m}. We use Taylor expansion of the smooth function $f(x) \in \mathbb{C}^{k+1}[a, b]$:

$$\begin{aligned} f(y_k) =& f(x_k) + f'(x_k)t(x_{k+m} - x_k) + \cdots + \frac{f^{(k)}(x_k)}{k!}t^k(x_{k+m} - x_k)^k \\ &+ O\left((x_{k+m} - x_k)^{k+1}\right). \end{aligned} \tag{1.44}$$

Neglecting small term $O\left((x_{k+m} - x_k)^{k+1}\right)$ in (1.44), we have

$$\begin{aligned} f(y_k) \approx& f(x_k) + f'(x_k)t(x_{k+m} - x_k) + \cdots + \frac{f^{(k)}(x_k)}{k!}t^k(x_{k+m} - x_k)^k \\ \equiv& P_k(t). \end{aligned} \tag{1.45}$$

From (1.45) it is clear that

$$f(x_k) = P_k(0).$$

We also require that

$$f(x_{k+m}) = P_k(1). \tag{1.46}$$

From (1.46) we find $f^{(k)}(x_k)/k!(x_{k+m} - x_k)^k$ and substituting it into (1.45), we get $P_k(t)$. From this we find $t > 1$ such that

$$f(y_k) \approx P_k(t) = 0. \tag{1.47}$$

Geometrically, (1.45) means that the graph (plot) of $f(x)$ in the vicinity of root x^* is replaced by curve $P_k(t)$ passing through the given points $(x_k, f(x_k))$ and $(x_{k+m}, f(x_{k+m}))$. Thus, Eq. (1.47) for $k = 1, 2$ and $k = 3$ gives us

$$P_0(t) = f(x_k) \approx f(y_k), \quad y_k = x_k, \tag{1.48a}$$

$$P_1(t) = f(x_k) + (f(x_{k+m}) - f(x_k))\,t, \tag{1.48b}$$

$$\begin{aligned} P_2(t) =& f(x_k) + f'(x_k)(x_{k+m} - x_k)t \\ &+ \left(f(x_{k+m}) - f(x_k) - f'(x_k)(x_{k+m} - x_k)\right)t^2, \end{aligned} \tag{1.48c}$$

$$P_3(t) = f(x_k) + f'(x_k)(x_{k+m} - x_k)t + \frac{f''(x_k)}{2}(x_{k+m} - x_k)^2 t^2$$

$$+\left(f(x_{k+m}) - f(x_k) - f'(x_k)(x_{k+m} - x_k) - \frac{f''(x_k)}{2}(x_{k+m} - x_k)^2\right)t^3, \quad (1.48d)$$

respectively. Thus, the parameter t in (1.43) is calculated as a root greater than 1 of Eq. (1.47). In particular, for $k = 1$, we have

$$t_{\text{opt}} = \frac{f(x_k)}{f(x_k) - f(x_{k+m})} > 1. \qquad (1.49)$$

Since $P_k(0)P_k(1) = f(x_k)f(x_{k+m}) > 0$ for $k > 1$, Eq. (1.47) may have at least one root satisfying the condition $t^* > 1$. From (1.42) it follows that

$$|f(x_{k+1})| < \frac{1}{4}|f(x_k)|. \qquad (1.50)$$

Therefore, it is desirable to choose k and m such that

$$|f(x_{k+m})| < \left(\frac{1}{4}\right)^m |f(x_k)| \ll 0.1. \qquad (1.51)$$

This inequality is written in term of $P_k(t)$ as

$$|P_k(1)| \le \left(\frac{1}{4}\right)^m |P_k(0)| < 0.1.$$

On the other hand, from (1.45) we see that $P_k'(1)$ is not equal to 0 under the Assumption 1.1, i.e., $t = 1$ is not a critical point of $P_k(t)$. Thus, $P_k(t)$ is decreasing around $t = 1$. Therefore, there exists $t_{\text{opt}} > 1$ such that $P_k(t_{\text{opt}}) = 0$.

Lemma 1.4 *Assume that*

$$|x^* - x_k| = \varepsilon_k < 1. \qquad (1.52)$$

Then the following holds

$$t_{\text{opt}} - 1 = O(\varepsilon_k). \qquad (1.53)$$

Proof First of all, let us note that the inequality (1.52) is equivalent to

$$|f(x_k)| = O(\varepsilon_k), \qquad (1.54)$$

which follows from the expansion

$$0 = f(x^*) = f(x_k) + f'(\xi)(x^* - x_k).$$

Of course, $|x^* - x_{k+m}| < \varepsilon_k$ and $|x_{k+m} - x_k| < \varepsilon_k$ under (1.52). We also use an analogous expansion for $P_k(t)$

$$0 = P_k(t_{\text{opt}}) = P_k(1) + P_k'(\eta)(t_{\text{opt}} - 1), \quad \eta \in (1, t_{\text{opt}}). \tag{1.55}$$

Since $P_k(t)$ is decreasing around $t = 1$, then $P_k'(\eta) \neq 0$. Hence, from (1.54), (1.55) and (1.51) we conclude that

$$t_{\text{opt}} - 1 = -\frac{f(x_{k+m})}{P_k'(\eta)} \approx O(\varepsilon_k). \tag{1.56}$$

The Lemma is proved. $\qquad\square$

Note that in [28] the iterations was proposed:

$$x_{2k+1} = x_{2k} - \tau_k \frac{f(x_{2k})}{f'(x_{2k})}, \tag{1.57a}$$

$$x_{2k+2} = x_{2k+1} - \frac{x_{2k+1} - x_{2k}}{f(x_{2k+1}) - f(x_{2k})} f(x_{2k+1}), \quad k = 0, 1, \ldots, \tag{1.57b}$$

which has a third order of convergence when $\tau_k = 1$ or τ_k tends to 1. It is easy to show that the iteration (1.57b) coincides fully with our acceleration procedure (1.43) and (1.49) with $m = 1$ and $k = 1$. Therefore, one can expect a high acceleration when $k = 2, 3$ for Newton's method.

How to accelerate the convergence rate of iteration (1.41)? The answer of this question gives the following theorem.

Theorem 1.10 *Assume $f(x) \in \mathbb{C}^{k+2}$ and the condition (1.52) is satisfied. Then for y_k with t_{opt} we have*

$$\frac{|x^* - y_k|}{|x^* - x_k|^{k+2}} \approx O(1), \tag{1.58}$$

where O is Landau symbol.

Proof Let

$$x^* = x_k + t^*(x_{k+m} - x_k), \quad t^* \geq 1,$$
$$y_k = x_k + t_{\text{opt}}(x_{k+m} - x_k). \tag{1.59}$$

We use Taylor expansions of $f(x) \in \mathbb{C}^{k+2}$

$$0 = f(x^*) = \sum_{p=0}^{k} \frac{f^{(p)}(x_k)}{p!} (t^*)^p (x_{k+m} - x_k)^p$$

$$+ \frac{f^{(k+1)}(\eta_k)}{(k+1)!} (t^*)^{(k+1)} (x_{k+m} - x_k)^{(k+1)}, \tag{1.60}$$

$$f(x_{k+m}) - \sum_{p=0}^{k-1} \frac{f^{(p)}(x_k)}{p!}(x_{k+m} - x_k)^p = \frac{f^{(k)}(x_k)}{k!}(x_{k+m} - x_k)^k$$

$$+ \frac{f^{(k+1)}(\xi_k)}{(k+1)!}(x_{k+m} - x_k)^{k+1}, \tag{1.61}$$

and

$$0 = P_k(t_{\mathrm{opt}}) = \sum_{p=0}^{k-1} \frac{f^{(p)}(x_k)}{p!}(t_{\mathrm{opt}})^p(x_{k+m} - x_k)^p$$

$$+ \left(f(x_{k+m}) - \sum_{p=0}^{k-1} \frac{f^{(p)}(x_k)}{p!}(x_{k+m} - x_k)^p \right)(t_{\mathrm{opt}})^k, \tag{1.62}$$

where $\eta_k \in (x_k, x^*)$ and $\xi_k \in (x_k, x_{k+m})$. Using (1.61) in (1.62) and subtracting (1.62) from (1.60) we get

$$\left(f'(x_k) + \frac{f''(x_k)}{2}(x_{k+m} - x_k)(t^* + t_{\mathrm{opt}}) \right.$$

$$+ \frac{f'''(x_k)}{6}(x_{k+m} - x_k)^2(t^{*2} + t^* t_{\mathrm{opt}} + t_{\mathrm{opt}}^2) + \ldots$$

$$\left. + \frac{f^{(k)}(x_k)}{k!}\left(t^{*k-1} + t^{*k-2} t_{\mathrm{opt}} + \ldots + t_{\mathrm{opt}}^{k-1} \right)(x_{k+m} - x_k)^{k-1} \right)(t^* - t_{\mathrm{opt}})$$

$$= -\frac{(x_{k+m} - x_k)^k}{(k+1)!}\left(f^{(k+1)}(\eta_k)t^{*k+1} - f^{(k+1)}(\xi_k)t_{\mathrm{opt}}^k \right). \tag{1.63}$$

Since $f'(x_k) \neq 0$, then from last expression we deduce that

$$t^* - t_{\mathrm{opt}} = O\left(\varepsilon_k^k \right). \tag{1.64}$$

It is possible to derive a more precise estimation than (1.64). Indeed, using (1.64) and $f \in \mathbb{C}^{k+2}$ we evaluate

$$A_k = f^{(k+1)}(\eta_k)t^{*k+1} - f^{(k+1)}(\xi_k)t_{\mathrm{opt}}^k$$

$$= f^{(k+1)}(\xi_k)(t^{*k+1} - t_{\mathrm{opt}}^k) + f^{(k+2)}(\omega_k)(\eta_k - \xi_k),$$

where $\omega_k = \theta \eta_k + (1 - \theta)\xi_k, \theta \in (0, 1)$.

By definition we have

$$|\eta_k - \xi_k| \leq |x^* - x_k| = \varepsilon_k.$$

Using (1.53) and (1.64) we have

$$t^{*^{k+1}} - t^k_{\text{opt}} = \left(t_{\text{opt}} + O\left(\varepsilon^k_k\right)\right)^{k+1} - t^k_{\text{opt}}$$
$$= t^k_{\text{opt}}\left(t_{\text{opt}}(1 + O\left(\varepsilon^k_k\right))^{k+1} - 1\right) = t^k_{\text{opt}}\left(t_{\text{opt}} + O\left(\varepsilon^k_k\right) - 1\right) = O(\varepsilon_k).$$

Then $A_k = O(\varepsilon_k)$ and thereby from (1.63) we get

$$t^* - t_{\text{opt}} = O\left(\varepsilon^{k+1}_k\right).$$

Hence, from (1.59) and (1.60) we find that

$$x^* - y_k = O\left(\varepsilon^{k+2}_k\right)$$

which proves (1.58). $\qquad\qquad\square$

The sequence $\{y_k\}$ given by formula (1.43) can be considered as new a iteration. For it we have the following.

Theorem 1.11 *Assume $f(x) \in \mathbb{C}^{k+1}$ and the convergence order of iterations (1.41) equal to 2, i.e., the following holds*

$$|x^* - x_k| \leq Mq^{2^k}|x^* - x_0|, \quad 0 < q < 1, \quad M = const. \tag{1.65}$$

If the Eq. (1.47) has at least one root t_{opt} greater than 1, then the convergence order of new iteration (1.43) is the same as (1.41) and we have

$$|x^* - y_k| \leq M_1 q_1^{2^k}|x^* - y_0|, \quad 0 < q_1 = q^{k+1} < 1, \quad M_1 = const. \tag{1.66}$$

Proof By virtue of (1.65) the condition (1.52) is satisfied for large k. Then by Theorem 1.10, the relation (1.58) holds. Using (1.65) in (1.58), we get

$$|x^* - y_k| \leq C\left(q^{d^k}\right)^{k+2}|x^* - x_0|^{k+2} = C\left(q^{k+2}\right)^{d^k}|x^* - x_0|^{k+2}$$
$$= Cq_1^{d^k}|x^* - x_0|^{k+2} \leq M_1 q_1^{d^k}|x^* - y_0|, \quad q_1 = q^{k+2} < q < 1.$$

The proof is completed. $\qquad\qquad\square$

Theorem 1.11 shows that the convergence order of iteration (1.43) is the same as iteration (1.41).

However, the speed of convergence of these iterations depends on the factor q_1 and q in (1.65) and (1.66), respectively. Since $q_1 = q^{k+2} < q$ for $k = 1, 2, 3$, one can expect a more rapid convergence of iteration (1.43). Of course, the higher is acceleration of iteration attained at $k = 3$.

From (1.65) and (1.66) it is clear that the iteration (1.43) converges to x^* more rapidly than iteration (1.41) by virtue of $q_1 = q^{k+1} < q$. This accelerating procedure is useful, especially at the beginning of iterations, but under condition (1.52). Of course, the acceleration procedure (1.43) can be continued further with $x_{k+m} := y_k$, $x_k := x_{k+m}$ and with $t > 1$ if y_k and x_{k+m} are located on side of x^* and with $t \in (0, 1)$ if y_k and x_{k+m} are located on two-sides of root. From Theorem 1.11, it is clear that the sequence $\{y_k\}$ given by (1.43) together with (1.41) can be considered as a new iteration process with a small factor compared to (1.41). The acceleration procedure is achieved without additional calculations, so that the iteration (1.43) possesses a high computational efficiency. However, despite the sequence x_k is monotone, the new iteration (1.43) may not be monotone. For instance, when $k = 1$ it is easy to show that

$$f(y_k) = \frac{f''(\xi_k)}{2}(x_{k+m} - x_k)^2.$$

From this it is clear that

$$f(y_k) > 0 \quad \text{if} \quad f''(x) > 0,$$

and

$$f(y_k) < 0 \quad \text{if} \quad f''(x) < 0.$$

Let us know two successive approximations x_k and x_{k+1}, for which

$$f(x_k)f(x_{k+1}) < 0 \tag{1.67}$$

holds. We consider

$$y_k = x_k + t(x_{k+1} - x_k), \quad 0 \le t \le 1.$$

The acceleration technique will be the same as a previous case with $m = 1$. In this case, according to (1.67) we have $P_k(0)P_k(1) = f(x_k)f(x_{k+1}) < 0$ for $k = 1, 2, 3$. Hence, Eq. (1.47) has a root $t_{\text{opt}} \in (0, 1)$. Obviously, the new approximation

$$y_k = x_k + t_{\text{opt}}(x_{k+1} - x_k), \quad 0 \le t \le 1,$$

will be situated more close to x^* as compared to x_k, x_{k+1} and Theorem 1.10 holds true for this case, too. It indicates that the two-sided approximations are useful not only for estimations of roots, but also for finding it approximately with a higher accuracy. Note that the accelerating procedure (1.43) is applicable not only for iterations (1.41), but also for any iteration, in particular, for the following iterations (1.68), (1.69), (1.72) and (1.75).

Now we consider the accelerated iteration

$$y_k = x_k - \frac{f(x_k)}{f'(x_k)}, \quad x_{k+1} = x_k + t_{\text{opt}}(y_k - x_k), \quad k = 0, 1, \ldots. \tag{1.68}$$

The iteration (1.68) is a damped Newton's method [1, 4] with optimal parameter $\tau_k = t_{\text{opt}}$. The first step y_k is used for finding the optimal parameter.

Theorem 1.12 *The assumptions of Theorem 1.10 are fulfilled. Then the convergence order of iteration (1.68) with optimal t_{opt} is $d = k + 2$, $k = 1, 2, 3$*

Proof If we compare (1.68) with (1.41) and (1.43), then $x_{k+m} := y_k$ and $y_k := x_{k+1}$. Therefore, the expression (1.58) in the Theorem 1.10 has a form

$$\frac{|x^* - x_{k+1}|}{|x^* - x_k|^{k+2}} = O(1) \iff |x^* - x_{k+1}| \le M|x^* - x_k|^{k+2},$$

which completes the proof of the Theorem 1.12. $\square$

Now let us consider another three-step iteration

$$y_k = x_k - \frac{f(x_k)}{f'(x_k)}, \quad z_k = y_k - \frac{f(y_k)}{f'(x_k)},$$
$$x_{k+1} = y_k + t(z_k - y_k), \quad k = 0, 1, \ldots . \tag{1.69}$$

Note that if $t \equiv 1$ in (1.69), then it leads to

$$y_k = x_k - \frac{f(x_k)}{f'(x_k)}, \quad x_{k+1} = y_k - \frac{f(y_k)}{f'(x_k)}, \quad k = 0, 1, \ldots . \tag{1.70}$$

The iteration (1.70) is a particular case of scheme (40) given in [28] with $\sigma = 0$ end $\tau = 1$ and has a third order of convergence. Therefore, the iteration (1.69) can be considered as improvement of iteration (1.70).

Theorem 1.13 *The assumptions of the Theorem 1.10 are fulfilled. Then the convergence order of iteration (1.69) with optimal t_{opt} equal to $d = 2k + 3$.*

Proof If we compare (1.69) with (1.43), then $x_k := y_k, x_{k+m} := z_k, y_k := x_{k+1}$. Then form (1.63) and (1.54) we get

$$t^* - t_{\text{opt}} = M A_k (z_k - y_k)^k \approx O\left(\varepsilon_k^{2k+1}\right), \tag{1.71}$$

where

$$x^* = y_k + t^*(z_k - y_k), \quad x_{k+1} = y_k + t_{\text{opt}}(z_k - y_k).$$

From this and from (1.71) we obtained

$$x^* - x_{k+1} = (t^* - t_{\text{opt}})(z_k - y_k) \approx O\left(\varepsilon_k^{2k+1}\right) O\left(\varepsilon_k^2\right) = O\left(\varepsilon_k^{2k+3}\right),$$

i.e., we have

$$|x^* - x_{k+1}| \le M_1 |x^* - x_k|^{2k+3},$$

which means that the convergence order of iteration (1.69) is equal to $d = 2k + 3$, $k = 1, 2, 3$. □

From the Theorem 1.13, we see that the convergence order of iteration (1.70) can be increased two or four units at the expense of only two additional evaluations of the function. So the order of convergence and the computational efficiency of the method are greatly improved. In [22] was constructed Algorithm 2:

$$z_k = x_k - \frac{f(x_k)}{f'(x_k)}, \quad x_{k+1} = z_k - H(x_k, y_k)\frac{f(z_k)}{f'(x_k)},$$

and it is proved that the order of convergence equals 5, 6, 7 depending on a suitable choice of two-variable function $H(x_k, y_k)$. For comparison purpose we can rewrite iteration (1.69) as

$$y_k = x_k - \frac{f(x_k)}{f'(x_k)}, \quad x_{k+1} = y_k - t\frac{f(y_k)}{f'(y_k)}.$$

We see that these two methods are different from one another only by chosen factors t and $H(x_k, y_k)$.

Now we consider the following iteration:

$$y_k = x_k - \frac{f(x_k)}{f'(x_k)}, \quad z_k = y_k - \frac{f(y_k)}{f'(y_k)},$$
$$x_{k+1} = y_k + t(z_k - y_k), \quad k = 0, 1, \dots. \tag{1.72}$$

The iteration (1.72) can be considered as improvement of iteration

$$y_k = x_k - \frac{f(x_k)}{f'(x_k)}, \quad x_{k+1} = y_k - \frac{f(y_k)}{f'(y_k)}, \quad k = 0, 1, \dots, \tag{1.73}$$

since if $t \equiv 1$ in (1.72), then it leads to (1.73).

In [28], it was proven that the convergence order of (1.73) is four.

Theorem 1.14 *The assumptions of Theorem 1.10 are fulfilled. Then the convergence order of iteration (1.72) with optimal t_{opt} equal to $d = 2(k + 2)$, $k = 1, 2, 3$.*

Proof If we compare (1.72) with (1.43), then $x_k := y_k$, $x_{k+m} := z_k$, $y_k := x_{k+1}$. Therefore, the expression (1.58) reads as

$$\frac{|x^* - x_{k+1}|}{|x^* - y_k|^{k+2}} = O(1) \iff |x^* - x_{k+1}| \le M|x^* - y_k|^{k+2}.$$

From (1.72), we find that

$$x^* - y_k = x^* - x_k + \frac{f(x_k) - f(x^*)}{f'(x_k)}.$$

Substituting here the expansion of $f(x^*)$

$$f(x^*) = f(x_k) + f'(x_k)(x^* - x_k) + \frac{f''(\xi_k)}{2}(x^* - x_k)^2,$$

we have

$$|x^* - y_k| \le \frac{|f''(\xi_k)|}{|f'(x_k)|}|x^* - x_k|^2.$$

Using the last estimate in (1.71), we obtain

$$|x^* - x_{k+1}| \le M_1 |x^* - x_k|^{2(k+2)},$$

which means that the convergence order of iteration (1.72) equals $d = 2(k + 2)$, $k = 1, 2, 3$.

Note that the iterations (1.68), (1.69) and (1.72) can be rewritten as a damped Newton's method [4]

$$x_{k+1} = x_k - \tau_k \frac{f(x_k)}{f'(x_k)}, \tag{1.74}$$

with

$$\tau_k = t_{\text{opt}},$$

and

$$\tau_k = 1 + t_{\text{opt}} \frac{f(y_k)}{f(x_k)},$$

and

$$\tau_k = 1 + t_{\text{opt}} \frac{f(y_k)}{f(x_k)} \frac{f'(x_k)}{f'(y_k)},$$

respectively. The unified representation (1.74) of different iterations shows that the choice of the damped parameter τ_k in (1.74) is essentially affected for the convergence order. Of course, the parameter τ_k in (1.74) is defined by different ways, but in all cases $\tau_k \to 1$ as $k \to \infty$.

The speed of convergence of sequence $\{\tau_k\}$ to unit is different for each iteration methods. In [1] the conjecture was proposed:

$$|1 - \tau_k| \le M q^{\rho^n}, \quad 0 < q < 1.$$

Now we consider the following three-point iterative method:

$$y_k = x_k - \frac{f(x_k)}{f'(x_k)}, \quad z_k = x_k + \bar{t}(y_k - x_k),$$

$$x_{k+1} = y_k + t(z_k - y_k), \quad k = 0, 1, \ldots, \tag{1.75}$$

where $\bar{t}$ and t in (1.75) are some parameters to be determined. We can formulate the following theorem.

Theorem 1.15 *Assume that $f(x) \in \mathbb{C}^4(a, b)$ and an initial approximation x_0 is sufficiently close to the zero $x^* \in (a, b)$ and the parameter $\bar{t}$ is chosen by as a root of equation*

$$\theta_k \bar{t}^2 - \bar{t} + 1 = 0, \quad \theta_k = \frac{f(y_k)}{f(x_k)}, \tag{1.76}$$

and t is a root of equation

$$\Psi(t, \alpha) = \alpha \Psi_1(t) + (1 - \alpha)\Psi_2(t) = 0,$$

where

$$\Psi_1(t) = at^2 - \left(a + \frac{f(x_k)}{f(y_k)}\left(f(z_k) - f(y_k)\right)\right)t - f(x_k),$$
$$a = -2f(z_k) - f(x_k)(1 - \bar{t})^2,$$

and

$$\Psi_2(t) = \left((1 - \bar{t})(2 - \bar{t})f(x_k) - (2 - 3\bar{t})f(z_k)\right)t$$
$$+(1 - \bar{t})\left(2f(z_k) - (2 - \bar{t})f(x_k)\right).$$

Then the three-point methods (1.75) is of eight order of convergence.

Proof Using $z_k - y_k = (1 - \bar{t})f(x_k)/f'(x_k)$ in Taylor expansion

$$f(x_{k+1}) = f(y_k) + f'(y_k)t(z_k - y_k) + \frac{f''(y_k)}{2}t^2(z_k - y_k)^2 + O\left((z_k - y_k)^3\right),$$

we get

$$f(x_{k+1}) = f(y_k) + t(1 - \bar{t})\frac{f'(y_k)}{f'(x_k)}f(x_k)$$
$$+ \frac{f''(y_k)}{2}t^2(1 - \bar{t})^2\frac{f^2(x_k)}{(f'(x_k))^2} + O\left(f^6(x_k)\right). \tag{1.77}$$

Analogously, the Taylor expansion of $f(x_{k+1})$ at point $x = z_k$ gives

$$f(x_{k+1}) = f(z_k) - (1 - t)(1 - \bar{t})\frac{f'(z_k)}{f'(x_k)}f(x_k)$$
$$+ \frac{f''(z_k)}{2}(1 - t)^2(1 - \bar{t})^2\frac{f^2(x_k)}{(f'(x_k))^2} + O\left((1 - t)^3 f^6(x_k)\right). \tag{1.78}$$

Using $f'(z_k) = f'(y_k) + f''(y_k)(z_k - y_k) + O\left((z_k - y_k)^2\right)$ in the last expansion, we have

$$f(x_{k+1}) = f(z_k) - (1 - t)(1 - \bar{t})\frac{f'(z_k)}{f'(x_k)}f(x_k)$$
$$+ \frac{f''(z_k)}{2}(1 - t)^2(1 - \bar{t})^2\frac{f^2(x_k)}{(f'(x_k))^2} + O\left((1 - t)^3 f^6(x_k)\right).$$

Using $f'(z_k) = f'(y_k) + f''(y_k)(z_k - y_k) + O\left((z_k - y_k)^2\right)$ in the last expansion, we have

$$f(x_{k+1}) = f(z_k) - (1 - t)(1 - \bar{t})\frac{f'(y_k)}{f'(x_k)}f(x_k)$$
$$- \frac{f''(y_k)f^2(x_k)}{2\left(f'(x_k)\right)^2}(1 - t)^2(1 - \bar{t})^2 + O\left(f^8(x_k)\right). \tag{1.79}$$

From (1.77) and (1.79) one can eliminate the term with $f'(y_k)/f'(x_k)f(x_k)$. As a result, we have

$$f(x_{k+1}) = tf(z_k) + (1 - t)f(y_k)$$
$$- \frac{f''(y_k)f^2(x_k)}{2\left(f'(x_k)\right)^2}(1 - \bar{t})^2 t(1 - t) + O\left(f^8(x_k)\right). \tag{1.80}$$

Note that in driving (1.80), we keep in mind that

$$1 - t = O\left(f^2(x_k)\right). \tag{1.81}$$

Further, using Taylor expansion of $f(x) \in \mathbb{C}^4(\mathfrak{D})$ at points y_k we obtain

$$f''(y_k) = \frac{2\left(f'(x_k)\right)^2}{f^2(x_k)\bar{t}(1 - \bar{t})}\left((1 - \bar{t})f(x_k) + \bar{t}f(y_k) - f(z_k)\right)$$
$$- \frac{f'''(y_k)}{3}\frac{f(x_k)}{f'(x_k)}(2 - \bar{t}) + O\left(f^2(x_k)\right). \tag{1.82}$$

The same technique gives us

$$f''(y_k) = \frac{2\left(f(z_k) - (1 - \bar{t})f(x_k)\right)}{\bar{t}^2 f^2(x_k)}\left(f'(x_k)\right)^2$$
$$- \frac{f'''(y_k)}{3}\frac{f(x_k)}{f'(x_k)}(3 - \bar{t}) + O\left(f^2(x_k)\right). \tag{1.83}$$

For (1.82) and (1.83) one can eliminate the term with $f'''(y_k)$. As a result, we obtain

$$\frac{f''(y_k)f^2(x_k)}{2\left(f'(x_k)\right)^2} = \frac{1}{\bar{t}^2(1-\bar{t})}\left((3-\bar{t})\bar{t}\left(-f(z_k)+\bar{t}f(y_k)+(1-\bar{t})f(x_k)\right)\right.$$
$$\left.-(2-\bar{t})(1-\bar{t})\left(f(z_k)-(1-\bar{t})f(x_k)\right)\right)+O\left(f^4(x_k)\right). \quad (1.84)$$

Substituting (1.84) into (1.79), we obtain

$$f(x_{k+1}) = \Psi_1(t) + O\left(f^8(x_k)\right), \quad (1.85)$$

where

$$\Psi_1(t) = at^2 - \left(a + \frac{f(x_k)}{f(y_k)}\left(f(z_k)-f(y_k)\right)\right)t - f(x_k)$$

with $a = -2f(z_k) - f(x_k)(1-\bar{t})^2$. On the other hand, by virtue of (1.75) we have

$$x_{k+1} - z_k = -(1-\bar{t})(1-t)\frac{f(x_k)}{f'(x_k)}. \quad (1.86)$$

If we take (1.76) and (1.81) into account in (1.86), from it we deduce

$$x_{k+1} - z_k = O\left(f^4(x_k)\right).$$

Then, from (1.78) we get

$$f(x_{k+1}) = f(z_k) + (1-\bar{t})(1-t)\frac{f'(z_k)}{f'(x_k)}f(x_k) + O\left(f^8(x_k)\right). \quad (1.87)$$

Now we approximate $f'(z_k)$ by the method of undetermined coefficient such that

$$f'(z_k) \approx a_k f(x_k) + b_k f(y_k) + c_k f(z_k) + d_k f'(x_k) + O\left(f^4(x_k)\right), \quad (1.88)$$

This can be done by means of Taylor expansion of $f(x) \in \mathbb{C}^4(a,b)$ at point z_k and we obtain the following linear system of equations

$$\begin{cases} a_k + b_k + c_k = 0, \\ a_k(x_k-z_k) + b_k(y_k-z_k) + d_k = 1, \\ a_k(x_k-z_k)^2 + b_k(y_k-z_k)^2 + 2d_k(x_k-z_k) = 0, \\ a_k(x_k-z_k)^3 + b_k(y_k-z_k)^3 + 2d_k(x_k-z_k)^2 = 0, \end{cases}$$

which has a unique solution

$$a_k = \frac{\beta_k(2\beta_k - 3\omega_k)}{\omega_k(\beta_k - \omega_k)^2}, \quad b_k = \frac{\omega_k^2}{\beta_k(\beta_k - \omega_k)^2},$$

$$c_k = -\frac{2\beta_k + \omega_k}{\beta_k\omega_k}, \quad d_k = -\frac{\beta_k}{\beta_k - \omega_k}, \tag{1.89}$$

where

$$\omega_k = x_k - z_k = \bar{t}\,\frac{f(x_k)}{f'(x_k)}, \quad \beta_k = y_k - z_k = (1 - \bar{t})\frac{f(x_k)}{f'(x_k)}.$$

Substituting (1.88) with coefficients defined by (1.89) into (1.87), we get

$$f(x_{k+1}) = \Psi_2(t) + O\left(f^8(x_k)\right), \tag{1.90}$$

where

$$\Psi_2(t) = \left((1 - \bar{t})(2 - \bar{t})f(x_k) - (2 - 3\bar{t})f(z_k)\right)t$$
$$+(1 - \bar{t})\left(2f(z_k) - (2 - \bar{t})f(x_k)\right).$$

The linear combination of (1.85) and (1.90) gives

$$f(x_{k+1}) = \alpha\Psi_1(t) + (1 - \alpha)\Psi_2(t) + O\left(f^8(x_k)\right).$$

Clearly, if we choose t as a root of quadratic equations

$$\Psi(t, \alpha) = \alpha\Psi_1(t) + (1 - \alpha)\Psi_2(t) = 0, \tag{1.91}$$

then we have

$$f(x_{k+1}) = O\left(f^8(x_k)\right)$$

which completes the proof. $\qquad\qquad\qquad\qquad\qquad\qquad\qquad\qquad\quad\square$

Remark 1.4 Since t is a root of Eq. (1.91), it depends on the parameter α, i.e., $t = t(\alpha)$. Therefore, (1.75) are one parameter family of iterations.

It is easy to show that

$$\Psi_2(\hat{t}) = 0, \quad \hat{t} \to 1, \quad \Psi_1(\check{t}) = 0, \quad \check{t} \to 1.$$

Then taking this into account and passing to the limit $t \to 1$ in Eq. (1.91), we get

$$\Psi(t, \alpha) \xrightarrow{t \to 1} \alpha\Psi_1(1) + (1 - \alpha)\Psi_2(1) \approx 0.$$

It means that Eq. (1.91) has a root tending to unit for any $\alpha \in [0, 1]$.

We recall that, according to Kung-Traub hypothesis, the order of convergence of any multi-point method without memory cannot exceed the bound 2^{n-1} (called optimal order), where n is the number of function evaluations per iteration. As is known, the efficiency index of iteration defined by formula $E = d^{1/m}$, where d is the convergence order and m is the number of function and its derivative evaluations per iteration. Therefore, the optimal efficiency index would be $2^{(n-1)/n}$.

According to the Theorem 1.12, the iteration (1.68) has the convergence order fourth for $k = 2$, requiring only three function evaluations ($f(x_k)$, $f(y_k)$ and $f'(x_k)$), whereas the Theorem 1.15 shows that the iteration (1.75) has the convergence order eight, requiring four function evaluations ($f(x_k)$, $f(y_k)$, $f(z_k)$, $f'(x_k)$).

Hence, this order of convergence is optimal in the above mentioned sense of the Kung-Traub conjecture. This efficiency index is $4^{1/3} \approx 1.587$ and $8^{1/4} \approx 1.681$, respectively.

Thus, we obtain the iterations (1.68) and (1.75) with the optimal order of convergence 4 and 8, accelerating Newton's method. Our procedure of accelerations gives a genuine improvement of Newton's method. One of the advantages of iterations (1.68) and (1.75) is that these methods work well for the system of nonlinear equations, whereas the optimal order methods in [19–21, 24, 25] do not extend to the system of equations.

For convenience we present the efficiency index of the proposed above methods (1.68), (1.69), (1.72) and (1.75) in Table 1.1. From Table 1.1 one can see that the efficiency index of the iterations (1.68), (1.69), (1.72) and (1.75) is better or much better than that of Newton's method $\sqrt{2} \approx 1.414$.

Table 1.1 The efficiency index of the methods (1.68), (1.69), (1.72) and (1.75)

Methods	n	d	E
(1.68)	1	3	$3^{\frac{1}{3}} \approx 1.442249$
	2	4	$4^{\frac{1}{3}} \approx 1.587401$
	3	5	$5^{\frac{1}{4}} \approx 1.495348$
(1.69)	1	5	$5^{\frac{1}{4}} \approx 1.495348$
	2	7	$7^{\frac{1}{5}} \approx 1.475773$
	3	9	$9^{\frac{1}{5}} \approx 1.551845$
(1.72)	1	6	$6^{\frac{1}{5}} \approx 1.430969$
	2	8	$8^{\frac{1}{5}} \approx 1.515716$
	3	10	$10^{\frac{1}{6}} \approx 1.467799$
(1.75)	–	8	$8^{\frac{1}{4}} \approx 1.681792$

1.3.3 *Finding Optimal Parameter*

Let $m = 1$ in (1.43). Then the root (1.49) can be written as

$$t_{\mathrm{opt}}^{(1)} = \frac{1}{1 - \theta_k}, \quad \theta_k = \frac{f(y_k)}{f(x_k)}.$$

For $k = 2$, from (1.48c) we obtain

$$P_2(t) \equiv f(y_k)t^2 - f(x_k)t + f(x_k) = 0, \tag{1.92a}$$

or

$$P_2(t) \equiv \theta_k t^2 - t + 1 = 0. \tag{1.92b}$$

By the well known assertion and (1.50) we have

$$t_1 + t_2 = \frac{f(x_k)}{f(y_k)} > 1, \quad t_1 t_2 = \frac{f(x_k)}{f(y_k)}.$$

Hence

$$t_1 + t_2 = t_1 t_2.$$

From this we obtain

$$t_1^2 - \frac{f(x_k)}{f(y_k)}t_1 + \frac{f(x_k)}{f(y_k)} = 0. \tag{1.93}$$

The root of (1.93) greater than 1 is

$$t_{\mathrm{opt}}^{(2)} = \frac{1 - \sqrt{1 - 4\theta_k}}{2\theta_k} = \frac{2}{1 + \sqrt{1 - 4\theta_k}}.$$

In a similar way, from (1.41) and (1.48d) we obtain

$$P_3(t) = (\theta_k - \omega_k)t^3 + \omega_k t^2 - t + 1 = 0, \tag{1.94}$$

where

$$\omega_k = \frac{f''(x_k)f(x_k)}{2\left(f'(x_k)\right)^2}.$$

Since in all iterations (1.68), (1.69) and (1.72) we have

$$f(y_k) = \frac{f''(x_k)}{2}\frac{f^2(x_k)}{(f'(x_k))^2} + O\left(f^3(x_k)\right),$$

then

$$\omega_k = \frac{f(y_k)}{f(x_k)} + O\left(f^2(x_k)\right). \tag{1.95}$$

Using (1.56) and (1.95) in (1.94) we obtain approximates equation

$$\left(\frac{f(z_k)}{f(y_k)} - \frac{f(y_k)}{f(x_k)}\right) t^3 + \frac{f(y_k)}{f(x_k)} t^2 - t + 1 = 0. \tag{1.96}$$

Equation (1.96) approximates (1.94) with accuracy $O\left(f^4(x_k)\right)$ in case of (1.68) and with accuracy $O\left(f^5(x_k)\right)$ in case of (1.69) and (1.72) since $t_{\mathrm{opt}} - 1 = O\left(\varepsilon_k^2\right)$ for (1.68) and $t_{\mathrm{opt}} - 1 = O\left(\varepsilon_k^3\right)$ for (1.69) and (1.72). Therefore, Eq. (1.96) may be useful especially for (1.68).

Above, we obtain formula (1.92) for finding optimal value t_{opt} for iteration (1.68). However, it may be changed for iteration (1.69) and (1.72). Since $x_{k+m} := z_k$, $x_k := y_k$ and $y_k := x_{k+1}$ for iteration (1.69), then according to (1.48c) we have

$$P_2(t) \equiv f(y_k) + f'(y_k)(z_k - y_k)t + \left(f(z_k) - f(y_k) - f'(y_k)(z_k - y_k)\right) t^2 = 0,$$

or

$$P_2(t) \equiv \left(\frac{f(z_k)}{f(y_k)} - 1 + \frac{f'(y_k)}{f'(x_k)}\right) t^2 - \frac{f'(y_k)}{f'(x_k)} t + 1 = 0. \tag{1.97}$$

We rewrite (1.97) as

$$P_2(t) = \frac{f(z_k)}{f(y_k)} t^2 - t + 1 + \left(1 - \frac{f'(y_k)}{f'(x_k)}\right) t(1 - t) = 0.$$

From the last equation, it is clear that if we take into account the following estimate

$$1 - \frac{f'(y_k)}{f'(x_k)} = 2\frac{f(y_k)}{f(x_k)} + O\left(f^2(x_k)\right) = O\left(f(x_k)\right) \tag{1.98}$$

and

$$1 - t = O\left(f(x_{k+m})\right) = O\left(f(z_k)\right) = O\left(f^3(x_k)\right), \tag{1.99}$$

which follows from (1.56), then the Eq. (1.92b) with $\theta_k = f(z_k)/f(y_k)$ holds within the accuracy $O\left(f^4(x_k)\right)$.

If we wish to include the precise correction to (1.97), one can replace $1 - f'(y_k)/f'(x_k)$ by $2f(y_k)/f(x_k)$, then we arrive at

$$\left(\theta_k - 2\frac{f(y_k)}{f(x_k)}\right) t^2 + \left(2\frac{f(y_k)}{f(x_k)} - 1\right) t + 1 = 0. \tag{1.100}$$

By virtue of (1.98) and (1.99), Eq. (1.100) approximates Eq. (1.97) with accuracy $O\left(f^5(x_k)\right)$.

With respect to the iteration (1.72), Eq. (1.92b) remains true with $\theta_k = f(z_k)f(y_k)$.

Note that in most cases the value of the iteration parameter of the damped Newton's method varies from zero to unit, whereas in our case the value of the optimal parameter may be greater than unit.

References

1. T. Zhanlav, O. Chuluunbaatar, Convergence of the continuous analogy of Newton's method for solving nonlinear equations. Numer. Methods Program. **10**, 402–407 (2009)
2. M.A. Hernandez, M.A. Salanova, Modification of the Kantorovich assumptions for semilocal convergence of the Chebyshev method. J. Comput. Appl. Math. **126**, 131–143 (2000)
3. P. Deuflhard, *Newton Method for Nonlinear Problems. Affine Invariance and Adaptive Algorithms* (Springer International, 2004)
4. T. Zhanlav, I.V. Puzynin, The convergence of iteration based on a continuous analogue of Newton's method. Comput. Math. Math. Phys. **32**, 729–737 (1992)
5. M.K. Gavurin, Nonlinear functional equations and continuous analogues of iterative methods. Izv. Vyssh. Uchebn. Zaved. Mat. **5**(6), 18–31 (1958). (in Russian)
6. L.V. Kantorovich, Functional analysis and applied mathematics. Usp. Mat. Nauk **3**(6)(28), 89–185 (1948). (in Russian)
7. J. Stoer, R. Bulirsch, *Introduction to Numerical Analysis*, 3rd edn. (Springer, 2002)
8. E. Zeidler, *Nonlinear Functional Analysis and its Applications, Part 1: Fixed-Point Theorems* (Springer-Verlag, 1986)
9. T. Zhanlav, O. Chuluunbaatar, The local and global convergence of the continuous analogy of Newton's method. Bulletin of PFUR. Series Mathematics, Information Sciences and Physics **1**, 34–43 (2012)
10. R.S. Dembo, S.C. Eisenstat, T. Steihang, Inexact Newton methods. SIAM J. Numer. Anal. **19**, 400–408 (1982)
11. E. Cătinaş, The inexact, inexact perturbed, and quasi-Newton methods are equivalent models. Math. Comput. **74**, 291–301 (2004)
12. H.-B. An, Z.-Y. Mo, X.-P. Liu, A choice of forcing terms in inexact Newton method. J. Comput. Appl. Math. **200**, 47–60 (2007)
13. X.-C. Cai, W.D. Gropp, D.E. Keyes, M.D. Tidriti, Newton-Krylov-Schwarz methods in CFD, in *Proceeding of the International Workshop on Numerical Methods for the Navier-Stokes Equations* (Vieweg, Braunschwieg, 1995), pp. 17–30
14. S.C. Eisenstat, M.F. Walker, Choosing the forcing term in an inexact Newton method. SIAM J. Sci. Comput. **17**, 16–32 (1996)
15. B.M. Podlevs'kii, On certain two-sided analogues of Newton's method for solving nonlinear eigenvalue problems. Comput. Math. Math. Phys. **47**, 1745–1755 (2007)
16. T. Zhanlav, O. Chuluunbaatar, V. Ulziibayar, Two-sided approximation for some Newton's type methods. Appl. Math. Comput. **236**, 239–246 (2014)
17. T. Zhanlav, D. Hongorzul, The behavior of the convergence of the combined iteration method for solving nonlinear equations. Zh. Vychisl. Mat. Mat. Fiz. **52**, 790–800 (2012)
18. J.A. Ezquerro, M.A. Hernandez, N. Romero, A.I. Velasco, Improving the domain of starting points for secant-like methods. Appl. Math. Comput. **219**, 3677–3692 (2012)
19. W. Bi, H. Ren, Q. Wu, Three-step iterative methods with eighth-order convergence for solving nonlinear equations. J. Comput. Appl. Math. **225**, 105–112 (2009)
20. A. Cordero, J.L. Hueso, E. Martínez, J.R. Torregrosa, New modifications of Potra-Pták's method with optimal fourth and eighth orders of convergence. J. Comput. Appl. Math. **234**, 2969–2976 (2010)

21. A. Cordero, J.R. Torregrosa, M.P. Vassileva, Three-step iterative methods with optimal eighth-order convergence. J. Comput. Appl. Math. **235**, 3189–3194 (2011)
22. L. Fang, G. He, Some modifications of Newton's method with higher-order convergence for solving nonlinear equations. J. Comput. Appl. Math. **228**, 296–303 (2009)
23. J.R. Sharma, H. Arora, On efficient weighted-Newton methods for solving systems of nonlinear equations. Appl. Math. Comput. **222**, 497–506 (2013)
24. H.I. Siyyam, M.T. Shatnawi, I.A. Al-Subaihi, A new one parameter family of iterative methods with eighth-order of convergence for solving nonlinear equations. Int. J. Pure Appl. Math. **84**, 451–461 (2013)
25. R. Thukral, M.S. Petković, A family of three-point methods of optimal order for solving nonlinear equations. J. Comput. Appl. Math. **233**, 2278–2284 (2010)
26. X. Wang, T. Zhang, A new family of Newton-type iterative methods with and without memory for solving nonlinear equations. Calcolo **51**, 1–15 (2014)
27. S. Weerakoon, T.G.I. Fernando, A variant of Newton's method with accelerated third-order convergence. Appl. Math. Lett. **13**(8), 87–93 (2000)
28. T. Zhanlav, O. Chuluunbaatar, Some iteration methods with high order convergence for nonlinear equations. Bulletin of PFUR. Series Mathematics, Information Sciences and Physics **4**, 47–55 (2009)

Chapter 2
Two-Sided Approximations

Abstract Monotone approximations are interesting from not only a theoretical but also a practical point of view. In particular, two-sided approximations can be efficiently used as a posteriori estimations for the desired solution, which means that one can control the error at each iteration step. The iteration parameter in Newton-type methods plays an essential role in the convergence. It can expand the convergence domain and control the convergence behaviour. In particular, we established that in some cases τ_k from the interval $(0,1)$ enables the monotonicity of iterations, and the τ-region of two-sided convergence is completely included in the interval $(1,2)$.

2.1 On Convergence Behavior of Combined Iterative Methods for Solving Nonlinear Equations

We consider the following nonlinear equation:

$$f(x) = 0, \tag{2.1}$$

where $f : R \to R$ twice continuously differentiable function. Let x^* be isolated simple root of the Eq. (2.1). As known [1, 2] for finding the root of the Eq. (2.1) are widely used method Newton and its various modifications. In [3] was proposed the iteration

$$x_{k+1} = \tilde{x}_k - \tau_k \frac{f(\tilde{x}_k)}{f'(\tilde{x}_k)}, \tag{2.2a}$$

$$\tilde{x}_{k+1} = x_{k+1} - \frac{f(x_{k+1})}{f'(x_{k+1})}, \quad k = 0, 1, \dots \tag{2.2b}$$

to find a solution to the Eq. (2.1). Here $\tilde{x}_0$ is the initial approximation. Separately, the formula (2.2a) is a continuous analog of Newton's method (CANM) and (2.2b) is Newton's method. In general, the iterative method (2.2) is combined iterative method CANM. In [3] proved that if the iterative method (2.2) converges, then it converges quadratically for $\tau_k \neq 1$. The parameter τ_k in (2.2) plays essential role for

T. Zhanlav and O. Chuluunbaatar, *New Developments of Newton-Type Iterations for Solving Nonlinear Problems*, Mathematical Engineering, https://doi.org/10.1007/978-3-031-63361-4_2

convergence. The optimal choice of the parameter τ_k in (2.2a) not only expands the convergence domain [4], but also controls the rate of convergence.

According to Theorem 1.9 the iterative method (2.2) converges for any $x_0 \in \Omega_0$, if

$$\tau_k \in I_k = \left(0, \frac{-1 + \sqrt{1 + 4a_k}}{a_k}\right) \subseteq (0, 2), \quad k = 0, 1, \ldots. \tag{2.3}$$

We will show that using τ_k you can control the behavior of the convergence of method (2.2).

Let the condition be fulfilled:

$$a_k < \frac{1}{2} \quad \text{for all} \quad k = 0, 1, \ldots, \tag{2.4}$$

then $I_k \subseteq (0, 2)$ holds true.

2.1.1 Monotone Convergence of Method

If we introduce the notation $x_{2m} := \tilde{x}_m$, $x_{2m-1} := x_m$ in (2.2), then the iterative method (2.2) is written as:

$$x_{2m+1} = x_{2m} - \tau_{2m} \frac{f(x_{2m})}{f'(x_{2m})}, \tag{2.5a}$$

$$x_{2m+2} = x_{2m+1} - \frac{f(x_{2m+1})}{f'(x_{2m+1})}, \quad m = 0, 1, \ldots. \tag{2.5b}$$

Let the Assumption 1.1 be true. Considering (2.5b) in the Taylor expansion

$$f(x_{2m+2}) = f(x_{2m+1}) + f'(x_{2m+1})(x_{2m+2} - x_{2m+1}) + \frac{f''(\eta_{2m})}{2}(x_{2m+2} - x_{2m+1})^2,$$

where $\eta_{2m} = \theta x_{2m+1} + (1 - \theta)x_{2m+2}$, $\theta \in (0, 1)$, we have

$$f(x_{2m+2}) = \frac{f''(\eta_{2m})}{2}\left(\frac{f(x_{2m+1})}{f'(x_{2m+1})}\right)^2, \quad m = 0, 1, \ldots. \tag{2.6}$$

From (2.6) it follows that

$$f(x_{2m+2}) \geq 0, \quad m = 0, 1, \ldots \text{ if } f''(x) > 0, \tag{2.7a}$$

$$f(x_{2m+2}) \leq 0, \quad m = 0, 1, \ldots \text{ if } f''(x) < 0. \tag{2.7b}$$

Note that in the derivation of the formula (2.6), only the condition $f'(x) \neq 0$, and the condition of the conservation of the sign $f''(x)$ used in deriving inequality (2.7).

Similarly, if we take (2.5a) into account the Taylor expansion

$$f(x_{2m+1}) = f(x_{2m}) + f'(x_{2m})(x_{2m+1} - x_{2m}) + \frac{f''(\xi_{2m})}{2}(x_{2m+1} - x_{2m})^2,$$

where $\xi_{2m} = \alpha x_{2m} + (1 - \alpha)x_{2m+1}$, $\alpha \in (0, 1)$, we have

$$f(x_{2m+1}) = (1 - \tau_{2m})f(x_{2m}) + \frac{f''(\xi_{2m})}{2}\left(\frac{f(x_{2m})}{f'(x_{2m})}\right)^2 \tau_{2m}^2, \quad m = 0, 1, \ldots.$$

$$(2.8)$$

Taking into account the (2.7a) and the conditions of Theorem 1.9, from (2.8) directly follow:

$$(1 - \tau_{2m})f(x_{2m}) \le f(x_{2m+1}) \le \left(1 - \tau_{2m} + \frac{a_{2m}}{2}\tau_{2m}^2\right)f(x_{2m}). \qquad (2.9)$$

Note that the left inequality in (2.9) is holds for all $m = 0, 1, \ldots$, at that time, the right inequality in (2.9) true for all $m = 1, 2, \ldots$. And if $f(x_0) > 0$, then it true for all $m = 0, 1, 2, \ldots$. The following theorem holds.

Theorem 2.1 *Let the condition (2.4) and Assumption 1.1 be fulfilled, and the parameter τ_{2m} be chosen from half interval $(0, 1]$, i.e.*

$$0 < \tau_{2m} \le 1, \quad m = 0, 1, \ldots.$$

Then the iterative method (2.5) converges monotonically to the solution x^* of the Eq. (2.1) for any initial approximation taken from the neighborhood $U_r(x^*)$.

Proof Let $f''(x) > 0$. According to (2.6), we have $f(x_{2m}) \ge 0$ for all $m = 1, 2, \ldots$, then from left inequality (2.9) implies that

$$f(x_{2m+1}) \ge 0, \quad m = 1, 2, \ldots.$$

Thus, all approximations starting from x_2 are one by one side of solution x^*. Further, the equality (2.6) implies:

$$0 \le f(x_{2m+2}) \le a_{2m+1}f(x_{2m+1}) < \frac{1}{2}f(x_{2m+1}), \quad m = 1, 2, \ldots. \qquad (2.10)$$

This means that x_{2m+2} is closer to x^* compared to x_{2m+1}, i.e., either $x_{2m+1} < x_{2m+2}$, or $x_{2m+2} < x_{2m+1}$. On the other hand, a positive function $1 - \tau_{2m} + a_{2m}\tau_{2m}^2/2$ has a minimum at the point $\tau^* = a_{2m}^{-1} > 2$, and therefore

$$\sup_{\tau \in (0, 1]} \left(1 - \tau + \frac{a_{2m}}{2}\tau^2\right) = q \le 1.$$

Then the right inequality in (2.9) implies the inequality

$$f(x_{2m+1}) < qf(x_{2m}), \quad m = 1, 2, \ldots, \tag{2.11}$$

which also confirms the conclusion about the proximity of x_{2m+1} to the root of x^*, than x_{2m}. From inequalities (2.10), (2.11) directly follow:

$$0 \le f(x_{2m+2}) < \left(\frac{q}{2}\right)^m f(x_2), \quad m = 1, 2, \ldots,$$

$$0 \le f(x_{2m+1}) < \left(\frac{q}{2}\right)^{m-1} qf(x_2), \quad m = 1, 2, \ldots.$$

Passing to the limit $m \to \infty$ in the last inequalities, we get $\lim_{m\to\infty} x_m = x^*$. Suppose $f(x_0) > 0$. Then from (2.9), (2.10) it follows that

$$0 \le f(x_1) \le qf(x_0) \quad \Rightarrow \quad f(x_2) \le \frac{1}{2}f(x_1) \le \frac{q}{2}f(x_0). \tag{2.12}$$

It follows that $x_2 < x_1 < x_0$, or $x_0 < x_1 < x_2$, and from (2.10), (2.11) and (2.12) follows:

$$x^* < \cdots < x_{2m+2} < x_{2m+1} < x_{2m} < \cdots < x_3 < x_2 < x_1 < x_0$$

or

$$x_0 < x_1 < x_2 < x_3 < \cdots < x_{2m} < x_{2m+1} < x_{2m+2} < \cdots < x^*,$$

i.e., the sequence $\{x_k\}$ monotonically converges to x^*. If $f(x_0) < 0$, then from (2.10), (2.11) it follows that

$$x^* < \cdots < x_{2m+2} < x_{2m+1} < x_{2m} < \cdots < x_3 < x_2$$

or

$$x_2 < x_3 < \cdots < x_{2m} < x_{2m+1} < x_{2m+2} < \cdots < x^*,$$

i.e., sequence $\{x_k\}$ starting from $k = 2$ monotone converges to x^*. Now let $f''(x) < 0$. According to (2.6), we have $f(x_{2m}) \le 0$ for all $m = 1, 2, \ldots$, then from the right inequalities (2.9) implies that

$$f(x_{2m+1}) \le 0, \quad m = 1, 2, \ldots.$$

Hence, if we introduce the notation $g(x) = -f(x)$, then it is easy to see that for $g(x)$ is repeated literally the above considered proof. If $\tau_{2m} \equiv 1$, then (2.5) turns into Newton's method, which converges monotonically in a small neighborhood of x^* see [5, 6]. $\square$

Remark 2.1 If $\sum_{k=0}^{\infty} \tau_k = const$, then according to the necessary criterion of convergence of the series must be $\tau_k \to 0$ with $k \to \infty$. It is known that CANM does not allow small parameter τ_k and usually set the lower threshold $\tau \geq \tau_0 \approx 0.1$ [4].

2.1.2 Two-Sided Monotone Approximations

Two-sided monotone approximations are attractive because they simultaneously give an estimate of the error. From (2.6), (2.7) it is clear that in order to construct such approximations, it suffices to find the conditions on τ_{2m} for which the following inequalities hold

$$f(x_{2m+1}) \leq 0, \quad m = 1, 2, \ldots, \text{ if } f''(x) > 0, \tag{2.13a}$$
$$f(x_{2m+1}) \geq 0, \quad m = 1, 2, \ldots, \text{ if } f''(x) < 0. \tag{2.13b}$$

Lemma 2.1 *Let the Assumption 1.1 be fulfilled. Then the necessary condition for the inequalities (2.13a), (2.13b) to be holds is*

$$\tau_{2m} > 1.$$

Proof Consider the case of $f''(x) > 0$. Suppose the contrary, i.e., $\tau_{2m} \leq 1$, then from (2.9) taking into account (2.7a), it follows that $f(x_{2m+1}) \geq 0$, which contradicts (2.13a). Similarly, it is proved statement of the lemma in the case when $f''(x) < 0$. $\square$

Lemma 2.2 *Suppose that the Assumption 1.1 and condition (2.4) are fulfilled. Then a sufficient condition for the inequalities (2.13) to be holds, is*

$$\tau_{2m} \in \left[\frac{1 - \sqrt{1 - 2a_{2m}}}{a_{2m}}, \frac{1 + \sqrt{1 - 2a_{2m}}}{a_{2m}} \right]. \tag{2.14}$$

Proof Consider the case of $f''(x) > 0$. Then the inequality (2.7a) holds. Therefore, $1 - \tau_{2m} + a_m \tau_{2m}^2 / 2 < 0$ under (2.14). Hence from the right inequality in (2.9) implies (2.13a). Let $f''(x) < 0$, then (2.7b) holds. When selecting (2.14) we have $1 - \tau_{2m} < 0$. Therefore, from the left inequality in (2.9) follows (2.13b). $\square$

On the other hand, the parameter τ_{2m} in (2.5) should be in the τ-region of convergence (2.3). Therefore, we must choose τ_{2m} from the intersection of the domains (2.3) and (2.14), i.e.,

$$\tau_{2m} \in I_{2m} = \left[\frac{1 - \sqrt{1 - 2a_{2m}}}{a_{2m}}, \frac{-1 + \sqrt{1 + 4a_{2m}}}{a_{2m}} \right) \subseteq (1, 2), \tag{2.15}$$

which is non-empty if

$$a_{2m} < \frac{4}{9}, \quad m = 0, 1, \ldots . \tag{2.16}$$

Interval I_{2m} is called τ-region of two-sided monotone convergence, which is completely included in the interval $(1; 2)$.

Remark 2.2 In (2.15) as a_k, in particular, we can take

$$a_k = \frac{k|f(x_k)|}{(f'(x_k))^2}, \quad k = \sup_{x \in U_r(x^*)} |f''(x)|. \tag{2.17}$$

Theorem 2.2 *Suppose that Assumption 1.1 and conditions (2.15) and (2.16) are fulfilled. Then the iterative method (2.5) converges to the solution x^* for any initial approximation $x_0 \in U_r(x^*)$. Moreover, all approximations $\{x_{2m}\}$ and $\{x_{2m+1}\}$ are on opposite sides of the solution x^*, and starting from $m = 1$, they form monotone sequences, i.e., takes place*

$$x_3 < x_5 < \cdots < x_{2m-1} < \cdots < x^* < \cdots < x_{2m} < \cdots < x_4 < x_2 \tag{2.18}$$

or

$$x_2 < x_4 < \cdots < x_{2m} < \cdots < x^* < \cdots < x_{2m-1} < \cdots < x_5 < x_3. \tag{2.19}$$

Proof First of all, we note that using formulas (2.6), (2.9) and conditions (2.15), (2.16) it is easy to prove by induction that all approximations of x_k belong to the neighborhood $U_r(x^*)$. Let $f''(x) > 0$ in a neighborhood of the solution x^* then (2.7a) holds, i.e., $f(x_{2m}) \geq 0$ for all $m = 1, 2, \ldots$. By Lemma 2.2 and by the assumption of the theorem we have $f(x_{2m+1}) \leq 0$ for all $m = 1, 2, \ldots$. This means that all $\{x_{2m}\}$, starting with $m = 1$, are on the same side of the solution x^*, and $\{x_{2m+1}\}$ starting with $m = 1$ are on the other sides of the solution x^*. If we introduce the notation $g(x_{2m+1}) = -f(x_{2m+1}) \geq 0$, then from (2.6) and (2.9) immediately follow:

$$f(x_{2m}) \leq a_{2m-1} g(x_{2m-1}), \quad m = 1, 2, \ldots \tag{2.20}$$

and

$$-\left(1 - \tau_{2m-2} + \frac{a_{2m-2}}{2}\tau_{2m-2}^2\right) g(x_{2m-2})$$
$$\leq g(x_{2m-1}) \leq (\tau_{2m-2} - 1)g(x_{2m-2}), \quad m = 1, 2, \ldots . \tag{2.21}$$

From here we easily get

$$0 \leq f(x_{2m}) \leq a_{2m-1}(\tau_{2m-2} - 1)f(x_{2m-2}) < \frac{\overline{q}}{2}f(x_{2m-2}), \tag{2.22}$$

$$0 \leq f(x_{2m}) \leq \left(\frac{\overline{q}}{2}\right)^{m-1} f(x_2), \tag{2.23}$$

where

$$\bar{q} = \max_{\tau_{2m} \in I_{2m}} (\tau_{2m} - 1) < 1.$$

From (2.22) it is clear that x_{2m} is closer to x^* than x_{2m-2}. From (2.23) it follows that $\lim_{m \to \infty} x_{2m} = x^*$. Similarly, from (2.20) and (2.21) we get

$$0 \le g(x_{2m+1}) \le \frac{\bar{q}}{2} g(x_{2m-1}) \le \left(\frac{\bar{q}}{2}\right)^m g(x_1), \quad m = 1, 2, \ldots.$$

Hence it is clear that x_{2m+1} is closer to x^* than x_{2m-1} and $\lim_{m \to \infty} x_{2m+1} = x^*$. Therefore, (2.18) and (2.19) are true. $\qquad\square$

The question arises how the situation will change if we replace a_{2m} in (2.15) by $\tilde{a}_{2m} = f''(x_{2m}) f(x_{2m})/(f'(x_{2m}))^2$, which can be easily calculated by compared to (2.17). Note that a_{2m} is a parameter, located in the half-interval $\tilde{a}_{2m} \le a_{2m} < 4/9$. As a result, the width of the region I_{2m} also changes. The function $(-1 + \sqrt{1 + 4x})/x$ is decreasing, and the function $(1 - \sqrt{1 - 2x})/x$ is increasing. Therefore, the interval I_{2m} will be the widest when $a_{2m} = \tilde{a}_{2m}$. We denote the resulting interval by $\tilde{I}_{2m}$, i.e.

$$\tilde{I}_{2m} = \left[\frac{1 - \sqrt{1 - 2\tilde{a}_{2m}}}{\tilde{a}_{2m}}, \frac{-1 + \sqrt{1 + 4\tilde{a}_{2m}}}{\tilde{a}_{2m}}\right). \tag{2.24}$$

There is an inclusion

$$I_{2m} \subseteq \tilde{I}_{2m} \subseteq (1, 2).$$

Now we consider the specific choices of the parameter τ_{2m} in (2.5). The natural choice of parameter is

$$\tau_{2m}^{opt} = \frac{1 - \sqrt{1 - 2\tilde{a}_{2m}}}{\tilde{a}_{2m}} > 1, \tag{2.25}$$

which tends to unity when $\tilde{a}_{2m} \to 0$. Using the expansion

$$(1 + x)^\alpha = 1 + \alpha x + \frac{\alpha(\alpha - 1)}{2}x^2 + \frac{\alpha(\alpha - 1)(\alpha - 2)}{3!}x^3 + \cdots$$

we can write τ_{2m}^{opt}, given by the formula (2.25), in the form

$$\tau_{2m}^{opt} = 1 + \frac{\tilde{a}_{2m}}{2} + \frac{\tilde{a}_{2m}^2}{3} + \cdots \tag{2.26}$$

two-sided approximations of the form were constructed in [7]

$$x_{2m+1} = x_{2m} - \frac{f'(x_{2m})f(x_{2m})}{(f'(x_{2m}))^2 - f(x_{2m})f''(x_{2m})},$$

$$x_{2m+2} = x_{2m+1} - \frac{f(x_{2m+1})}{f'(x_{2m+1})}, \quad m = 0, 1, \ldots. \tag{2.27}$$

The iteration process (2.27) is a special case of iteration (2.5) with parameter

$$\tau_{2m}^* = \frac{1}{1 - \tilde{a}_{2m}} \in (1, 2). \tag{2.28}$$

It is easy to verify that $\tau_{2m}^* \in \tilde{I}_{2m}$. As seen from (2.28), the possible choice is the following

$$\tau_{2m}^{**} = 1 + \alpha \tilde{a}_{2m}, \quad \alpha > 0. \tag{2.29}$$

It is easy to see that $\tau_{2m}^{**} \in \tilde{I}_{2m}$ provided that

$$0 < \tilde{a}_{2m} < \frac{-1 + \sqrt{2\alpha}}{\alpha}, \quad \frac{1}{2} < \alpha < 2.$$

It is obvious that τ_{2m}^{**} with $\alpha = 1/2$ is not included in $\tilde{I}_{2m}$. Iterative process (2.5) with parameter (2.29) and $\alpha = 1/2$ becomes the famous Chebyshev method. So, the combination of the Chebyshev method with the Newton method does not give two-sided approximations. Unlike [7], we found the whole domain $\tilde{I}_{2m}$ of monotone two-sided convergence of the iteration (2.5). In particular, we propose concrete elections. of parameter τ from the domain $\tilde{I}_{2m}$. It should be noted that for $\tau_k \equiv 1$ the iteration process (2.5) leads to a known iteration process, which has the fourth order of convergence [3]. Therefore, for $\tau_m \to 1$ the order of convergence of method (2.5) increases from two to fourth. From the convergence speed of view, the most appropriate choice is (2.25), which is close to unity compared to all possible choices (2.24), including, (2.28), (2.29), i.e.,

$$1 < \tau_{2m}^{opt} \leq \tau_{2m}, \quad \tau_{2m} \in \tilde{I}_{2m}.$$

This is evident from (2.26), (2.28).

2.1.3 One- and Two-Sided Approximations for Systems of Nonlinear Equations

Let $F : D \subseteq R^n \to R^n$ be a twice continuously differentiable function of n variables. Let the second derivative $F''(x)$ is bounded in norm, i.e.,

$$\|F''(x)\| \leq K \quad x \in D.$$

In [8] proved the monotonicity of Newton's method provided that

$$(F'(z))^{-1} \geq 0, \quad z \in D, \tag{2.30}$$

which is understood by the elements, i.e., every element of the inverse matrix $(F'(z))^{-1}$ is non-negative. Suppose that one of the following three equivalent inequalities [9] holds

$$F(\lambda x + (1 - \lambda)y) \leq \lambda F(x) + (1 - \lambda)F(y), \tag{2.31}$$

$$F(y) - F(x) \geq F'(x)(y - x), \tag{2.32}$$

$$(F'(y) - F'(x))(y - x) \geq 0, \tag{2.33}$$

which are understood component wise by component. The combination of conditions (2.31)–(2.33) with (2.30) also gives three equivalent affine-covariant conditions [9]:

$$(F'(z))^{-1}F(\lambda x + (1 - \lambda)y) \leq (F'(z))^{-1}(\lambda F(x) + (1 - \lambda)F(y)), \tag{2.34}$$

$$(F'(x))^{-1}(F(y) - F(x)) \geq y - x, \tag{2.35}$$

$$(F'(z))^{-1}(F'(y) - F'(x))(y - x) \geq 0. \tag{2.36}$$

The iteration process (2.5) for systems of nonlinear equations looks like:

$$x_{2m+1} = x_{2m} - \tau_{2m}(F'(x_{2m}))^{-1}F(x_{2m}), \tag{2.37a}$$

$$x_{2m+2} = x_{2m+1} - (F'(x_{2m+1}))^{-1}F(x_{2m+1}), \quad m = 0, 1 \dots . \tag{2.37b}$$

It is known that (2.30) is necessary and sufficient condition [1, 5] for the monotonicity of the matrix $F'(z)$, i.e., from $F'(z)\upsilon \geq 0$ follows $\upsilon \geq 0$, and from $F'(z)\upsilon \leq 0$ follows $\upsilon \leq 0$. Thus, the matrix $F'(z)$ is a monotone type for all $z \in D$.

Theorem 2.3 *Let one of the equivalent inequalities (2.34)–(2.36) be holds and* $0 < \tau_{2m} \leq 1$ *in (2.37). Then the iterative process (2.37) converges monotonically, i.e., converges to* x^* *on one side.*

Proof Let the condition (2.35) be satisfied, then we have

$$(F'(x_{2m-1}))^{-1}(F(x_{2m}) - F(x_{2m-1})) \geq x_{2m} - x_{2m-1},$$

which is equivalent to

$$F(x_{2m}) - F(x_{2m-1}) \geq F'(x_{2m-1})(x_{2m} - x_{2m-1}) = -F(x_{2m-1}).$$

It follows that

$$F(x_{2m}) \geq 0, \quad m = 1, 2, \dots . \tag{2.38}$$

From (2.37a), (2.38) it follows that

$$F'(x_{2m})(x_{2m+1} - x_{2m}) = -\tau_{2m} F(x_{2m}) \leq 0.$$

Since $F'(z)$ is a monotone type matrix for all $z \in D$, including $z = x_{2m}$, then the last inequality implies:

$$x_{2m+1} \leq x_{2m}, \quad m = 1, 2, \ldots. \tag{2.39}$$

According to (2.37b) and (2.35) we have

$$x_{2m} - x^* = x_{2m-1} - (F'(x_{2m-1}))^{-1}[F(x_{2m-1}) - F(x^*)] - x^* \geq$$
$$\geq x_{2m-1} - x_{2m-1} + x^* - x^* = 0,$$

i.e.,

$$x^* \leq x_{2m}. \tag{2.40}$$

Similarly, from (2.37a) and (2.35) we have

$$x_{2m+1} - x^* = x_{2m} - x^* + \tau_{2m}(F'(x_{2m}))^{-1}[F(x^*) - F(x_{2m})] \geq$$
$$\geq x_{2m} - x^* + \tau_{2m}(x^* - x_{2m}) = (1 - \tau_{2m})(x_{2m} - x^*).$$

Then it is clear that

$$x^* \leq x_{2m+1} \quad \text{for} \quad 0 < \tau_{2m} \leq 1.$$

On the other hand, we have

$$F(x_{2m+1}) - F(x_{2m}) \geq F'(x_{2m})(x_{2m+1} - x_{2m}) = -\tau_{2m} F(x_{2m})$$

and

$$F(x_{2m+1}) \geq (1 - \tau_{2m}) F(x_{2m}). \tag{2.41}$$

From here we have

$$F(x_{2m+1}) \geq 0, \quad m = 1, 2, \ldots.$$

Then from (2.37b), in turn, it follows that $x_{2m+2} \leq x_{2m+1}$. As a result, we have

$$x^* \leq \cdots \leq x_{2m+2} \leq x_{2m+1} \leq x_{2m} \leq \cdots \leq x_3 \leq x_2,$$

i.e., the iterative process converges to the solution x^* of the equation $F(x) = 0$ monotonically from right side. $\square$

Lemma 2.3 *Let all the conditions of Theorem 2.3 be satisfied, then $(F'(x_{2m}))^{-1}$ is a matrix of monotone-type.*

Proof Indeed, according to (2.37a), we have

$$x_{2m+1} - x_{2m} = -\tau_{2m}(F'(x_{2m}))^{-1}F(x_{2m}).$$

Hence, according to (2.39), we have

$$(F'(x_{2m}))^{-1}F(x_{2m}) \geq 0.$$

On the other hand, $F(x_{2m}) \geq 0$, which means monotony matrices $\Gamma_{2m} = (F'(x_{2m}))^{-1}$.
$\square$

Lemma 2.4 *Let the conditions (2.30) and (2.32) be fulfilled. Then F is a monotone operator.*

Proof Previously, we showed that the matrix $F'(z)$ is monotone type under condition (2.30). Let $F(y) - F(x) \leq 0$, then it follows from (2.32) that $y - x \leq 0$ due to the monotonicity of $F'(x)$, that is, from $F(y) \leq F(x)$ follows $y \leq x$. $\square$

Lemma 2.5 *Let one of the inequalities (2.34)–(2.36) holds. Then the condition*

$$\tau_{2m} > 1$$

is necessary for inequality $F(x_{2m+1}) < 0$ to be holds.

Proof The proof follows from (2.38) and (2.41). $\square$

Theorem 2.4 *Let conditions (2.30), and one of the equivalent inequalities (2.31)–(2.33) (or (2.34)–(2.36)) be fulfilled and iteration parameter selected from interval*

$$\tau_{2m} \in I_{2m} = \left[\frac{1 - \sqrt{1 - 2a_{2m}}}{a_{2m}}, \frac{-1 + \sqrt{1 + 4a_{2m}}}{a_{2m}}\right),$$

where $\|F''(x)\|\,\|\Gamma_{2m}\|\,\|\Gamma_{2m}F(x_{2m})\| \leq a_{2m} < 4/9$. Then the iterative process (2.37) produces two-sided approximations, namely

$$x_3 \leq \cdots \leq x_{2m+1} \leq x^* \leq x_{2m+2} \leq x_{2m} \leq \cdots \leq x_2. \tag{2.42}$$

Proof Under the conditions (2.30) and (2.34)–(2.36) above we proved the inequalities (2.38), (2.40), i.e.,

$$x^* \leq x_{2m}; \qquad m = 1, 2, \ldots.$$

Using the Taylor expansion, and taking into account we get (2.37b), gives

$$F(x_{2m+2}) = \frac{F''(\eta)}{2}\left((F'(x_{2m+1}))^{-1}F(x_{2m+1})\right)^2,$$

$$F(x_{2m+1}) = (1 - \tau_{2m})F(x_{2m}) + \frac{F''(\xi)}{2}\tau_{2m}^2(\Gamma_{2m}F(x_{2m}))^2,$$

$$\xi = \theta x_{2m} + (1 - \theta)x_{2m+1}, \quad \eta = \bar{\theta}x_{2m+1} + (1 - \bar{\theta})x_{2m+2}, \quad \theta, \bar{\theta} \in (0, 1).$$

From this it follows that

$$\|F(x_{2m+2})\| \le \frac{a_{2m+1}}{2} \|F(x_{2m+1})\|,$$

$$\|F(x_{2m+1})\| \le \left|1 - \tau_{2m} + \frac{a_{2m}}{2} \tau_{2m}^2\right| \|F(x_{2m})\| < \|F(x_{2m})\|.$$

Thus, we have

$$\|F(x_{2m+2})\| < \frac{1}{4} \|F(x_{2m})\|,$$

which indicates a greater proximity of x_{2m+2} to x^* than x_{2m}, i.e.,

$$x^* \le x_{2m+2} \le x_{2m}, \qquad m = 1, 2, \ldots.$$

This proves the right-hand side of the inequality (2.42). As above, considering the expansion of $F(x_{2m+3})$ at the point x_{2m+2}, it is easy to verify that x_{2m+3} is close to x^* than x_{2m+1}. This proves the left side of the inequality (2.42). $\qquad\square$

Taking the proportionality of values a_{2m} and $\|F(x_{2m})\|$ into account as τ_{2m} can be taken value

$$\tau_{2m} = \frac{2}{1 + \sqrt{1 - 2\|F(x_{2m})\|}}, \qquad m = 0, 1, \ldots,$$

under conditions $\|F(x_{2m})\| \le 0.5$.

2.2 Two-Sided Approximations for Some Newton-Type Methods

Now we consider another way to construct two-sided approximations. Let $a, b \in R$, $a < b$, $f : [a, b] \to R$ and consider the following nonlinear equation

$$f(x) = 0. \tag{2.43}$$

Assume that $f(x) \in C^3[a, b]$, $f'(x) \ne 0, x \in [a, b]$ and Eq. (2.43) has a unique root $x^* \in [a, b]$. For a numerical solution of Eq. (2.43) we propose the following iterations

$$x_{2k+1} = x_{2k} - \tau_k \frac{f(x_{2k})}{f'(x_{2k})}, \tag{2.44}$$

$$x_{2k+2} = x_{2k+1} - \omega_k f(x_{2k+1}), \quad k = 0, 1, \ldots. \tag{2.45}$$

Here $\tau_k > 0$ and ω_k are the iteration parameters to be determined properly. It should be mentioned that the first iteration (2.44) is a continuous analogue of Newton's method (or damped Newton's method), while the second one (2.45) is a simple iteration. In [10] it is shown that the iterations (2.44) and (2.45) with

$$\omega_k = \frac{1}{f'(x_{2k+1})}$$

have a two-sided approximation behavior, and was proved the convergence rate of these iterations is 4 when $\tau_k \to 1$ as $k \to \infty$.

On the other hand, the iterations (2.44) and (2.45) can be considered as simple iterations

$$x_{2k+1} = p(x_{2k}), \quad x_{2k+2} = q(x_{2k+1}), \quad k = 0, 1, \ldots, \tag{2.46}$$

for two equations

$$x - p(x) = 0, \quad x - q(x) = 0,$$

which are equivalent to the above Eq. (2.43) and with functions

$$p(x) = x - \tau \frac{f(x)}{f'(x)}, \quad q(x) = x - \omega f(x). \tag{2.47}$$

Suppose that [10]

$$\left| \frac{f''(x)}{(f'(x))^2} f(x) \right| \le M_2 \left| \frac{f(x)}{(f'(x))^2} \right| \le a(x) < \frac{4}{9}, \quad x \in [a, b], \tag{2.48}$$

where $M_2 = \max_{x \in [a,b]} |f''(x)|$. Then it is easy to show that the function $p(x)$ satisfies

$$0 < p'(x_{2k}) < 1, \quad k = 0, 1, \ldots \tag{2.49}$$

under condition

$$\tau_k \in \left(0, \frac{1}{1 - a_{2k}} \right), \quad a_{2k} = M_2 \left| \frac{f(x_{2k})}{(f'(x_{2k}))^2} \right|.$$

A sufficient condition for $q(x)$ to be decreasing is

$$\omega_k f'(x_{2k+1}) > 1, \quad k = 0, 1, \ldots. \tag{2.50}$$

It should be noted that the conditions (2.49) and (2.50) were used first in [11, 12] for bilateral approximations of Aitken-Steffensen-Hermite type methods.

Using Taylor expansion of $f(x_{2k+2})$ at point x_{2k+1}, and (2.45), we obtain

$$\frac{f(x_{2k+2})}{f(x_{2k+1})} = 1 - \omega_k f'(x_{2k+1}) + \frac{f''(\xi_{2k})}{2} f(x_{2k+1})\omega_k^2, \qquad (2.51)$$

where $\xi_{2k} = \theta x_{2k+2} + (1-\theta)x_{2k+1}$, $\theta \in (0,1)$.

Lemma 2.6 *Suppose that*

$$f''(\xi_{2k})f(x_{2k+1}) < 0, \quad k = 0, 1, \ldots, \qquad (2.52)$$

and the inequality (2.50) holds. Then

$$\frac{f(x_{2k+2})}{f(x_{2k+1})} < 0, \quad k = 0, 1, \ldots. \qquad (2.53)$$

Proof If take into account (2.52) and (2.50), then from formula (2.51) we get

$$\frac{f(x_{2k+2})}{f(x_{2k+1})} < 1 - \omega_k f'(x_{2k+1}) < 0.$$

The Lemma is proved. $\qquad\square$

Analogously, using Taylor expansion of $f(x_{2k+1})$ at point x_{2k}, and (2.44) we obtain

$$\frac{f(x_{2k+1})}{f(x_{2k})} = 1 - \tau_k + \frac{f''(\eta_{2k})}{2}\frac{f(x_{2k})}{(f'(x_{2k}))^2}\tau_k^2, \quad k = 0, 1, \ldots, \qquad (2.54)$$

where $\eta_{2k} = \alpha x_{2k+1} + (1-\alpha)x_{2k}$, $\alpha \in (0,1)$.

Lemma 2.7 *Suppose that the inequality (2.48) holds. Then*

$$\frac{f(x_{2k+1})}{f(x_{2k})} < 0, \quad k = 0, 1, \ldots \qquad (2.55)$$

under condition

$$\tau_k \in I_{2k} = \left[\frac{1 - \sqrt{1 - 2a_{2k}}}{a_{2k}}, \frac{-1 + \sqrt{1 + 4a_{2k}}}{a_{2k}}\right) \subseteq [1, 2).$$

Proof From (2.54) we obtain

$$\frac{f(x_{2k+1})}{f(x_{2k})} < 1 - \tau_k + \frac{a_{2k}}{2}\tau_k^2 \le 0.$$

From this it follows the condition (2.55) holds if

$$\tau_k \in \left[\frac{1 - \sqrt{1 - 2a_{2k}}}{a_{2k}}, \frac{1 + \sqrt{1 - 2a_{2k}}}{a_{2k}}\right].$$

On the other hand, as shown in Sect. 1.1.2, the iteration parameter τ_k must be taken from the τ-region of convergence of iteration (2.44)

$$\tau_k \in \left(0, \frac{-1 + \sqrt{1 + 4a_{2k}}}{a_{2k}}\right) \subseteq (0, 2).$$

Hence, the inequality (2.55) is valid for $\tau_k \in I_{2k}$, and I_{2k} is not an empty interval because of (2.48). The Lemma is proved. $\qquad\square$

We obtained the following results when $f(x)$ is increasing and convex on the interval $x \in [a, b]$.

Theorem 2.5 *Let* $x_0 \in (x^*, b]$, *and* $f(x)$ *satisfies the following conditions:*

(i_1) $f'(x) > 0$, $x \in [a, b]$,
(ii_1) $f''(x) > 0$, $x \in [a, b]$,
(iii_1) *the inequality (2.48) holds.*

If the parameters τ_k *and* ω_k *are chosen such that*

$$\tau_k \in \left(0, \frac{1}{1 - a_{2k}}\right) \cap I_{2k}, \quad k = 0, 1, \ldots, \tag{2.56}$$

and

$$\omega_k f'(a) \geq 1, \quad k = 0, 1, \ldots, \tag{2.57}$$

then the following relations hold:

(j_1) $x_1 < x_3 < \cdots < x_{2k+1} < x^* < x_{2k} < \cdots < x_2 < x_0,$
(jj_1) $\lim_{k \to \infty} x_{2k+1} = \lim_{k \to \infty} x_{2k} = x^*.$

Proof By (i_1) it follows that $x^* \in (a, b)$ is the unique solution of Eq. (2.43). From (2.49), (2.50) the functions $p(x)$ and $q(x)$ have no extremum on the own domain of definition, and $p(a) > a$, $p(b) < b$, $q(a) > a$, $q(b) < b$ because of $f(a) < 0$ and $f(b) > 0$. Therefore, all the approximations generated by (2.44) and (2.45) belong to $[a, b]$, i.e., $x_{2k+1}, x_{2k+2} \in (a, b)$, $k = 0, 1, \ldots$. By assumption of theorem $f(x_0) > 0$. Then, according to Lemma 2.7, from (2.55) it follows that $f(x_1) < 0$. By (ii_1), $f'(x)$ is increasing on the interval $[a, b]$, i.e.,

$$0 < f'(a) < f'(x) < f'(b), \quad x \in (a, b).$$

Therefore,

$$\frac{1}{f'(a)} > \frac{1}{f'(x)} > \frac{1}{f'(b)}, \quad x \in (a, b).$$

According to (2.57), we have

$$\omega_0 \geq \frac{1}{f'(a)} > \frac{1}{f'(x)}, \quad x \in (a, b),$$

which holds, for instance, for x_1, i.e., the condition (2.50) is fulfilled for $k = 0$. By virtue of (ii_1) and $f(x_1) < 0$, the assumption (2.52) is valid for $k = 0$. Then, according to Lemma 2.6, from (2.53) we obtain $f(x_2) > 0$. By induction on k from (2.53) and (2.55) one can show that

$$f(x_{2k}) > 0, \quad f(x_{2k+1}) < 0,$$

and also can prove (2.49) and (2.50) for all $k = 0, 1, \dots$.

Thus, the sequence $\{x_{2k+1}\}$ generated by (2.44) (or (2.46)) is increasing and the sequence $\{x_{2k+2}\}$ generated by (2.45) (or (2.46)) is decreasing. Consequently we have

$$x_1 < x_3 < \cdots < x_{2k+1} < x^* < x_{2k} < \cdots < x_2 < x_0,$$

i.e., (j_1) is proved. The (jj_1) follows from (j_1) passing to the limit $k \to \infty$. Since $f(x)$ is increasing function on the interval $[a, b]$, then we have

$$f(x_1) < f(x_3) < \cdots < f(x_{2k+1}) < f(x^*) < f(x_{2k}) < \cdots < f(x_2) < f(x_0),$$

and

$$\lim_{k \to \infty} f(x_{2k+1}) = \lim_{k \to \infty} f(x_{2k}) = f(x^*) = 0.$$

It should be pointed out that the intersection of two intervals in (2.56) is not an empty set because of (2.48). The Theorem is proved. □

If $f(x)$ is decreasing and concave on the interval $x \in [a, b]$, instead of Eq. (2.43), we consider the equation

$$-f(x) = 0.$$

The function $-f(x)$ satisfies the conditions of Theorem 2.5. From here we have:

Corollary 2.1 *Let $x_0 \in (x^*, b]$, and $f(x)$ satisfies the following conditions:*

(i_2) $f'(x) < 0, x \in [a, b]$,
(ii_2) $f''(x) < 0, x \in [a, b]$,
(iii_2) *the inequality (2.48) holds.*

If the parameter τ_k is taken from the interval (2.56) and parameter ω_k is chosen such that (2.57), then the relations (j_1) and (jj_1) hold too.

When

$$f''(x)f'(x) < 0, \quad x \in [a, b],$$

we consider instead of iterations (2.44) and (2.45) the following iterations

$$x_{2k+1} = x_{2k} - \omega_k f(x_{2k}), \tag{2.58}$$

$$x_{2k+2} = x_{2k+1} - \tau_k \frac{f(x_{2k+1})}{f'(x_{2k+1})}, \quad k = 0, 1, \ldots. \tag{2.59}$$

As above, the iterations (2.58) and (2.59) can be considered as

$$x_{2k+1} = q(x_{2k}), \quad x_{2k+2} = p(x_{2k+1}),$$

with functions $q(x)$ and $p(x)$ given by (2.47).

Using Taylor expansion of $f(x_{2k+2})$ at point x_{2k+1}, and (2.59), we obtain

$$\frac{f(x_{2k+2})}{f(x_{2k+1})} = 1 - \tau_k + \frac{f''(\xi_{2k+2}^*)}{2} \frac{f(x_{2k+1})}{(f'(x_{2k+1}))^2} \tau_k^2, \tag{2.60}$$

where $\xi_{2k+2}^* = \theta^* x_{2k+2} + (1 - \theta^*) x_{2k+1}, \theta^* \in (0, 1)$.

Lemma 2.8 *Suppose that*

$$f''(\xi_{2k+2}^*) f(x_{2k+1}) < 0, \quad k = 0, 1, \ldots. \tag{2.61}$$

Then the inequality

$$\frac{f(x_{2k+2})}{f(x_{2k+1})} < 0, \quad k = 0, 1, \ldots \tag{2.62}$$

holds for any $\tau_k \geq 1$.

Proof The inequality (2.62) immediately follows from (2.60), if we take into account (2.61) and $\tau_k \geq 1$. The Lemma is proved. $\qquad\square$

Analogously, using Taylor expansion of $f(x_{2k+1})$ at point x_{2k}, and (2.58), we obtain

$$\frac{f(x_{2k+1})}{f(x_{2k})} = 1 - \omega_k f'(x_{2k}) + \frac{f''(\eta_{2k}^*)}{2} f(x_{2k}) \omega_k^2,$$

where $\eta_{2k}^* = \alpha^* x_{2k+1} + (1 - \alpha^*) x_{2k}, \alpha^* \in (0, 1)$.

Lemma 2.9 *Suppose that the inequality (2.48) holds and*

$$f''(\eta_{2k}^*) f(x_{2k}) > 0, \quad k = 0, 1, \ldots.$$

Then

$$\frac{f(x_{2k+1})}{f(x_{2k})} < 0, \quad k = 0, 1, \ldots, \tag{2.63}$$

under condition

$$\omega_k f'(x_{2k}) \in I_{2k}, \quad k = 0, 1, \ldots.$$

Proof The proof of Lemma 2.9 is the same as the proof of Lemma 2.7.   □

Now we are ready to prove the following theorem when $f(x)$ is increasing and concave on the interval $x \in [a, b]$.

Theorem 2.6 *Let $x_0 \in [a, x^*)$, and $f(x)$ satisfies the following conditions*

(i_3) $f'(x) > 0$, $x \in [a, b]$,
(ii_3) $f''(x) < 0$, $x \in [a, b]$,
(iii_3) the inequality (2.48) holds.

If the parameters τ_k and ω_k are chosen such that

$$\tau_k \in \left[1, \frac{1}{1 - a_{2k+1}}\right), \quad a_{2k+1} = M_2 \left|\frac{f(x_{2k+1})}{(f'(x_{2k+1}))^2}\right|, \quad k = 0, 1, \ldots, \tag{2.64}$$

and

$$\omega_k f'(x_{2k}) \in I_{2k}, \quad k = 0, 1, \ldots, \tag{2.65}$$

then the following relations hold:

(j_3) $x_0 < x_2 < \cdots < x_{2k} < x^* < x_{2k+1} < \cdots < x_3 < x_1$,
(jj_3) $\lim_{k\to\infty} x_{2k} = \lim_{k\to\infty} x_{2k+1} = x^*$.

Proof By (i_3) it follows that $x^* \in (a, b)$ is the unique solution of Eq. (2.43) and $f(x_0) < 0$. By (ii_3) and $f(x_0) < 0$ the assumption of Lemma 2.9 is fulfilled for $k = 0$. Then from (2.63) it follows that $f(x_1) > 0$ under condition (2.65). The assumption (2.61) is valid for $k = 1$. Then by Lemma 2.8 the inequality (2.62) is valid for $k = 1$, i.e., $f(x_2) < 0$ under condition (2.64). By induction on k from (2.62) and (2.63) one can prove that

$$f(x_{2k}) < 0, \quad f(x_{2k+1}) > 0,$$

and

$$q'(x_{2k}) = 1 - \omega_k f'(x_{2k}) \leq 0, \quad 0 < p'(x_{2k+1}) < 1.$$

Therefore, we have

$$x_0 < x_2 < \cdots < x_{2k} < x^* < x_{2k+1} < \cdots < x_3 < x_1.$$

The (j_3) is proved. The (jj_3) follows from (j_3) passing to the limit $k \to \infty$. Since $f(x)$ is increasing on $[a, b]$, then from (j_3) it follows that

$$f(x_0) < f(x_2) < \cdots < f(x_{2k}) < f(x^*) < f(x_{2k+1}) < \cdots < f(x_3) < f(x_1)$$

and

$$\lim_{k \to \infty} f(x_{2k}) = \lim_{k \to \infty} f(x_{2k+1}) = f(x^*) = 0.$$

The Theorem is proved. $\square$

If $f(x)$ is decreasing and convex on the interval $x \in [a, b]$, the function $-f(x)$ satisfies the conditions of Theorem 2.6. From here we have:

Corollary 2.2 *Let $x_0 \in [a, x^*)$, and $f(x)$ satisfies the following conditions*

(i_4) $f'(x) < 0$, $x \in [a, b]$,
(ii_4) $f''(x) > 0$, $x \in [a, b]$,
(iii_4) the inequality (2.48) holds.

If the parameter τ_k is taken from the interval (2.64) and parameter ω_k is chosen such that (2.65), then the relations (j_3) and (jj_3) hold, too.

2.2.1 Convergence Order of Proposed Iterations

The convergence order of proposed iterations (2.44), (2.45) and (2.58), (2.59) is given in the following results.

Theorem 2.7 *Assume that $f(x) \in C^4[a, b]$, $f'(x) \neq 0$, $x \in [a, b]$, and there exists a unique solution $x^* \in [a, b]$ of Eq. (2.43). Then the q-convergence (or p-convergence) order of the sequence $\{x_k\}$ generated by iterations (2.44) and (2.45) (or (2.58) and (2.59)) is at least 2, when $\tau_k \to 1$ as $k \to \infty$.*

Proof First we consider the convergence order of iterations (2.44) and (2.45). Let $e_k = x_k - x^*$. From (2.44) and (2.45) it follows that

$$e_{2k+1} = e_{2k} - \tau_k \frac{f(x_{2k}) - f(x^*)}{f'(x_{2k})}, \tag{2.66}$$

$$e_{2k+2} = e_{2k+1} - \omega_k(f(x_{2k+1}) - f(x^*)). \tag{2.67}$$

Using the Taylor expansions for $f(x^*)$ at points x_{2k} and x_{2k+1} in (2.66) and (2.67), respectively, we obtain

$$e_{2k+1} = e_{2k}\left(1 - \tau_k + \tau_k \frac{f''(x_{2k})}{2f'(x_{2k})}e_{2k} - \tau_k \frac{f'''(x_{2k})}{6f'(x_{2k})}e_{2k}^2\right) + O\left(e_{2k}^4\right), \quad (2.68)$$

$$e_{2k+2} = e_{2k+1}\left(1 - \omega_k f'(x_{2k+1}) + \omega_k \frac{f''(x_{2k+1})}{2}e_{2k+1} -\right.$$
$$\left. -\omega_k \frac{f'''(x_{2k+1})}{6}e_{2k+1}^2\right) + O\left(e_{2k+1}^4\right). \tag{2.69}$$

Substituting e_{2k+1} from (2.68) into (2.69), we have

$$e_{2k+2} = (1 - \tau_k)(1 - \omega_k f'(x_{2k+1}))e_{2k}$$
$$+ \left((1 - \tau_k)^2 \omega_k \frac{f''(x_{2k+1})}{2} + \tau_k \frac{f''(x_{2k})}{2f'(x_{2k})}(1 - \omega_k f'(x_{2k+1}))\right)e_{2k}^2$$
$$+ \left(-\tau_k \frac{f'''(x_{2k})}{6f'(x_{2k})}(1 - \omega_k f'(x_{2k+1})) + (1 - \tau_k)\tau_k \omega_k \frac{f''(x_{2k+1})f''(x_{2k})}{2f'(x_{2k})}\right.$$
$$\left. -(1 - \tau_k)^3 \omega_k \frac{f'''(x_{2k+1})}{6}\right)e_{2k}^3 + O\left(e_{2k}^4\right) + O\left(e_{2k+1}^4\right). \tag{2.70}$$

When $\tau_k \to 1$ as $k \to \infty$, from (2.70) we conclude that

$$e_{2k+2} = O\left(e_{2k}^2\right), \quad \text{if } \omega_k = const,$$
$$e_{2k+2} = O\left(e_{2k}^3\right), \quad \text{if } \omega_k - (f'(x_{2k+1}))^{-1} = O\left(e_{2k}\right),$$
$$e_{2k+2} = O\left(e_{2k}^4\right), \quad \text{if } \omega_k = (f'(x_{2k+1}))^{-1}.$$

Using similar calculations for iterations (2.58) and (2.59), we obtain

$$e_{2k+2} = (1 - \tau_k)(1 - \omega_k f'(x_{2k}))e_{2k}$$
$$+ \left((1 - \tau_k)\omega_k \frac{f''(x_{2k})}{2} + \tau_k \frac{f''(x_{2k+1})}{2f'(x_{2k+1})}(1 - \omega_k f'(x_{2k}))^2\right)e_{2k}^2$$
$$+ \left(-\tau_k \frac{f'''(x_{2k+1})}{6f'(x_{2k+1})}(1 - \omega_k f'(x_{2k}))^3 + \tau_k \omega_k \frac{f''(x_{2k+1})f''(x_{2k})}{2f'(x_{2k+1})}(1 - \omega_k f'(x_{2k}))\right.$$
$$\left. -(1 - \tau_k)\omega_k \frac{f'''(x_{2k})}{6}\right)e_{2k}^3 + O\left(e_{2k}^4\right) + O\left(e_{2k+1}^4\right). \tag{2.71}$$

When $\tau_k \to 1$ as $k \to \infty$, from (2.71) we have

$$e_{2k+2} = O\left(e_{2k}^2\right), \quad \text{if } \omega_k = const,$$
$$e_{2k+2} = O\left(e_{2k}^4\right), \quad \text{if } \omega_k - (f'(x_{2k}))^{-1} = O\left(e_{2k}\right),$$
$$e_{2k+2} = O\left(e_{2k}^4\right), \quad \text{if } \omega_k = (f'(x_{2k}))^{-1}.$$

Thus, the q-convergence (or p-convergence) order of iterations (2.44) and (2.45) (or (2.58) and (2.59)) is at least 2, which completes the proof of theorem. $\qquad \square$

From Theorems 2.5–2.7 and Corollaries 2.1, 2.2 it is clear that the best choice of parameters are

$$\tau_k = \left.\frac{1 - \sqrt{1 - 2a_{2k}}}{a_{2k}}\right|_{k\to\infty} \to 1, \quad \omega_k = \frac{1}{f'(x_{2k+1})}, \tag{2.72}$$

for iterations (2.44) and (2.45) and

$$\tau_k = 1, \quad \omega_k = \left.\frac{1 - \sqrt{1 - 2a_{2k}}}{a_{2k}f'(x_{2k})}\right|_{k\to\infty} \to \frac{1}{f'(x_{2k})}. \tag{2.73}$$

for iterations (2.58) and (2.59), respectively.

Remark 2.3 Note that in general cases, it is difficult to find $M_2 = \max\limits_{x\in[a,b]}\left|f''(x)\right|$. Therefore, instead of (2.72), (2.73) we can use the following parameters

$$\tau_k = \left.\frac{1 - \sqrt{1 - 2\tilde{a}_{2k}}}{\tilde{a}_{2k}}\right|_{k\to\infty} \to 1, \quad \omega_k = \frac{1}{f'(x_{2k+1})},$$

$$\tau_k = 1, \quad \omega_k = \left.\frac{1 - \sqrt{1 - 2\tilde{a}_{2k}}}{\tilde{a}_{2k}f'(x_{2k})}\right|_{k\to\infty} \to \frac{1}{f'(x_{2k})},$$

respectively. Here

$$\tilde{a}_{2k} = \left|\frac{f''(x_{2k})f(x_{2k})}{(f'(x_{2k}))^2}\right| \leq a_{2k},$$

and as shown in [10], holds the inclusion

$$I_{2k} \subseteq \tilde{I}_{2k} = \left[\frac{1 - \sqrt{1 - 2\tilde{a}_{2k}}}{\tilde{a}_{2k}}, \frac{-1 + \sqrt{1 + 4\tilde{a}_{2k}}}{\tilde{a}_{2k}}\right) \subseteq [1, 2).$$

References

1. L.V. Kantorovich, G.P. Akilov, *Functional Analysis in Normed Spaces* (Fizmatgiz, Moscow, 1959). (in Russian)
2. M.A. Krasnoselsky, G.M. Vainikko, P.P. Zabreiko, *Approximate Solution of Operator Equations* (Nauka, Moscow, 1969). (in Russian)
3. T. Zhanlav, O. Chuluunbaatar, Some iteration methods with high order convergence for nonlinear equations. Bulletin of PFUR, Series Mathematics, Information Sciences and Physics **4**, 47–55 (2009)
4. T. Zhanlav, I.V. Puzynin, The convergence of iteration based on a continuous analogue of Newton's method. Comput. Math. Math. Phys. **32**, 729–737 (1992)
5. L. Collatz, *Functional Analysis and Computational Mathematics* (Nauka, Moscow, 1969)
6. H.H. Kalitkin, *Numerical Methods* (Nauka, Moscow, 1978). (in Russian)

7. B.M. Podlevs'kii, An approach to the construction of two-sided iteration method for nonlinear equations. Dop. NAN. Ukr. **5**, 37–41 (1998)
8. J.M. Ortega, W.C. Rheinboldt, *Iterative Solution of Nonlinear Equations in Several Variables* (Academic Press, New-York, 1970)
9. P. Deuflhard, *Newton Method for Nonlinear Problems. Affine Invariance and Adaptive Algorithms* (Springer International, 2004)
10. T. Zhanlav, D. Hongorzul, The behavior of the convergence of the combined iteration method for solving nonlinear equations. Zh. Vychisl. Mat. Mat. Fiz. **52**, 790–800 (2012)
11. I. Păvăloiu, E. Cătinaş, Bilateral approximations for some Aitken-Steffensen-Hermite type methods of order three. Appl. Math. Comput. **217**, 5838–5846 (2011)
12. I. Păvăloiu, E. Cătinaş, On a Steffensen-Hermite method of order three. Appl. Math. Comput. **215**, 2663–2672 (2009)

Chapter 3
New Developments and Extensions of Newton-Type Methods

Abstract Two- and three-point iteration methods for solving nonlinear equations are presented. They contain iteration parameters, and the suitable choice of these parameters allows increasing the convergence order. Necessary and sufficient conditions for the convergence of Newton-type iterations are suggested. The convergence order of any existing methods can be established using the proposed sufficient condition. Moreover, these conditions allow constructing new iterations. In particular, we propose a generating function method for constructing new two- and three-point iterations. Based on this approach, we design a new family of optimal two-point iterations that includes well-known methods as special cases. New developments and extensions of some optimal three-point iterations are also driven.

3.1 Necessary and Sufficient Conditions for Convergence of Two- and Three-Point Newton-Type Iterations

In recent years, many modified iterative methods for solving nonlinear equations have been developed to improve the local order of convergence of some methods such as Newton, Ostrowski's or King's methods. The most efficient methods studied on the literature are the optimal eighth-order methods with efficiency index $8^{1/4} \approx 1.682$ see, for example [1–9] and references therein. Also was made attempt to compare them from the convergence behaviour point of view [10, 11]. Many of these methods involve the optimal Ostrowski's or King's methods in the first step and use arbitrary real parameters and weight functions not always to determine. The increased order of convergence is usually achieved at the expense of additional function or derivative evaluations to carry out iterations, which may affect the efficiency of the method.

So, nowadays, obtaining new optimal methods is still important in spite of the fact that there are many optimal high order ones. It is known, that usually the Taylor expansions used to find error equations or to establish the convergence order. But these are cumbersome and lead to tedious calculations. To overcome these difficulties and to find good candidates for weighted functions and parameters in last years often used symbolic computation in programming packages Mathematica and Maple [3, 12].

In this chapter we propose a new technique or acceleration procedure to obtain higher and optimal order methods. We shall show that the optimal choices of parameters involving in considered method allow us to increase the convergence order.

3.1.1 Two-Point Iterative Methods

Let x^* be a simple zero of a real function $f(x) : I \subset \mathbb{R} \to \mathbb{R}$ in an open interval I. We consider two-point iterative method

$$y_k = x_k - \frac{f(x_k)}{f'(x_k)}, \quad x_{k+1} = x_k - \tau_k \frac{f(x_k)}{f'(x_k)}, \quad k = 0, 1, \ldots, \tag{3.1}$$

where τ_k is an iteration parameter. We formulate the following definition of the convergence order of the iteration methods.

Definition 3.1 Let $f(x)$ be a real function with simple root $x^* \in I$ and let $\{x_k\}$ be a sequence of real numbers that converges towards x^*. The order of convergence p is given by

$$|x_{k+1} - x^*| \le M|x_k - x^*|^p, \quad p \in \mathbb{R}^+, \quad 0 < M < \infty. \tag{3.2}$$

It is easy to show that the condition (3.2) is equivalent to

$$|f(x_{k+1})| \le C|f(x_k)|^p, \quad 0 < C < \infty, \tag{3.3}$$

provided that $f'(x) \ne 0$ in the small vicinity of x^*.

In what follows we shall use (3.3) in order to study the convergence of the different iterations. We will find the suitable choice of parameter τ_k to improve the local order of convergence of iteration (3.1). To this end we rewrite (3.1) as

$$x_{k+1} = y_k + (\tau_k - 1)(y_k - x_k).$$

By using Taylor expansion of $f(x_{k+1})$ around y_k, we obtain

$$f(x_{k+1}) = f(y_k) + f'(y_k)(\tau_k - 1)(y_k - x_k) + O\left(\tau_k^2(y_k - x_k)^2\right). \tag{3.4}$$

We approximate $f'(y_k)$ using already computed function values. This can be done by the method of undetermined coefficients, such that

$$f'(y_k) = a_k f(x_k) + b_k f(y_k) + c_k f'(x_k) + O\left(f^2(x_k)\right). \tag{3.5}$$

By using the Taylor expansions of $f(x_k)$, $f'(x_k)$ around y_k, we obtain the following linear system of equations

$$
\begin{aligned}
a_k + b_k &= 0, \\
a_k(x_k - y_k) + c_k &= 1, \\
a_k(x_k - y_k)^2 + 2c_k(x_k - y_k) &= 0,
\end{aligned}
$$

which has a unique solution:

$$
a_k = -\frac{2}{y_k - x_k}, \quad b_k = \frac{2}{y_k - x_k}, \quad c_k = -1. \tag{3.6}
$$

Substituting (3.6) into (3.5), we have

$$
f'(y_k) = A_k + O\left(f^2(x_k)\right), \tag{3.7}
$$

where

$$
A_k = 2\frac{f(y_k) - f(x_k)}{y_k - x_k} - f'(x_k).
$$

Replacing $f'(y_k)$ in (3.4) by (3.7), we get

$$
f(x_{k+1}) = f(x_k)\left(\theta_k + (\tau_k - 1)(2\theta_k - 1)\right) + O\left((\tau_k - 1)^2(y_k - x_k)^2\right), \tag{3.8}
$$
$$
\theta_k = \frac{f(y_k)}{f(x_k)}.
$$

We choose τ_k such that the first term in the right hand side of (3.8) vanishes, i.e.,

$$
\tau_k = \frac{1 - \theta_k}{1 - 2\theta_k}. \tag{3.9}
$$

From (3.1) it clears that

$$
f(y_k) = O\left(f^2(x_k)\right), \quad y_k - x_k = O\left(f(x_k)\right). \tag{3.10}
$$

So $\theta_k = O\left(f(x_k)\right)$ and $\tau_k - 1 = O\left(f(x_k)\right)$. Then from (3.8) it follows that

$$
|f(x_{k+1})| \leq M|f(x_k)|^4,
$$

under condition (3.9). Thus, we obtain optimal fourth-order method (3.1), (3.9). We call the value of parameter defined by (3.9) the optimal one. This optimal choice allows us to increase the convergence order of method (3.1). Using the well known expansion:

$$\frac{1}{1-x} = 1 + x + x^2 + x^3 + \cdots, \quad |x| < 1, \tag{3.11}$$

we write (3.9) as

$$\tau_k = 1 + \theta_k + 2\theta_k^2 + O\left(\theta_k^3\right). \tag{3.12}$$

We can formulate more strong result as follows.

Theorem 3.1 *Assume that the function f is sufficiently differentiable and has a simple zero $x^* \in I$. The initial approximation x_0 is sufficiently close to x^*. Then the iterative method (3.1) has a fourth-order of convergence if and only if the parameter τ_k satisfies (3.12).*

Proof Suppose that τ_k in (3.1) satisfies (3.12). According to (3.12) and (3.10) the right hand side of (3.8) is of order $O\left(f^4(x_k)\right)$ i.e.,

$$f(x_{k+1}) = O\left(f^4(x_k)\right).$$

Conversely, let the iteration (3.1) be a fourth-order convergence. Then from (3.8) it follows that

$$\theta_k + (\tau_k - 1)(2\theta_k - 1) = O\left(f^3(x_k)\right).$$

From last expression we deduce

$$\tau_k = 1 + \frac{\theta_k}{1 - 2\theta_k} + O\left(f^3(x_k)\right) = 1 + \theta_k + 2\theta_k^2 + O\left(\theta_k^3\right).$$

The Theorem is proved. $\qquad\qquad\square$

Method (3.1) uses the values $f(x_k)$, $f(y_k)$, $f'(x_k)$ and has the optimal fourth order of convergence. Its efficiency index is $4^{1/3} \approx 1.587$, which accord with the conjecture of Kung-Truab. The well known fourth-order methods can be rewritten as (3.1). Some of these we list in Table 3.1. It is easy to show that all the parameters τ_k in Table 3.1 satisfy the condition (3.12). The convergence of any existing method can be established using the sufficient convergence criterion (3.12). In our opinion,

Table 3.1 Choices of parameter τ_k

Methods	τ_k
[13]	$1 + \dfrac{\theta_k(1+b\theta_k)}{1+(b-2)\theta_k}$
[14]	$\dfrac{1+\theta_k^2}{1-\theta_k} + \beta\theta_k \dfrac{f^2(x_k)}{f^2(x_k)+(f'(x_k))^2}$
[3]	$1 + \theta_k + 2\theta_k^2 + \theta_k^3$
[15]	$\dfrac{1}{1-\theta_k} + \theta_k^2$

the method (3.1), (3.12) can be represents the class of two-point and fourth-order convergence methods completely.

Analogously we can state the following.

Theorem 3.2 *Assume the assumptions of Theorem 3.1 are fulfilled. Then the iterative method (3.1) has a third-order convergence if and only if the parameter τ_k satisfies*

$$\tau_k = 1 + \theta_k + O\left(\theta_k^2\right). \tag{3.13}$$

Proof Proof is immediately follows from (3.8) and (3.13).   □

It is evident that the iterative method (3.1) has quadratically convergent if $\tau_k \equiv 1 + O\left(\theta_k\right)$. For comparison purposes, consider the iterative method [16]

$$x_{k+1} = \phi(x_k), \tag{3.14}$$

where $\phi(x) = x - a_1\omega_1(x) - a_2\omega_2(x) - a_3\omega_3(x)$,

$$\omega_1(x) = \frac{f(x)}{f'(x)}, \quad \omega_2(x) = \frac{f(x + \beta\omega_1(x))}{f'(x)}, \quad \omega_3(x) = \frac{f(x + \gamma\omega_1(x) + \delta\omega_2(x))}{f'(x)}.$$

In particular, in the case $a_1 = a_2 = 1$, $\beta = -1$, $a_3 = 0$, we can rewrite (3.14) in the form (3.1) with $\tau_k = 1 + \theta_k$. Hence, according to Theorem 3.2, this method converges with the order $p = 3$, which is in agreement with the results obtained in [16] (see Table 9.1 in that paper).

3.1.2 Three-Point Iterative Newton-Type Methods

Now we consider three-point iterative methods

$$y_k = x_k - \frac{f(x_k)}{f'(x_k)}, \quad z_k = x_k - \tau_k\frac{f(x_k)}{f'(x_k)},$$
$$x_{k+1} = y_k + t(z_k - y_k), \quad k = 0, 1, \ldots, \tag{3.15}$$

where τ_k defined by (3.9) and t is iteration parameter to be determined in a suitable way. As preceding subsection we rewrite x_{k+1} in (3.10) as

$$x_{k+1} = z_k + (t - 1)(z_k - y_k).$$

We use Taylor expansion of $f(x) \in \mathbb{C}^2(I)$ around z_k:

$$f(x_{k+1}) = f(z_k) + f'(z_k)(t - 1)(z_k - y_k) + O\left((t - 1)^2(z_k - y_k)^2\right), \tag{3.16}$$

and approximate $f'(z_k)$ using already computed function values as in [17]:

$$f'(z_k) = a_k f(x_k) + b_k f(y_k) + c_k f(z_k) + d_k f'(x_k) + O\left(f^4(x_k)\right), \quad (3.17)$$

where

$$a_k = \frac{\gamma_k(2\gamma_k - 3\omega_k)}{\omega_k(\gamma_k - \omega_k)^2}, \quad b_k = \frac{\omega_k^2}{\gamma_k(\gamma_k - \omega_k)^2}, \quad (3.18)$$

$$c_k = -\frac{2\gamma_k + \omega_k}{\gamma_k \omega_k}, \quad d_k = \frac{\gamma_k}{\omega_k - \gamma_k}, \quad \omega_k = x_k - z_k, \quad \gamma_k = y_k - z_k.$$

From (3.15) it clears that

$$\omega_k = \tau_k \frac{f(x_k)}{f'(x_k)}, \quad \gamma_k = \frac{f(x_k)}{f'(x_k)}(\tau_k - 1), \quad \gamma_k - \omega_k = -\frac{f(x_k)}{f'(x_k)}. \quad (3.19)$$

Substituting (3.19) into (3.18), we get

$$a_k = -\frac{(\tau_k + 2)(\tau_k - 1)}{\tau_k}\frac{f'(x_k)}{f(x_k)}, \quad b_k = \frac{\tau_k^2}{\tau_k - 1}\frac{f'(x_k)}{f(x_k)}, \quad (3.20)$$

$$c_k = \frac{2 - 3\tau_k}{\tau_k(\tau_k - 1)}\frac{f'(x_k)}{f(x_k)}, \quad d_k = \tau_k - 1.$$

Substituting (3.20) into (3.17), we obtain

$$f'(z_k) = A_k + O\left(f^4(x_k)\right), \quad (3.21)$$

where

$$A_k = f'(x_k)\left(-\frac{2(\tau_k - 1)}{\tau_k} + \theta_k\frac{\tau_k^2}{\tau_k - 1} + \theta_k\frac{2 - 3\tau_k}{\tau_k(\tau_k - 1)}\frac{f(z_k)}{f(y_k)}\right) \neq 0. \quad (3.22)$$

Replacing $f'(z_k)$ in (3.16) by A_k, we have

$$f(x_{k+1}) = (f(z_k) + A_k(t - 1)(z_k - y_k)) + O\left(f^6(x_k)(t - 1)\right)$$
$$+ O\left(f^4(x_k)(t - 1)^4\right). \quad (3.23)$$

In deriving (3.23) have been used the following

$$\tau_k - 1 = O\left(f(x_k)\right), \quad z_k - y_k = O\left(f^2(x_k)\right). \quad (3.24)$$

Now we choose the parameter t so that the first term in the right hand side of (3.23) vanishes, i.e.,

$$t - 1 = -\frac{f(z_k)}{A_k(z_k - y_k)}. \tag{3.25}$$

Taking into account (3.22), (3.24) and Theorem 3.1, from (3.25) it follows that

$$t - 1 = O\left(f^2(x_k)\right).$$

Hence, from (3.23) we conclude that

$$f(x_{k+1}) = O\left(f^8(x_k)\right),$$

under conditions (3.9) and (3.25).

Typically, the three-point iterative methods has a form

$$y_k = x_k - \frac{f(x_k)}{f'(x_k)}, \quad z_k = y_k - \tilde{\tau}_k \frac{f(y_k)}{f'(x_k)},$$

$$x_{k+1} = z_k - \alpha_k \frac{f(z_k)}{f'(x_k)}, \quad k = 0, 1, \ldots. \tag{3.26}$$

In particular, the iterative method (3.15) can be rewritten as (3.26) with parameters

$$\tilde{\tau}_k = \frac{\tau_k - 1}{\theta_k}, \quad \alpha_k = \frac{f'(x_k)}{A_k} = \frac{\tau_k(\tau_k - 1)}{\theta_k \tau_k^3 - 2(\tau_k - 1)^2 + \theta_k(2 - 3\tau_k)\frac{f(z_k)}{f(y_k)}}. \tag{3.27}$$

To analyze the convergence order of iterations (3.26) we make an assumption

$$\tilde{\tau}_k = 1 + 2\theta_k + \beta\theta_k^2 + \gamma\theta_k^3 + \cdots. \tag{3.28}$$

Then

$$\tau_k = 1 + \tilde{\tau}_k\theta_k = 1 + \theta_k + 2\theta_k^2 + \beta\theta_k^3 + \gamma\theta_k^4 + \cdots. \tag{3.29}$$

Substituting (3.29) into (3.27) and using expansion (3.11), we obtain

$$\alpha_k = 1 + 2\theta_k + (\beta + 1)\theta_k^2 + (2\beta + \gamma - 4)\theta_k^3 + (1 + 4\theta_k)\frac{f(z_k)}{f(y_k)} + O\left(\theta_k^4\right) \tag{3.30}$$

With these preparation we are now ready to state our main result.

Theorem 3.3 *Let the assumptions of Theorem 3.1 be fulfilled. Then the iterative method (3.26) has a eighth-order of convergence if and only if the parameters $\tilde{\tau}_k$ and α_k satisfy (3.28) and (3.30), respectively.*

Proof We use Taylor expansion of $f(x_{k+1})$ around z_k:

$$f(x_{k+1}) = f(y_k) - f'(z_k)\frac{f(z_k)}{f'(x_k)}\alpha_k + O\left(f^2(z_k)\right)$$

$$= \left(1 - \frac{A_k}{f'(x_k)}\alpha_k\right) f(z_k) + O\left(f^2(z_k)\right), \tag{3.31}$$

in which we have used (3.21) and (3.27). Suppose that $\tilde{\tau}_k$ and α_k are given by (3.28) and (3.30). Then by virtue of (3.29) the condition (3.12) satisfied. Then by Theorem 3.1 we have

$$f(z_k) = O\left(f^4(x_k)\right).$$

Hence, from (3.31) it follows

$$f(x_{k+1}) = O\left(f^8(x_k)\right).$$

Conversely, assume that the convergence order of iterations (3.26) equal to eight. Then from (3.31) it follows

$$f(z_k) = O\left(f^4(x_k)\right), \tag{3.32}$$

and

$$1 - \frac{A_k}{f'(x_k)}\alpha_k = O\left(f^4(x_k)\right). \tag{3.33}$$

From (3.32) and from Theorem 3.1 it follows that τ_k satisfies the condition (3.12) or (3.29). From (3.33) it follows

$$\alpha_k = \frac{f'(x_k)}{A_k}\left(1 + O\left(f^4(x_k)\right)\right). \tag{3.34}$$

Substituting (3.29) into (3.34) we arrive at (3.30). $\square$

Remark 3.1 The method (3.26) use four values $f(x_k)$, $f(y_k)$, $f(z_k)$ and $f'(x_k)$ at each iteration step and have the optimal order. Its efficiency index is $8^{1/4} \approx 1.682$.

The some known three-point iterative methods can be written as (3.26) in a unified way. This allows us to compare them from the consistency point of view. It is not difficult to show that all the parameters $\tilde{\tau}_k$ and α_k in these eighth-order methods have a general asymptotic behavior like (3.28) and (3.30), although they defined by different formulas. Therefore, the iterative methods (3.26) with parameters given by (3.28) and (3.30) can be represent completely the class of three-point eighth-order methods. Thus, the convergence of any existing method can be established using the sufficient convergence criteria (3.28), (3.30) without using Taylor expansions many

Table 3.2 The parameters $\tilde{\tau}_k$ and α_k for the eighth-order known iterative methods

Methods	$\tilde{\tau}_k$	α_k
TP [6]	$\frac{1+b\theta_k}{1+(b-2)\theta_k}$	$\varphi(\theta_k) + \frac{f(z_k)}{f(y_k)-af(z_k)} + 4\frac{f(z_k)}{f(x_k)},$ $\varphi(0)=1,\ \varphi'(0)=2,\ \varphi''(0)=10-2b,$ $\varphi'''(0)=12b^2-72b+12$
ZU [17]	$\frac{1+b\theta_k}{1+(b-2)\theta_k}$	$\dfrac{\tilde{\tau}_k(1+\theta_k\tilde{\tau}_k)}{(1+\theta_k\tilde{\tau}_k)^3-2\theta_k\tilde{\tau}_k^2-(1+3\theta_k\tilde{\tau}_k)\frac{f(z_k)}{f(y_k)}}$
BRW [1]	The same with $b=-\frac{1}{2}$	$H(\mu_k)\frac{f'(x_k)}{f[z_k,y_k]+f[z_k,x_k,x_k](z_k-y_k)},\ \mu_k=\frac{f(z_k)}{f(x_k)},$ $H(0)=1,\ H'(0)=2,\ |H''(0)|<\infty$
BWR [2]	a. $1+\frac{4\theta_k}{2-5\theta_k}$ b. $1+2\theta_k+5\theta_k^2+\theta_k^3$ c. $\frac{1}{1-2\theta_k-\theta_k^2+\theta_k^3}$	$\frac{f(x_k)+\beta f(z_k)}{f(x_k)+(\beta-2)f(z_k)}\ \frac{f'(x_k)}{f[z_k,y_k]+f[z_k,x_k,x_k](z_k-y_k)}$
WL [8]	$\frac{1}{1-2\theta_k}$	a. $\frac{1}{2}+\frac{5+8\theta_k+2\theta_k^2}{5-12\theta_k}\left(\frac{1}{2}+\frac{f(z_k)}{f(y_k)}\right)$ b. $\frac{5-2\theta_k+\theta_k^2}{5-12\theta_k}+(1+4\theta_k)\frac{f(z_k)}{f(y_k)}$ c. $1+\frac{4f(z_k)}{f(x_k)+af(z_k)}\left(\frac{1}{1-2\theta_k-\theta_k^2}+\frac{f(z_k)}{f(y_k)}\right)$
CTV [4]	$\frac{1}{1-2\theta_k}$	$\left(\frac{f(x_k)-f(y_k)}{f(x_k)-2f(y_k)}+\frac{f(z_k)}{2(f(y_k)-2f(z_k))}\right)^2+$ $\frac{\alpha_1(u_k-z_k)+\alpha_2(y_k-x_k)+\alpha_3(z_k-x_k)}{\beta_1(u_k-z_k)+\beta_2(y_k-x_k)+\beta_3(z_k-x_k)}$ $\alpha_2=\alpha_3,\ \alpha_1=3(\beta_2+\beta_3)\neq 0$
SS [5]	$\frac{1}{1-2\theta_k}$	$\frac{f'(x_k)f[x_k,y_k]}{f[x_k,z_k]f[y_k,z_k]}w(\mu_k),\ w(0)=1,$ $w'(0)=1,\ \mu_k=\frac{f(z_k)}{f(x_k)}$

times. This discover a new and simple way to establish the convergence order of iterative methods and is key significance of our approach.

The well known eighth-order iterative methods can be rewritten as (3.26) with parameters $\tilde{\tau}_k$ and α_k. We list in Table 3.2 them for some methods.

It is easy to show that all the parameters $\tilde{\tau}_k$ and α_k given in Table 3.2 satisfy the conditions (3.28) and (3.30). It means that the convergence of any existing eighth-order method can be established using the sufficient convergence conditions (3.28) and (3.30). It is worthy to mention that the assumption for $\tilde{\tau}_k$ usually holds. Analogously can be proved the following.

Theorem 3.4 *Let the assumptions of Theorem 3.1 be fulfilled. Then, the convergence order of the iterative method (3.26) is p if and only if the parameters $\tilde{\tau}_k$ and α_k are given by Table 3.3.*

Table 3.3 The parameters and for $p = 5, 6$ and 7

p	τ_k	α_k
7	$1 + \theta_k + 2\theta_k^2 + O\left(\theta_k^3\right)$	(3.35)
6	$1 + \theta_k + 2\theta_k^2 + O\left(\theta_k^3\right)$	(3.36)
	$1 + \theta_k + O\left(\theta_k^2\right)$	(3.35)
5	$1 + \theta_k + 2\theta_k^2 + \beta\theta_k^3 + O\left(\theta_k^4\right)$	(3.37)
	$1 + \theta_k + O\left(\theta_k^2\right)$	(3.36)

Table 3.4 The parameters $\tilde{\tau}_k$ and α_k for the sixth- and seventh-order known iterative methods

Methods	$\tilde{\tau}_k$	α_k	Satisfies	Order
KLW [18]	$\frac{1}{1-2\theta_k}$	$\left(\frac{1-\theta_k}{1-2\theta_k}\right)^2 + \frac{f(z_k)}{f(y_k)-\alpha f(z_k)}$	(3.35)	7
CH [19]	$\frac{1}{1-2\theta_k}$	$1 + 2\theta_k + O\left(\theta_k^2\right)$	(3.36)	6
BRW [2]	$\frac{1+b\theta_k}{1+(b-2)\theta_k}$, $b \in \mathbb{R}$	$H(\mu_k)\frac{f'(x_k)}{f[z_k,y_k]+f[z_k,x_k,x_k](z_k-y_k)}$, $\mu_k = \frac{f(z_k)}{f(x_k)}$, $H(0) = 1$	(3.35) with $\beta = 4 - 2b$	7
CN [20]	$\frac{1}{(1-\theta_k)^2}$	$\frac{1}{\left(1-\theta_k\left(1+\frac{f(z_k)}{f(y_k)}\right)\right)^2}$	(3.36)	6

Proof The proof immediately follows from (3.31). $\qquad\square$

In the Table 3.3, the following formulas were used:

$$\alpha_k = 1 + 2\theta_k + (\beta + 1)\theta_k^2 + \frac{f(z_k)}{f(y_k)} + O\left(\theta_k^3\right), \tag{3.35}$$

$$\alpha_k = 1 + 2\theta_k + O\left(\theta_k^2\right), \tag{3.36}$$

and

$$\alpha_k = 1 + O\left(\theta_k\right). \tag{3.37}$$

Table 3.4 shows the list of the parameters $\tilde{\tau}_k$ and α_k for some known sixth- and seventh-order methods. The simplest choice (3.30) is called optimal. It is clear that the iterative methods (3.26) with the optimal order are the best ones among the methods with different parameters in terms of the computational cost (see Tables 3.2 and 3.3). In Tables 3.2 and 3.4, the first and the second divided differences were used

$$f[x, y] = \frac{f(y) - f(x)}{y - x}, \quad f[x, y, z] = \frac{f[y, z] - f[x, y]}{z - x},$$

and $H(x)$ is a generating function (see the Sect. 3.2).

3.2 Generating Function Method for Constructing New Iterations

In this section we propose a generating function method for constructing new two and three-point iterations with p ($3 \leq p \leq 8$) order of convergence. This approach allows us to derive a new family of optimal order iterative methods that includes well known methods as a special cases. The necessary and sufficient conditions for p-th order convergence of proposed iterations are given in term of parameters τ_k and α_k. We also propose some generating functions for τ_k and α_k. We give the extension of class of optimal fourth-order Jarratt's type iterations with $a \neq 2/3$. We develop a unified representation of all optimal eighth-order methods. Several numerical results are given to demonstrate the efficiency and performance of the presented methods and compare them with some other existing methods.

Solving a nonlinear equations is one of the most important problems. In recent years, a number of higher-order iterative methods have been developed and analysed for solving nonlinear equations, see [1, 3–5, 8, 13, 15, 21–24] and references therein.

We show that Theorems 3.1 and 3.3 not only give sufficient conditions for iterations to be $p = 4, 8$ order of convergence, but also allow us to construct new iterations with p order of convergence. Obtaining new optimal methods of order four is still important, because they combine higher-order of convergence and low computational cost. We consider the following choice of parameter τ_k

$$\tau_k = H(\theta_k), \tag{3.38}$$

where $H(\theta)$ is a real valued function to be determined properly. Obviously τ_k will satisfy the condition (3.12) if

$$H(0) = 1, \quad H'(0) = 1, \quad H''(0) = 4. \tag{3.39}$$

We call the function $H(\theta)$ satisfying conditions (3.39) generating one for iteration (3.1).

Construction of generating function allows us to derive new optimal order family of iterations. The following theorem is a consequence of Theorem 3.1.

Theorem 3.5 *Assume that all assumptions of Theorem 3.1 are fulfilled. Then the optimal fourth-order two-point iterations (3.1) are obtained by generating function (3.38) satisfying the conditions (3.39).*

Of course, many different variants of generating function $H(x)$, satisfying condition (3.39) are possible. We cite here one simple form of them

$$H(x) = \frac{1 + (1 - m\alpha)x + \left(2 - m\alpha + \frac{m(m-1)}{2}\alpha^2\right)x^2 + \omega x^3}{(1 - \alpha x)^m},$$
$$\alpha, m, \omega \in \mathbb{R}. \tag{3.40}$$

The optimal two-point iterations (3.1) with $\tau_k = H(\theta_k)$ given by (3.40) include many well-known iterations as a special cases. For example, if $\omega = 0$, $m = 1$ and $\alpha = 2 - b$, $b \in \mathbb{R}$ then (3.1) leads to King's one [13]. If $\alpha = 0$, $m = 1$ and $\omega = 1$ in (3.40) then (3.1) leads to new modification of Potra-Pták's one [3]. If $\alpha = m = 1$ and $\omega = -1$ in (3.40) then (3.1) leads to Maheshwari's one [15]. If $\alpha = 1$, $m = 2$ and $\omega = 0$ in (3.40) then (3.1) leads to Chun and Lee's one [21] and so on (see Table 3.5). More recently, Behl et al. [25] proposed a general class of fourth-order optimal methods that includes the well-known Ostrowski's and King's family as a special cases. We note that this general class of optimal fourth-order iterations also includes in our methods with $\tau_k = H(\theta_k)$ given by (3.40) as a special cases. Namely, if $m = 3$, α replaced by $-\alpha$ and

$$\omega = \alpha^2 + \frac{5}{3}\alpha + \frac{4}{3} \text{ or } \omega = \left(1 - \frac{\beta}{6}\right)\alpha^3 + \alpha^2 - 2\alpha$$

then the iterations (3.1) with $\tau_k = H(\theta_k)$ given by (3.40) reduce to (3.8) and (3.10) in [25], respectively. This show that our class of optimal fourth-order methods is wider than that of [25]. So, we have obtained an optimal fourth-order convergence family of iterative methods with three degrees of freedom based on the generating function method.

Analogously, one can construct generating function (3.38) for the third-order iterations (3.1). Indeed, the generating function for it has a form

$$\tau_k = H(\theta_k) \equiv \frac{1 + (1 + \eta)\theta_k}{1 + \eta\theta_k}, \quad \eta \in \mathbb{R}. \tag{3.41}$$

3.2.1   Two-Point Iterations with Two Parameters

Now we consider the following two-point iterative method

$$y_k = x_k - a\frac{f(x_k)}{f'(x_k)}, \quad x_{k+1} = x_k - \tau_k\frac{f(x_k)}{f'(x_k)}, \quad k = 0, 1, \ldots, \tag{3.42}$$

where τ_k is iteration parameter to be determined properly.

Note that when $a = 1$ the iteration (3.42) leads to (3.1). But in some case the iteration (3.42) is applied with $a \neq 1$. Therefore, study of convergence of iteration (3.42) is also interesting and we prove the following.

Theorem 3.6 *Assume that all assumptions of Theorem 3.1 are fulfilled. Then the two-point iteration (3.42) has a fourth-order of convergence if and only if the parameter τ_k is given by*

Table 3.5 Some choices of generating function $H(x)$

No	m	α	ω	$H(x)$	β	γ	Name of iterations with function $H(x)$
1	0	$\forall$	0	$1+x+2x^2$	0	0	Bi et al. [1]
2	$\forall$	0	0				
3	0	0	1	$1+x+2x^2+x^3$	1	0	Optimal Potra-Pták (OP4)
4	1	2	0	$\frac{1-x}{1-2x}$	4	8	Wang and Liu [8] Cordero et al. [4]
5	1	$\frac{9}{4}$	0	$\frac{1-\frac{5}{4}x-\frac{1}{4}x^2}{1-\frac{9}{4}x}$	$\frac{9}{2}$	$\frac{81}{8}$	Wang and Liu [8]
6	1	1	0	$\frac{1+x^2}{1-x}$	2	2	Chun and Lee [21] $\beta=0$
7	2	1	0	$\frac{1-x+x^2}{(1-x)^2}$	3	5	Chun and Lee [21]
8	1	$\forall$	$\forall$	$1+\frac{x(1+(2-\alpha)x+\omega x^2)}{1-\alpha x}$	$2\alpha+\omega$	$\alpha(2\alpha+\omega)^2$	
9	1	0	$\forall$	$1+x+2x^2+\omega x^3$	ω	0	
10	1	$2-b$	0	$1+\frac{x(1+bx)}{1+(b-2)x}$	$2(2-b)$	$2(2-b)^2$	Zhanlav and Ulziibayar [17], Thukral and Petković [6] Sharma and Sharma [5]
11	1	2.5	0	$\frac{1-1.5x-0.5x^2}{1-2.5x}$	5	12.5	Bi et al. [1]
12	1	1	-1	$1+\frac{x(1+x-x^2)}{1-x}$	1	1	Maheshwari [15]

$$\tau_k = 1 + \frac{\theta_k - (1-a)}{a^2} + 2\left(\frac{\theta_k - (1-a)}{a^2}\right)^2$$

$$+ \frac{(1-a)(a+4)}{a^3}\left(1 - \frac{f'(y_k)}{f'(x_k)} - \frac{2(\theta_k - (1-a))}{a}\right) + O\left(f^3(x_k)\right). \qquad (3.43)$$

Proof We use Taylor expansion of $f(x_{k+1})$ at point x_k.

$$f(x_{k+1}) = (1 - \tau_k)f(x_k) + \frac{f''(x_k)}{2}\tau_k^2\left(\frac{f(x_k)}{f'(x_k)}\right)^2$$

$$- \frac{f'''(x_k)}{6}\tau_k^3\left(\frac{f(x_k)}{f'(x_k)}\right)^3 + O\left((f^4(x_k))\right).$$

Multiplying by $(f'(x_k))^{-1}$ two-sides of last expression, we get

$$(f'(x_k))^{-1}f(x_{k+1})$$

$$= \left(1 - \tau_k + \tau_k^2\bar{\theta}_k - \frac{\tau_k^3}{6}(f'(x_k))^{-1}f'''(x_k)\left(\frac{f(x_k)}{f'(x_k)}\right)^2\right)(f'(x_k))^{-1}f(x_k)$$

$$+ O\left(f^4(x_k)\right), \qquad (3.44)$$

where

$$\bar{\theta}_k = \frac{(f'(x_k))^{-1}f''(x_k)}{2}\frac{f(x_k)}{f'(x_k)} = O\left(f(x_k)\right). \qquad (3.45)$$

We seek for τ_k in the form:

$$\tau_k = 1 + \bar{\theta}_k + c\bar{\theta}_k^2 + d_k + O\left(f^3(x_k)\right). \qquad (3.46)$$

Substituting (3.46) into (3.44) we obtain

$$(f'(x_k))^{-1}f(x_{k+1})$$

$$= \left(-\bar{\theta}_k - c\bar{\theta}_k^2 - d_k + \bar{\theta}_k + 2\bar{\theta}_k^2 - \frac{f'''(x_k)}{6f'(x_k)}\left(\frac{f(x_k)}{f'(x_k)}\right)^2\right)(f'(x_k))^{-1}f(x_k)$$

$$+ O\left(f^4(x_k)\right),$$

because of $\tau_k^2 = 1 + 2\bar{\theta}_k + O\left(f^2(x_k)\right)$, $\tau_k^3 = 1 + O\left(f(x_k)\right)$.

From this we conclude that

$$f(x_{k+1}) = O\left(f^4(x_k)\right),$$

under condition

$$c = 2, \quad d_k = -\frac{1}{6}(f'(x_k))^{-1} f'''(x_k) \left(\frac{f(x_k)}{f'(x_k)}\right)^2. \tag{3.47}$$

Hence

$$\tau_k = 1 + \bar{\theta}_k + 2\bar{\theta}_k^2 - \frac{1}{6}(f'(x_k))^{-1} f'''(x_k) \left(\frac{f(x_k)}{f'(x_k)}\right)^2 + O\left(f^3(x_k)\right). \tag{3.48}$$

It means that the iteration (3.42) has fourth-order convergence if and only if τ_k is given by (3.48).

Next, we using Taylor expansions of $f(y_k)$ and $f'(y_k)$ at point x_k we get

$$\theta_k = \frac{f(y_k)}{f(x_k)} = (1 - a) + a^2 \bar{\theta}_k + a^3 d_k + O\left(f^3(x_k)\right), \tag{3.49a}$$

$$(f'(x_k))^{-1} f'(y_k) = 1 - 2a\bar{\theta}_k - 3a^2 d_k + O\left(f^3(x_k)\right). \tag{3.49b}$$

From this we find $\bar{\theta}_k$

$$\bar{\theta}_k = \frac{1}{a^2}\left(\theta_k - (1 - a) - a^3 d_k\right) + O\left(f^3(x_k)\right). \tag{3.50}$$

Substituting $\bar{\theta}_k$ given by (3.50) into (3.48) we get

$$\tau_k = 1 + \frac{\theta_k - (1 - a)}{a^2} + 2\left(\frac{\theta_k - (1 - a)}{a^2}\right)^2$$
$$+ (1 - a)\left(1 + \frac{4}{a}\right) d_k + O\left(f^3(x_k)\right). \tag{3.51}$$

Elimination $\bar{\theta}_k$ from (3.49a) and (3.49b) gives us

$$d_k = \frac{1}{a^2}\left(1 - (f'(x_k))^{-1} f'(y_k) - \frac{2(\theta_k - (1 - a))}{a}\right) + O\left(f^3(x_k)\right). \tag{3.52}$$

If we take into account (3.52) in (3.51), we get (3.43). The converse is evident from (3.44) and (3.46). $\qquad\qquad\square$

Note that Theorem 3.6 is converted to Theorem 3.3 when $a = 1$. The iteration (3.42) with $a \neq 1$ is not optimal although its convergence order is four, because it requires four function evalutions $f(x_k)$, $f(y_k)$, $f'(x_k)$ and $f'(y_k)$ per iteration.

From (3.45) and (3.47) we find that

$$d_k = -\omega_2 \bar{\theta}_k^2, \quad \omega_2 = \frac{2}{3}\frac{f'''(x_k)f'(x_k)}{(f''(x_k))^2}.$$

Hence the expression (3.48) can be written as

$$\tau_k = 1 + \bar{\theta}_k + (2 - \omega_2)\bar{\theta}_k^2 + O\left(f^3(x_k)\right).$$

3.2.2 Optimal Jarratt-Type Iterations

Another class of optimal fourth-order methods appears when τ_k depends not only on θ_k, but also on $f'(y_k)/f'(x_k)$. The well-known Jarratt optimal fourth-order method [26] and method found in [27] has a form (3.42) with

$$a = \frac{2}{3}, \quad \tau_k = \frac{1}{2}\left(1 + \frac{1}{1 + \frac{3}{2}b_k}\right)$$

and

$$\tau_k = 1 - \frac{3}{4}b_k + \frac{9}{8}b_k^2, \quad b_k = \frac{f'(y_k)}{f'(x_k)} - 1,$$

respectively.

It is easy to show that τ_k in these methods satisfies the sufficient fourth-order convergence condition (3.48).

We consider another family of two-point iterative methods

$$y_k = x_k - a\frac{f(x_k)}{f'(x_k)}, \quad x_{k+1} = x_k - \tau_k \frac{f(x_k)}{f'(x_k)}, \tag{3.53}$$

where iteration parameter τ_k is given by

$$\tau_k = \frac{1 - \left(\frac{1}{2a} + m\alpha\right)(\xi_k - 1) + \left(\frac{1}{2a^2} + \frac{m\alpha}{2a} + \frac{m(m-1)}{2}\alpha^2\right)(\xi_k - 1)^2}{(1 + \alpha(\xi_k - 1))^m}$$

$$+ \frac{1}{a^2}\left(1 - \frac{3a}{2}\right)\left(\frac{2(1 - a - \theta_k)}{a} - (\xi_k - 1)\right), \quad \alpha, m \in \mathbb{R}, \tag{3.54}$$

with

$$\theta_k = \frac{f(y_k)}{f(x_k)} \quad \text{and} \quad \xi_k = \frac{f'(y_k)}{f'(x_k)}.$$

Above, we prove that the necessary and sufficient condition for two-point iteration to be fourth-order convergence is (3.48). Using (3.49b) and (3.52) it is easy to show that the parameter τ_k given (3.54) satisfies the sufficient fourth-order convergence condition (3.48).

From (3.54) we see that τ_k depends in generally not only on ξ_k but also on θ_k. The exception is the case $a = 2/3$. In this case the second term in (3.54) disappears and

Table 3.6 Some particular cases of (3.55)

No	Choices of parameters	τ_k	Methods
1	$m = 1,\ \alpha = -\frac{3}{2}$	$\frac{1}{2}\left(1 + \frac{1}{1+\frac{3}{2}(\xi_k-1)}\right)$	Jarratt [26]
2	$m = 0$ or $\alpha = 0$	$1 - \frac{3}{4}(\xi_k - 1) + \frac{9}{8}(\xi_k - 1)^2$	Petković et al. [27], Chun [28]
3	$m = 1,\ \alpha = -\frac{3}{4}$	$-1 - \frac{3}{4}(1 - \xi_k) + \frac{8}{1+3\xi_k}$	Chun [28]
4	$m = 2,\ \alpha = -\frac{3}{4}$	$1 - \frac{4}{1+3\xi_k} + \left(\frac{4}{1+3\xi_k}\right)^2$	Chun [28]
5	$m = 1,\ \alpha = -\frac{1}{2}$	$\frac{2g(\xi_k)}{\xi_k},\ g(1) = 1,$ $g'(1) = -\frac{1}{4},\ g''(1) = \frac{3}{2}$	Lotfi [29]
6	$m = 0$ or $\alpha = 0$	$\frac{1}{2}(3 - \xi_k)G(\xi_k),\ G(1) = 1,$ $G'(1) = \frac{3}{4},\ G''(1) = 2$	Jaisway [30]
7	$m = 0$ or $\alpha = 0$	$1 + \sum_{j=1}^{4} \alpha_j \xi_k^j,\ \alpha_1 = \frac{21}{8} - \alpha_4,$ $\alpha_2 = -\frac{9}{2} + 3\alpha_4,$ $\alpha_3 = \frac{15}{8} - 3\alpha_4$	Kattri and Abbasbandy [31]

τ_k depends only on ratio ξ_k and per iteration it requires evaluations of $f(x_k)$, $f'(x_k)$ and $f'(y_k)$. So the two parameter family of iterations (3.53) is optimal one. When $a = 2/3$ the formula (3.54) leads to

$$\tau_k = \frac{1 - \left(\frac{3}{4} + m\alpha\right)(\xi_k - 1) + \left(\frac{8}{9} + \frac{3}{4}m\alpha + \frac{m(m-1)}{2}\alpha^2\right)(\xi - 1)^2}{(1 + \alpha(\xi_k - 1))^m}, \quad \alpha, m \in \mathbb{R}. \tag{3.55}$$

The iterations (3.53) with generating function given by (3.55) include many well-known iterations as a particular cases. For example, when $m = 0$, $\alpha = -3/2$ the iteration (3.53) with (3.55) leads to Jarratt's one [26]. When $m = 0$ or $\alpha = 0$ then the iteration (3.53) with (3.55) leads to method in Chun [28] and Petković et al. [27]. When $m = 1$, $\alpha = -3/4$ and $m = 2$, $\alpha = -3/4$ the iterations (3.53) with (3.55) leads to methods given in Chun [28]. When $m = 1$ and $\alpha = -1/2$ the iteration (3.53) with (3.55) leads to one found in Lotfi [29] (see Table 3.6). So our iterations (3.53), (3.55) can be considered as generalization of many optimal fourth-order convergence iterations. The fourth order convergent iterations (3.53), (3.54) are not optimal when $a \neq 2/3$. Per iteration it requires evaluations of $f(x_k)$, $f(y_k)$, $f'(x_k)$ and $f'(y_k)$.

Remark 3.2 If, instead of a in (3.53) and (3.54) we take

$$a_k = \frac{2}{3} + cf(x_k) \neq 0, \quad c \in \mathbb{R}, \tag{3.56}$$

then the influence of second term in (3.54) is neglectible because it has a form $O\left(f^3(x_k)\right)$.

Thus, the iterations (3.53) remain in the class of optimal fourth-order convergence methods under choice

$$\tau_k = \frac{1 - \left(\frac{1}{2a_k} + m\alpha\right)(\xi_k - 1) + \left(\frac{1}{2a_k^2} + \frac{m\alpha}{2a_k} + \frac{m(m-1)}{2}\alpha^2\right)(\xi_k - 1)^2}{(1 - \alpha(\xi_k - 1))^m}, \quad (3.57)$$

where a_k is determined by (3.56). Thus, we extend significantly the class of optimal fourth-order methods due to (3.56), (3.57).

3.2.3 Proper Representation of Three-Point Iterative Methods

Recently, based on optimal fourth-order methods some higher-order, in particular eighth order three-point methods have been proposed for solving nonlinear equations (see Table 3.5). It is easy to show that $\tau_k = H(\theta_k)$ given by (3.40) satisfies the condition (3.29) with constants

$$\beta = \omega + 2m\alpha - \frac{m(m-1)}{2}\alpha^2 + \frac{m(m-1)(m-2)}{6}\alpha^3,$$

$$\gamma = \omega m\alpha + m(m+1)\alpha^2 - \frac{(m-1)m(m+1)}{3}\alpha^3 + \frac{(m-2)(m-1)m(m+1)}{8}\alpha^4.$$

In Table 3.5 we present some function $H(\theta_k)$, satisfying the condition (3.29).

The following is a consequence of Theorem 3.3.

Theorem 3.7 *Assume that all assumptions of Theorem 3.1 are fulfilled. Then the family of three-point iterative methods (3.26) has a eighth-order of convergence if and only if the parameters τ_k and α_k are given by (3.38), (3.40) and*

$$\alpha_k = \left(H(\theta_k) + \theta_k + (\beta - 1)\theta_k^2 + (\beta + \gamma - 4)\theta_k^3\right) + (1 + 4\theta_k)\frac{f(z_k)}{f(y_k)}. \quad (3.58)$$

Proof It is easy to show that $\tau_k = H(\theta_k)$ satisfies the condition (3.29) with parameters β and γ given in Table 3.5 and α_k given by (3.58) satisfies (3.30). Then by Theorem 3.3 the family of methods (3.26) has eighth-order of convergence. $\square$

Thus, we propose the families of three-point iterative methods (3.26) with generating function $\tau_k = H(\theta_k)$. They include many well-known eighth-order methods, as a particular cases (see Table 3.5). Above mentioned methods differ from one to another only by α_k. Moreover, they have the same asymptotic (3.30), although these are determined by different formulas. Our approach proposed in [24] is constructive in the sense that it discover new way to obtain optimal eighth-order iterations.

From (3.41) and (3.58) we see that the parameters τ_k and α_k are expressed through generating function $H(\theta_k)$. It should be pointed out that to prove the convergence order of iterations usually applied computer algebraic systems such as Maple, Mathematica and so on, whereas in our approach use only easily verifiable sufficient conditions (3.29) and (3.30). The expression in brackets in (3.58) can be approximated by simple rational function without loss of generality. Then α_k can be represented as

$$\alpha_k = \frac{1 - (2 - mq)\theta_k + c\theta_k^2 + \omega\theta_k^3}{\left(1 - \theta_k(d\theta_k^2 + p\theta_k + q)\right)^m} + (1 + 4\theta_k)\frac{f(z_k)}{f(y_k)}, \quad q, p, d, m \in \mathbb{R}, \qquad (3.59)$$

where

$$c = \beta + 1 - m\left(p + 2q + \frac{1}{2}(m - 1)q^2\right),$$

$$\omega = (2\beta + \gamma - 4)$$

$$-m\left(d + 2p + (\beta + 1 + (1 - m)p)q - (m - 1)q^2 + \frac{(m - 1)(m - 2)}{6}q^3\right).$$

The optimal order three-point iterative methods (3.26) with parameters τ_k and α_k given by (3.38), (3.40) and (3.59), respectively, we call these proper representation. It is easy to show that all the well-known optimal order three-point iterative methods can be represented in the proper form uniquely (see [1, 4–6, 8, 10, 17, 21–24, 32, 33] and references therein). It should be mentioned that Wu and Lee in [8] first used proper representation of (3.26). Thus, by means of (3.38), (3.40) and (3.59) we find a unified representation of all optimal order three-point iterations.

Our families of three-point iterative method (3.26) with parameters τ_k and α_k given by (3.38), (3.40) and (3.59) include the well-known optimal order methods as a particular cases (see Table 3.7).

Theorem 3.8 *Assume that all assumptions of Theorem 3.1 are fulfilled. Then the iterations (3.26) has a p-th order of convergence if and only if the parameters τ_k and α_k are given by formulas presented in Table 3.8.*

In Table 3.8 we used following formulae:

$$\alpha_k = \frac{1 + (2 - \alpha)\theta_k + (\beta + 1 - 2\alpha)\theta_k^2}{1 - \alpha\theta_k} + \frac{f(z_k)}{f(y_k)}, \quad \alpha, \beta \in \mathbb{R}, \qquad (3.60)$$

$$\alpha_k = \frac{1 + (2 - \xi)\theta_k}{1 - \xi\theta_k}, \quad \xi \in \mathbb{R}, \qquad (3.61)$$

$$\alpha_k = \frac{1 + \varepsilon_1\theta_k}{1 - \varepsilon_2\theta_k}, \quad \varepsilon_1, \varepsilon_2 \in \mathbb{R}. \qquad (3.62)$$

Proof The parameter τ_k and α_k given by (3.38), (3.40), (3.41) and (3.60), (3.61), (3.62) will satisfy conditions for τ_k and α_k of Theorem 3.7 in [24]. Hence, by Theorem 3.7 in [24], the order of iterations (3.26) is p. $\qquad\square$

Table 3.7 The optimal order three-point iterative methods

No	m	$\alpha_k - (1 + 4\theta_k)\frac{f(z_k)}{f(y_k)}$	Methods
1	0	$1 + 2\theta_k + (\beta + 1)\theta_k^2 + (2\beta + \gamma - 4)\theta_k^3$	$\beta = \gamma = 1$, [6] Maheshwari-based
			$\beta = 4$, $\gamma = 8$, Method 1 in [6] see [24]
			$\beta = 3$, $\gamma = 4$, Chun and Lee [21]
2	1	$\dfrac{\beta+1+(6-\gamma)\theta_k+((\beta+1)^2-2(2\beta+\gamma-4))\theta_k^2}{\beta+1-(2\beta+\gamma-4)\theta_k}$	$p = d = \omega = 0$, [1, 4–6, 8, 21–23]
			$\beta = 4$, $\gamma = 8$, Method 2 in [6]
3	2	$\dfrac{1+2(1-q)\theta_k+\left(\frac{6-\gamma}{2}+(\beta-3)q\right)\theta_k^2}{(1-q\theta_k)^2}$	$p = d = \omega = b = 0$,
			$\beta = 4$, $\gamma = 8$, $q = 2$ Methods 3 in [6]
4	1	$\dfrac{2-\theta_k}{6\theta_k^2-5\theta_k+2}$	$p = -3$, $d = 0$, $q = \frac{5}{2}$, Maheshwari-based optimal methods [32]
5	1	$\dfrac{2+(5+2\beta)\theta_k+2(3+\beta)\theta_k^2}{2+(1+2\beta)\theta_k-2\beta\theta_k^2}$	$d = 0$, $q = -\frac{1+2p}{2}$, [32]
6	1	$\dfrac{2\beta-1+2\beta(\beta-2)\theta_k}{2\beta-1+2(\beta^2-4\beta+1)\theta_k+(1+4\beta)\theta_k^2}$	$d = 0$, $q = \frac{2(\beta^2-4\beta+1)}{1-2\beta}$, $p = \frac{1+4\beta}{1-2\beta}$, King-based optimal methods [33]
7	1	$\dfrac{1}{1-2\theta_k-\theta_k^2}$	$\beta = 4$, $\gamma = 8$, $q = 2$, $p = 1$, $d = \omega = 0$, method 4 in [6]

Table 3.8 The p-order iterative methods

p	τ_k	α_k
7	(3.38), (3.40)	(3.60)
6	(3.38), (3.40)	(3.61)
	(3.41)	(3.60)
5	(3.38), (3.40)	(3.62)
	(3.41)	(3.61)

3.3 On Developments and Extensions of Some Classes of Optimal Three-Point Iterations for Solving Nonlinear Equations

We develop a new families of optimal eight-order methods for solving nonlinear equations. We also extend some classes of optimal methods for any suitable choice of iteration parameter. Such development and extension was made using sufficient convergence conditions (3.28) and (3.30). Numerical examples are given in [34] to check the convergence order of new families and extensions of some well-known methods.

3.3.1 Developments of New Families of Optimal Three-Point Methods

First, we consider iterations (3.26) with parameter α_k given by

$$\alpha_k = \frac{f'(x_k)}{f[y_k, z_k] + 2(f[x_k, z_k] - f[x_k, y_k]) + (y_k - z_k)f[y_k, x_k, x_k]}, \quad (3.63)$$

where

$$f[y_k, x_k, x_k] = \frac{f[y_k, x_k] - f'(x_k)}{y_k - x_k}.$$

To show the convergence analysis of methods (3.26), (3.63), the following results is proven.

Theorem 3.9 *Let the function $f(x)$ be sufficiently smooth and have a simple root x^* on the open interval $I \subset R$. Furthermore, let the initial approximation x_0 be sufficiently close to x^* and the parameter $\bar{\tau}_k$ in (3.26) satisfies the condition (3.28). Then the order of convergence of the methods (3.26), (3.63) is eight.*

Proof Using the relations

$$f[x_k, y_k] = f'(x_k)(1 - \theta_k),$$
$$f[y_k, z_k] = f'(x_k)\frac{1 - \upsilon_k}{\bar{\tau}_k}, \quad (3.64)$$
$$f[x_k, z_k] = f'(x_k)\frac{1 - \theta_k \upsilon_k}{1 + \bar{\tau}_k \theta_k},$$

in (3.63), we obtain

$$\alpha_k = \frac{\bar{\tau}_k}{1 - \upsilon_k + 2\frac{\bar{\tau}_k \theta_k}{1 + \bar{\tau}_k \theta_k}(1 - \upsilon_k - \bar{\tau}_k + \bar{\tau}_k \theta_k) + \bar{\tau}_k^2 \theta_k^2}. \quad (3.65)$$

Using (3.28) and well-known expansion (3.11) in (3.65) we obtain

$$\alpha_k = \bar{\tau}_k(1 + \theta_k^2 - (6 - 2\tilde{\beta})\theta_k^3 + (1 + 2\theta_k)\upsilon_k) + O\left(f^4(x_k)\right). \quad (3.66)$$

From (3.66), it follows that α_k satisfies the condition (3.30) that completes the proof of theorem. $\square$

Of course, there are many possibility for choice $\bar{\tau}_k$ satisfying the condition (3.28). In particular, we give $\bar{\tau}_k$ as

$$\bar{\tau}_k = \frac{c + (2c + d)\theta_k + \omega\theta_k^2}{c + d\theta_k + b\theta_k^2}, \quad c + d + b \neq 0, \quad (3.67)$$

Table 3.9 Choices of parameters

Cases	Methods	Special case of (3.67)	$\bar{\tau}_k$	$\widetilde{\beta}$	$\widetilde{\gamma}$
I	Potra-Ptack's	$c = \omega = 1,$ $d = b = 0$	$1 + 2\theta_k + \theta_k^2$	1	0
II	Maheshwari's	$c = 1,\ b = 0,$ $d = \omega = -1$	$\frac{1+\theta_k-\theta_k^2}{1-\theta_k}$	1	1
III	Kung-Traub's	$c = b = 1,$ $d = -2,\ \omega = 0$	$\frac{1}{(1-\theta_k)^2}$	3	4
IV	King's type	$c = 1, \omega = b = 0,$ $d = \beta - 2$	$\frac{1+\beta\theta_k}{1+(\beta-2)\theta_k}$	$2(2-\beta)$	$2(2-\beta)^2$

that includes four free parameters. In Table 3.9, we list some well-known choices.

Note that similar theorem for iteration (3.26), (3.63) based on Kung-Traub's iteration were proved by Petković et al. [35] and by Zhanlav et al. [17] for Kings type iteration and by Wang et al. [7] for Ostrowski's type method. Thus, theorem 3.9 extend essentially the class of families of optimal eight-order iterations (3.26), (3.63). Now we consider the iterations (3.26) with α_k given by

$$\alpha_k = \frac{f'(x_k)(1 + A\theta_k + B\theta_k^2 + C\theta_k^3 + (\delta + \Delta\theta_k)\upsilon_k)}{\omega_1 f[x_k, z_k] + \omega_2 f[z_k, y_k] + \omega_3 f[x_k, y_k]}, \tag{3.68}$$

where $\omega_1 + \omega_2 + \omega_3 = 1$ and $A, B, C, \delta, \Delta, \omega_1, \omega_2$ and ω_3 are free parameters to be determined such that the iterations (3.26) with α_k given by (3.68) has optimal eight-order of convergence. Namely we can prove.

Theorem 3.10 *Let all assumptions of Theorem 3.9 be fulfilled. Then the order of convergence of the iterations (3.26), (3.68) is eight when*

$$A = \delta = 1 - \omega_2, \quad B = (\widetilde{\beta} - 2)(1 - \omega_2) + 1 - \omega_1, \quad \Delta = 3 - \omega_1 - \omega_2,$$
$$C = \widetilde{\gamma}(1 - \omega_2) + \widetilde{\beta}(1 + \omega_2 - \omega_1) + \omega_1 - \omega_2 - 5. \tag{3.69}$$

Proof The proof is the same as that if Theorem 3.9. For convenience, here we only give the main step of proof. As before, using (3.11) and (3.64) after some manipulations we obtain

$$\begin{aligned}
\alpha_k &= \frac{f'(x_k)}{\omega_1 f[x_k, z_k] + \omega_2 f[z_k, y_k] + \omega_3 f[x_k, y_k]}\\
&= 1 + (1 + \omega_2)\theta_k + (\omega_1 + \widetilde{\beta}\omega_2 + (1 - \omega_2)^2)\theta_k^2\\
&\quad + (\widetilde{\gamma}\omega_2 + \widetilde{\beta}(2(1 - \omega_2)^2 - \omega_1 - 2\omega_3) + \omega_1\\
&\quad + 2(1 - \omega_2)(2 - \omega_1) - 2(1 - \omega_2)^2 - (1 - \omega_2)^3)\theta_k^3\\
&\quad + (\omega_2 + (\omega_1 + 2\omega_2^2)\theta_k)\upsilon_k + O\left(f^4(x_k)\right).
\end{aligned} \tag{3.70}$$

Substituting (3.70) into (3.68) and comparing (3.68) with (3.30) we arrive at (3.69). $\square$

The expression in the numerator of (3.68) can be expressed through first order divided differences $f[x_k, y_k]$, $f[x_k, z_k]$ and $f[z_k, y_k]$ within accuracy $O\left(f^4(x_k)\right)$. Indeed using the iterations

$$f[x_k, y_k] - f[x_k, z_k] = f'(x_k)(\theta_k^2 + (\widetilde{\beta} - 3)\theta_k^3 + \theta_k v_k),$$

and

$$f[z_k, x_k] - f[y_k, z_k] = f'(x_k)(\theta_k + (\widetilde{\beta} - 5)\theta_k^2 + (\widetilde{\gamma} - 5\widetilde{\beta} + 11)\theta_k^3 + (1 - 3\theta_k)v_k),$$

in (3.68) we obtain

$$\alpha_k = \frac{(3\omega_2 + \omega_1 - 5)(f[x_k, z_k] - f[x_k, y_k]) + F_k + Q_k}{\omega_1 f[x_k, z_k] + \omega_2 f[z_k, y_k] + \omega_3 f[x_k, y_k]}, \qquad (3.71)$$

where $Q_k = f'(x_k)(1 + (\omega_2 - 2)\theta_k^2 + 2(1 - \omega_1 - \omega_2)\theta_k^3)$ and $F_k = (1 - \omega_2)$ $(f[x_k, y_k] - f[y_k, z_k])$. From (3.68), (3.69) we see that α_k includes two free parameters ω_1 and ω_2. Thus, we develop the class of optimal eight-order iterations (3.26) with α_k given by (3.68), (3.69). We consider some choices of parameters ω_1 and ω_2.

1. Let $\omega_1 = \omega_3 = 0$, $\omega_2 = 1$. Then (3.71) converted to

$$\alpha_k = \frac{(3 + \theta_k)f[x_k, y_k] - 2f[z_k, x_k]}{f[y_k, z_k]}.$$

2. Let $\omega_1 = 1$, $\omega_2 = \omega_3 = 0$. Then (3.71) converted to

$$\alpha_k = \frac{5f[x_k, y_k] - 4f[z_k, x_k] - f[y_k, z_k] + f'(x_k)(1 - 2\theta_k^2)}{f[x_k, z_k]},$$

3. Let $\omega_1 = \omega_2 = 0$, $\omega_3 = 1$. Then (3.71) converted to

$$\alpha_k = \frac{6f[x_k, y_k] - 5f[z_k, x_k] - f[y_k, z_k] + f'(x_k)(1 - 2\theta_k^2 + 2\theta_k^3)}{f[y_k, x_k]},$$

4. Let $\omega_1 = -1$, $\omega_2 = 2$, $\omega_3 = 0$. Then (3.71) converted to

$$\alpha_k = \frac{f[z_k, y_k] - f[x_k, y_k] + f'(x_k)}{2f[z_k, y_k] - f[z_k, x_k]}. \qquad (3.72)$$

The iteration (3.26), (3.72) can be considered as another variant of iterations given by Sharma et al. in [5, 36, 37] and given by Zhanlav et al. in [38].

5. Let $\omega_1 = \omega_2 = 1$, $\omega_3 = -1$. Then (3.71) converted to

$$\alpha_k = \frac{f[y_k, x_k] - f[z_k, x_k] + f'(x_k)(1 - \theta_k^2 - 2\theta_k^3)}{f[z_k, x_k] + f[z_k, y_k] - f[y_k, x_k]}.$$

It can be rewritten as:

$$\alpha_k \approx \frac{1}{\left(1 - \frac{f(z_k)}{f(x_k)}\right)\left(1 + (5 - \widetilde{\beta})\left(\frac{f(y_k)}{f(x_k)}\right)^3\right)} \frac{f'(x_k)}{f[z_k, x_k] + f[z_k, y_k] - f[y_k, x_k]}.$$

It is worthy to note that similar results for derivative-free case and for some choices of $\bar{\tau}_k$ were obtained by Thukral in [39] and by Khattri et al. in [40]. We also note that the iteration [35, 41] for $\widetilde{\beta} = 4$ was considered by Sharma et al. in [41].

6. Let $\omega_1 = -1$, $\omega_2 = \omega_3 = 1$. Then (3.71) converted to

$$\alpha_k = \frac{3f[x_k, y_k] - 3f[z_k, x_k] + f'(x_k)(1 - \theta_k^2 + 2\theta_k^3)}{f[y_k, x_k] + f[z_k, y_k] - f[z_k, x_k]}.$$

In each iteration step the methods (3.26), (3.63) and (3.26), (3.68) require three function evaluations and one evaluation of first derivative. Based on the conjecture of Kung and Traub, they reached the optimality with higher efficiency index $E = 8^{1/4} = 1.682$. One of main advantageous of the proposed iterative methods (3.26), (3.63) and (3.26), (3.68) is that they work well for any choice of parameter $\bar{\tau}_k$ satisfying the condition (3.28).

3.3.2 Extensions of Some Classes of Optimal Eighth-Order Methods

Now we consider the iterations (3.26) with parameter α_k given by

$$\alpha_k = (p(t_k) + \hat{\gamma}\theta_k^3)\frac{f'(x_k)}{f[z_k, y_k] + (z_k - y_k)f[z_k, x_k, x_k]}, \tag{3.73}$$

where $t_k = f(z_k)/f(x_k) = \theta_k v_k$ and $\hat{\gamma}$ is such that (3.73) satisfies (3.30).

Namely, we have.

Theorem 3.11 *Let all assumptions of Theorem 3.9 be fulfilled. Then the order of convergence of the iterations (3.26) and (3.73) is eight when*

$$p(0) = 1, \quad p'(0) = 2, \quad \hat{\gamma} = 2(\widetilde{\beta} - 5). \tag{3.74}$$

Proof We denote the second factor in (3.73) by $\hat{\alpha}_k$. That is

$$\hat{\alpha}_k = \frac{f'(x_k)}{f[z_k, y_k] + (z_k - y_k)f[z_k, x_k, x_k]}.$$

Then using the relations (3.64) we obtain

$$\hat{\alpha}_k = \frac{\bar{\tau}_k}{1 - \upsilon_k - \bar{\tau}_k \left(\frac{\bar{\tau}_k \theta_k}{1 + \bar{\tau}_k \theta_k}\right)^2}. \tag{3.75}$$

Using the expansion (3.11) in (3.75) and taking into account $\upsilon_k = O\left(f(x_k)^2\right), \theta_k = O\left(f(x_k)\right)$ under condition (3.28) we obtain

$$\hat{\alpha}_k = \bar{\tau}_k \left(1 + \upsilon_k + \bar{\tau}_k \left(\frac{\bar{\tau}_k \theta_k}{1 + \bar{\tau}_k \theta_k}\right)^2\right) + O\left(f^4(x_k)\right). \tag{3.76}$$

By using (3.11) and (3.28) it is easy to show that

$$\left(\frac{\bar{\tau}_k \theta_k}{1 + \bar{\tau}_k \theta_k}\right)^2 = \theta_k^2 + 2\theta_k^3 + O\left(f^4(x_k)\right). \tag{3.77}$$

Substituting (3.28) and (3.77) into (3.76) we get

$$\hat{\alpha}_k = 1 + 2\theta_k + (\tilde{\beta} + 1)\theta_k^2 + (\tilde{\gamma} + 6)\theta_k^3 + (1 + 2\theta_k)\upsilon_k + O\left(f^4(x_k)\right).$$

Then (3.73) is written as

$$\begin{aligned}
\alpha_k =\,&(p(t_k) + \hat{\gamma}\theta_k^3)(1 + 2\theta_k + (\tilde{\beta} + 1)\theta_k^2 \\
&+ (\tilde{\gamma} + 6)\theta_k^3 + (1 + 2\theta_k)\upsilon_k) + O\left(f^4(x_k)\right),
\end{aligned}$$

which satisfies the condition (3.30) provided that (3.74). $\qquad\qquad\square$

Thus, we develop the family of optimal three-point iterative methods (3.26) with α_k given by

$$\alpha_k = (p(t_k) + 2(\tilde{\beta} - 5)\theta_k^3)\frac{f'(x_k)}{f[z_k, y_k] + (z_k - y_k)f[z_k, x_k, x_k]}. \tag{3.78}$$

Similar results were obtained in [2, 42] for the iterations (3.26) with

$$\bar{\tau}_k = \frac{1 - \theta_k/2}{1 - 5\theta_k/2}, \tag{3.79}$$

and

Table 3.10 Choices of parameters

N	Methods	$\bar{\tau}_n$	$\widetilde{\beta}$	$\widetilde{\gamma}$	α_n	Extension factor ξ_n
1	Sharma et al. [5]	$\dfrac{1}{1-2\theta_n}$	4	8	$\dfrac{W(t_n)f[x_n,y_n]f'(x_n)}{f[x_n,z_n]f[y_n,z_n]}$ $W(0)=1,\,W'(0)=1$	$1+(\widetilde{\beta}-4)\theta_n^3$
2	DP8 [35]	$\dfrac{1}{1-2\theta_n}$	4	8	$\dfrac{1}{(1-2\theta_n-\theta_n^2)(1-\upsilon_n)(1-2\theta_n\upsilon_n)}$	$1+(\widetilde{\beta}-4)\theta_n^2+(\widetilde{\gamma}-8)\theta_n^3$
3	GK8 [43]	$\dfrac{1+\beta\theta_n+\lambda\theta_n^2}{1+(\beta-2)\theta_n+\mu\theta_n^2}$ $\mu=-\frac{3\beta}{2},\,\lambda=-1+\frac{\beta}{2}$	3	6	$\dfrac{1}{1-2\theta_n-\upsilon_n}$	$1+(\widetilde{\beta}-3)\theta_n^2+(\widetilde{\gamma}-6)\theta_n^3$
4	Chun et al. [21]	$\dfrac{1}{(1-\theta_n)^2}$	3	4	$\dfrac{1}{(1-H(\theta_n)-J(t_n)-P(\upsilon_n))^2}$	$1+(\widetilde{\beta}-3)\theta_n^2+(\widetilde{\gamma}-4)\theta_n^3$
5	Thukral-Petković [6]	$\dfrac{1+\beta\theta_n}{1+(\beta-2)\theta_n}$	$2(2-\beta)$	$2(2-\beta)^2$	$\phi(\theta_n)+\dfrac{t_n}{\theta_n-at_n}+4t_n$	$1+(\widetilde{\beta}-2(2-\beta))\theta_n^2$ $+(\widetilde{\gamma}-2(2-\beta)^2)\theta_n^3$
6	Lotfi et al. [22]	$\dfrac{1}{1-2\theta_n}$	4	8	$\dfrac{H(\theta_n)+K(t_n)}{G(\upsilon_n)}$	$1+(\widetilde{\beta}-4)\theta_n^2+(\widetilde{\gamma}-8)\theta_n^3$

$$\bar{\tau}_k = \frac{1}{1 - 2\theta_k - \theta_k^2 - \theta_k^3/2}, \tag{3.80}$$

respectively. The parameters $\bar{\tau}_k$ given by (3.79) and (3.80) satisfy the condition (3.28) with $\widetilde{\beta} = 5$. In this case $\hat{\gamma} = 0$ in (3.74) and the α_k given by (3.73) leads to

$$\alpha_k = p(t_k) \frac{f'(x_k)}{f[z_k, y_k] + (z_k - y_k)f[z_k, x_k, x_k]}.$$

This means that our iterations (3.26) and (3.78) include the iterations proposed by Bi et al. [2] and by Cordero et al. [42] as particular cases.

Now we consider the expression

$$\tilde{\alpha}_k = \frac{f'(x_k)}{f[z_k, y_k] + (z_k - y_k)f[y_k, x_k, x_k]}. \tag{3.81}$$

As before, using the relations (3.64) in (3.81) we obtain

$$\tilde{\alpha}_k = \frac{\bar{\tau}_k}{1 - \upsilon_k - \bar{\tau}_k^2 \theta_k^2} = \bar{\tau}_k(1 + \upsilon_k + \bar{\tau}_k^2 \theta_k^2) + O\left(f^4(x_k)\right).$$

By (3.28) one can easily to check that

$$\tilde{\alpha}_k = \hat{\alpha}_k + O\left(f^4(x_k)\right).$$

It means that instead of (3.78) one can also use

$$\alpha_k = (p(t_k) + \hat{\gamma}\theta_k^3)\frac{f'(x_k)}{f[z_k, y_k] + (z_k - y_k)f[y_k, x_k, x_k]}, \tag{3.82}$$

Therefore, Theorem 3.11 holds true for iterations (3.26), (3.82). Similar extension can be done for all optimal eight-order iterations. As examples, we present in Table 3.10 some of methods and their extension $\tilde{\alpha}_k = \xi_k \alpha_k$ with extension factor ξ_k.

Thus, we obtain extensions of some well-known optimal methods that work well for any suitable parameter $\bar{\tau}_k$, satisfying the condition (3.28). This allows us to expand the applicability of the original methods.

The main contributions of this section are:

1. The development of wide class of optimal eight-order iterative methods and extensions of some optimal methods that work well for any suitable choice of parameter $\bar{\tau}_k$ satisfying the condition (3.28).
2. The proposed iterative methods can be regarded as an advancement in the topic and can compete with other well-known methods.

References

1. W. Bi, H. Ren, Q. Wu, Three-step iterative methods with eighth-order convergence for solving nonlinear equations. J. Comput. Appl. Math. **225**, 105–112 (2009)
2. W. Bi, Q. Wu, H. Ren, A new family of eighth-order iterative methods for solving nonlinear equations. Appl. Math. Comput. **211**, 236–245 (2009)
3. A. Cordero, J.L. Hueso, E. Martínez, J.R. Torregrosa, New modifications of Potra-Pták's method with optimal fourth and eighth orders of convergence. J. Comput. Appl. Math. **234**, 2969–2976 (2010)
4. A. Cordero, J.R. Torregrosa, M.P. Vassileva, Three-step iterative methods with optimal eighth-order convergence. J. Comput. Appl. Math. **235**, 3189–3194 (2011)
5. J.R. Sharma, R. Sharma, A new family of modified Ostrowski's methods with accelerated eighth-order convergence. Numer. Algorithms **54**, 445–458 (2010)
6. R. Thukral, M.S. Petković, A family of three-point methods of optimal order for solving nonlinear equations. J. Comput. Appl. Math. **233**, 2278–2284 (2010)
7. X. Wang, L. Liu, Modified Ostrowski's method with eighth-order convergence and high efficiency index. Appl. Math. Lett. **23**, 549–554 (2010)
8. X. Wang, L. Liu, New eighth-order iterative methods for solving nonlinear equations. J. Comput. Appl. Math. **234**, 1611–1620 (2010)
9. T. Zhanlav, O. Chuluunbaatar, Convergence of the continuous analogy of Newton's method for solving nonlinear equations. Numer. Methods Prog. **10**, 402–407 (2009)
10. C. Chun, B. Neta, An analysis of a new family of eighth-order optimal methods. Appl. Math. Comput. **245**, 86–107 (2014)
11. C. Chun, B. Neta, Comparison of several families of optimal eighth order methods. Appl. Math. Comput. **274**, 762–773 (2016)
12. Y. Ham, C. Chun, S. Lee, Some higher-order modifications of Newton's method for solving nonlinear equations. J. Comput. Appl. Math. **222**, 477–486 (2008)
13. R. King, A family of fourth order methods for nonlinear equations. SIAM J. Numer. Anal. **10**, 876–879 (1973)
14. C. Chun, A family of composite fourth-order iterative methods for solving nonlinear equations. Appl. Math. Comput. **187**, 951–956 (2007)
15. A.K. Maheshwari, A fourth-order iterative methods for solving nonlinear equations. Appl. Math. Comput. **211**, 383–391 (2009)
16. J.F. Traub, *Iterative methods for the solution of equations* (Englewood Cliffs, NJ, Prentice-Hall, 1964)
17. T. Zhanlav, V. Ulziibayar, Modified King's methods with optimal eighth-order convergence and high efficiency index. Am. J. Comput. Appl. Math. **6**(5), 177–181 (2016)
18. J. Kou, Y. Li, X. Wang, Some variants of Ostrowski's method with seventh-order convergence. J. Comput. Appl. Math. **209**, 153–159 (2007)
19. C. Chun, Y. Ham, Some sixth-order variants of Ostrowski's methods with seventh-order convergence. Appl. Math. Comput. **193**, 389–394 (2007)
20. C. Chun, B. Neta, A new sixth-order scheme for nonlinear equations. Appl. Math. Lett. **25**, 185–189 (2012)
21. C. Chun, M.Y. Lee, A new optimal eighth-order family of iterative methods for the solution of nonlinear equations. Appl. Math. Comput. **223**, 506–519 (2013)
22. T. Lotfi, S. Sharifi, M. Salimi, S. Siegmund, A new class of three-point methods with optimal convergence order eight and its dynamics. Numer. Algorithms **68**, 261–288 (2015)
23. S. Sarifi, M. Ferrara, M. Salimi, S. Siegmant, New modification of Maheshwari's method with optimal eighth-order convergence for solving nonlinear equations. Open Math. **14**, 443–451 (2016)
24. T. Zhanlav, V. Ulziibayar, O. Chuluunbaatar, Necessary and sufficient conditions for the convergence of two- and three-point Newton-type iterations. Comput. Math. Math. Phys. **57**, 1090–1100 (2017)

25. R. Behl, A. Cordero, S.S. Motsa, J.R. Torregrosa, Construction of fourth-order optimal families of iterative methods and their dynamics. Appl. Math. Comput. **271**, 89–101 (2015)
26. P. Jarratt, Some fourth-order multipoint methods for solving equations. Math. Comput. **20**, 434–437 (1966)
27. M.S. Petković, B. Neta, L.D. Petković, J. Džunić, Multipoint methods for solving nonlinear equations. Appl. Math. Comput. **226**, 635–660 (2014)
28. C. Chun, M.Y. Lee, B. Neta, J. Džunić, On optimal fourth-order iterative methods free from second derivative and their dynamics. Appl. Math. Comput. **218**, 6427–6438 (2012)
29. T. Lotfi, A new optimal method of fourth-order convergence for solving nonlinear equations. Int. J. Ind. Math. **6**, 121–124 (2014)
30. J.P. Jaiswal, A class of iterative methods for solving nonlinear equations with optimal fourth-order convergence. Univers. J. Math. Appl. **2**, 283–289 (2014)
31. S.K. Khattri, S. Abbasbandy, Optimal fourth-order family of iterative methods. Mat. Vesn. **63**, 67–72 (2011)
32. C. Chun, B. Neta, An analysis of a family of Maheshwari-based optimal eighth-order methods. Appl. Math. Comput. **253**, 294–307 (2015)
33. C. Chun, B. Neta, An analysis of a King-based family of optimal eighth-order methods. Am. J. Algorithm Comput. **2**(1), 1–17 (2015)
34. T. Zhanlav, K. Otgondorj, On the development and extensions of some classes of optimal three-point iterations for solving nonlinear equations. J. Numer. Anal. Approx. Theory **50**, 180–193 (2021)
35. M.S. Petković, B. Neta, L.D. Petković, J. Džunić, *Multipoint Methods for Solving Nonlinear Equations* (Elsevier, 2013)
36. J.R. Sharma, H. Arora, A new family of optimal eighth order methods with dynamics for nonlinear equations. Appl. Math. Comput. **273**, 924–933 (2016)
37. J.R. Sharma, H. Arora, An efficient family of weighted-Newton methods with optimal eighth order convergence. Appl. Math. Lett. **29**, 1–6 (2014)
38. T. Zhanlav, K. Otgondorj, A new family of optimal eighth-order methods for solving nonlinear equations. Am. J. Comput. Appl. Math. **8**, 15–19 (2018)
39. R. Thukral, Eighth-order iterative methods without derivatives for solving nonlinear equations. ISRN Appl. Math. **2011**(1), 12 (2011). Article ID: 693787
40. S.K. Khattri, R.P. Agarwal, Derivative-free optimal iterative methods. Comput. Methods Appl. Math. **10**, 368–375 (2010)
41. R. Sharma, A. Bahl, Optimal eighth order convergent iteration scheme based on Lagrange interpolation. Acta Math. Sin. Engl. Ser. **33**, 1093–1102 (2017)
42. A. Cordero, M. Fardi, M. Ghasemi, J.R. Torregrosa, Accelerated iterative methods for finding solutions of nonlinear equations and their dynamical behavior. Calcolo **51**, 17–30 (2014)
43. Y.H. Geum, Y.I. Kim, A uniparametric family of three-step eighth-order multipoint iterative methods for simple roots. Appl. Math. Lett. **24**, 929–935 (2011)

Chapter 4
Derivative-Free Iterative Methods

Abstract Necessary and sufficient conditions for derivative-free two- and three-point iterations are obtained. In particular, we propose the finite-difference versions of well-known optimal fourth-order methods. We applied the generating function method for constructing new derivative-free methods and made the comparison of methods based on the dynamical behaviour. Two new parametric sixth- and seventh-order iterative methods with memory are presented, and the optimal values of additional parameters that increase the convergence order are found.

4.1 Derivative-Free Two and Three-Point Iterations

Families of optimal derivative-free two-point iterative methods for solving nonlinear equations are suggested. We obtain necessary and sufficient conditions for these methods to have the fourth-order of convergence. The convergence order and accuracy of method are tested on smooth and non-smooth functions. Moreover, we compare the performance of our methods to existing methods [1, 2].

Consider the derivative-free two-point iterative method

$$y_k = x_k - \frac{f(x_k)}{\phi(x_k)}, \tag{4.1a}$$

$$x_{k+1} = y_k - \bar{\tau}_k \frac{f(y_k)}{\phi(x_k)}, \tag{4.1b}$$

where

$$f'(x) \approx \phi(x) = \frac{f(x + \gamma f(x)) - f(x)}{\gamma f(x)}, \quad \gamma \in R, \tag{4.2}$$

γ is a free nonzero parameter, and $\bar{\tau}_k$ is parameter to be determined properly. Here the function $\phi(x) \equiv \phi(x, \gamma)$ depends not only on x but also on the parameter γ. By the definition of derivative, we have

$$f'(x) = \phi(x, \gamma), \quad \gamma \to 0.$$

To determine the convergence order of iterations (4.1a), (4.1b) we consider the following ratio:

$$\varsigma_k = \frac{f'(x_k)}{\phi(x_k)} \neq 0. \tag{4.3}$$

4.1.1 Convergence Analysis

Let $f(x) \in C^3(D)$, where D is an interval containing the root of the equation $f(x) = 0$. Then, the Taylor expansions of $f(y_k)$ and $f(x_k + \gamma f(x_k))$ give

$$f(y_k) = (1 - \varsigma_k) f(x_k) + \frac{f''(x_k)}{2} \varsigma_k^2 \left(\frac{f(x_k)}{f'(x_k)} \right)^2 + O\left(f^3(x_k)\right), \tag{4.4}$$

and

$$\phi(x_k) = f'(x_k) \left(1 + \gamma \frac{f''(x_k)}{2} \frac{f(x_k)}{f'(x_k)} \right) + O\left(f^2(x_k)\right). \tag{4.5}$$

Substituting (4.5) into (4.3) we obtain

$$\varsigma_k = \frac{1}{1 + \gamma \frac{f''(x_k)}{2} \frac{f(x_k)}{f'(x_k)}} + O\left(f^2(x_k)\right) = 1 - \gamma \frac{f''(x_k)}{2} \frac{f(x_k)}{f'(x_k)} + O\left(f^2(x_k)\right), \tag{4.6}$$

or

$$\varsigma_k = 1 + O\left(f(x_k)\right). \tag{4.7}$$

Taking into account (4.7) in (4.4) we have

$$f(y_k) = O\left(f^2(x_k)\right). \tag{4.8}$$

As in Chap. 3, we use the notation

$$\theta_k = \frac{f(y_k)}{f(x_k)}. \tag{4.9}$$

From (4.8), (4.9) it follows that $\theta_k = O\left(f(x_k)\right)$. Using (4.4) in (4.9) we obtain

$$\theta_k = 1 - \varsigma_k + \frac{1}{2} \varsigma_k \frac{f''(x_k) f(x_k)}{f'(x_k) \phi(x_k)} + O\left(f^2(x_k)\right). \tag{4.10}$$

Elimination of $\frac{f''(x_k) f(x_k)}{f'(x_k)}$ from (4.6) and (4.10) gives

$$\varsigma_k^2 - (1 - \gamma\phi(x_k))\varsigma_k - (1 - \theta_k)\gamma\phi(x_k) = O\left(f^2(x_k)\right). \tag{4.11}$$

From (4.11) it clear that ς_k depends on θ_k. By virtue of (4.7) one can seek for ς_k in the form

$$\varsigma_k = 1 - a_k\theta_k + O\left(f^2(x_k)\right). \tag{4.12}$$

Substituting (4.12) into (4.11) we have

$$\theta_k\left(\gamma\phi_k - a_k(1 + \gamma\phi_k)\right) = O\left(f^2(x_k)\right), \quad \phi_k = \phi(x_k). \tag{4.13}$$

From (4.13) it follows that

$$a_k = \frac{\gamma\phi_k}{1 + \gamma\phi_k} + O\left(f(x_k)\right). $$

Thereby, we have

$$\varsigma_k = 1 - \frac{\gamma\phi_k}{1 + \gamma\phi_k}\theta_k + O\left(f^2(x_k)\right). \tag{4.14}$$

On the other hand, the Taylor expansion of $f(x_{k+1})$ gives us

$$f(x_{k+1}) = \left(1 - \frac{f'(y_k)}{\phi_k}\bar{\tau}_k\right)f(y_k) + O\left(f^2(y_k)\right). \tag{4.15}$$

Due to (4.1a) we have

$$f'(y_k) = f'(x_k)\left(1 - \frac{f''(x_k)f(x_k)}{f'(x_k)\phi(x_k)}\right) + O\left(f^2(x_k)\right). \tag{4.16}$$

Elimination of term $(f''(x_k)f(x_k))(f'(x_k)\phi(x_k))$ from (4.10) and (4.16) gives

$$f'(y_k) = -f'(x_k)\frac{\varsigma_k + 2(\theta_k - 1)}{\varsigma_k} + O\left(f^2(x_k)\right). \tag{4.17}$$

Substituting ς_k given by (4.14) into (4.17) and using well-known expansion (3.11), we get

$$f'(y_k) = f'(x_k)\left(1 - \frac{2}{1 + \gamma\phi_k}\theta_k\right) + O\left(f^2(x_k)\right). \tag{4.18}$$

Using (4.18) in (4.15) we obtain

$$f(x_{k+1}) = \left(1 - (1 - \hat{d}_k\theta_k)\bar{\tau}_k\right)f(y_k) + O\left(f^2(y_k)\right), \quad \hat{d}_k = \frac{2 + \gamma\phi_k}{1 + \gamma\phi_k}. \tag{4.19}$$

Now we are ready to prove the following results.

Theorem 4.1 *Let $f(x) \in C^3(D)$ and the initial approximation x_0 be sufficiently close to simple zero $x^* \in D$ of function $f(x)$. Then convergence order of the iterative method (4.1) is four if and only if the parameter $\bar{\tau}_k$ satisfies condition*

$$\bar{\tau}_k = \frac{1}{1 - \hat{d}_k \theta_k} + O\left(f^2(x_k)\right) = 1 + \hat{d}_k \theta_k + O\left(f^2(x_k)\right). \tag{4.20}$$

Proof Assume that $\bar{\tau}_k$ in (4.1) satisfies condition (4.20). Then

$$1 - (1 - \hat{d}_k \theta_k)\bar{\tau}_k = O\left(f^2(x_k)\right)$$

and $f(y_k) = O\left(f^2(x_k)\right)$ because of (4.7). Hence, by virtue of (4.19), we have

$$f(x_{k+1}) = O\left(f^4(x_k)\right), \tag{4.21}$$

i.e., the convergence order of (4.1) is four under (4.20). Conversely, let the method (4.1) have the fourth-order of convergence, i.e., (4.21) is true. Then from (4.21) and (4.19) it follows that $f(y_k) = O\left(f^2(x_k)\right)$ and $1 - (1 - \hat{d}_k \theta_k)\bar{\tau}_k = O\left(f^2(x_k)\right)$, i.e., $\bar{\tau}_k$ satisfies the convergence (4.20). $\qquad\square$

The iterative method (4.1) uses the values $f(x_k)$, $f(y_k)$ and $\phi(x_k)$ at each iteration, so it is optimal in sense of Kung-Traub's conjecture [3]. The second step in (4.1) can be rewritten as

$$x_{k+1} = x_k - \tau_k \frac{f(x_k)}{\phi(x_k)},$$

where

$$\tau_k = 1 + \bar{\tau}_k \theta_k = 1 + \theta_k + \hat{d}_k \theta_k^2 + O\left(f^3(x_k)\right). \tag{4.22}$$

If $\phi(x_k, \gamma) = f'(x_k)$ as $\gamma \to 0$ then $\varsigma_k = 1$ and the formula (4.20) and (4.22) leads to

$$\bar{\tau}_k = 1 + 2\theta_k + O\left(f^2(x_k)\right),$$

and

$$\tau_k = 1 + \theta_k + 2\theta_k^2 + O\left(f^3(x_k)\right),$$

respectively. So the iteration (4.1) leads to optimal fourth-order two-point iteration (3.1)

$$y_k = x_k - \frac{f(x_k)}{f'(x_k)},$$

$$x_{k+1} = x_k - \tau_k \frac{f(x_k)}{f'(x_k)}.$$

Note that generating function method considered in Sect. 3.2 is also applicable for constructing new iterations (4.1). Of course many different variants of generating function $\bar{\tau}_k = H(\theta_k)$, satisfying the conditions

$$H(0) = 1, \quad H'(0) = \hat{d}_k, \tag{4.23}$$

are possible. We site here one simple form of them

$$H(x) = \frac{c + (H'(0)c + d)x + \omega x^2}{c + dx + bx^2}, \quad c + d + b \neq 0, \quad c, d, b, \omega \in R. \tag{4.24}$$

Now we consider some interesting cases of H.

1. Let $c = 1, d = \beta - 2, b = \omega = 0$ in (4.24). Then we obtain

$$H(x) = \frac{1 + \left(\beta - \frac{\gamma \phi_k}{1 + \gamma \phi_k}\right) x}{1 + (\beta - 2)x}.$$

The iteration (4.1) with choice $\bar{\tau}_k = H(\theta_k)$ has a form

$$y_k = x_k - \frac{f(x_k)}{\phi(x_k)}, \tag{4.25}$$

$$x_{k+1} = y_k - \frac{1 + \left(\beta - \frac{\gamma \phi_k}{1 + \gamma \phi_k}\right) \theta_k}{1 + (\beta - 2)\theta_k} \frac{f(y_k)}{\phi(x_k)}.$$

As $\gamma \to 0$, the iteration (4.25) gives the well-known King's one. We call this iteration (4.25) the finite-difference (FD) variant of King's method.

2. Let $c = b = 1, \quad d = -2$ and $\omega = 0$ in (4.24). Then we obtain

$$H(x) = \frac{1 - \frac{\gamma \phi_k}{1 + \gamma \phi_k} x}{(1 - x)^2}.$$

The iteration (4.1) with such choice $\bar{\tau}_k = H(\theta_k)$ has a form

$$y_k = x_k - \frac{f(x_k)}{\phi(x_k)}, \tag{4.26}$$

$$x_{k+1} = y_k - \frac{1 - \frac{\gamma \phi_k}{1 + \gamma \phi_k} \theta_k}{(1 - \theta_k)^2} \frac{f(y_k)}{\phi(x_k)}.$$

As $\gamma \to 0$, the iteration (4.26) produces the well-known Kung and Traub's optimal fourth-order method. By this reason, we call the iteration (4.26) the (FD) variant of Kung and Traub's one.

3. Let $c = 1$, $\omega = d = -1$ and $b = 0$ in (4.24). Then we obtain

$$H(x) = \frac{1 + \frac{1}{1+\gamma\phi_k}x - x^2}{1 - x}.$$

The iteration (4.1) with such choice $\bar{\tau}_k = H(\theta_k)$ has a form

$$y_k = x_k - \frac{f(x_k)}{\phi(x_k)},\tag{4.27}$$

$$x_{k+1} = y_k - \frac{1 + \frac{1}{1+\gamma\phi_k}\theta_k - \theta_k^2}{1 - \theta_k}\frac{f(y_k)}{\phi(x_k)}.$$

As $\gamma \to 0$, the iteration (4.27) gives the Maheshwari's one [4]. By this reason, we call the iteration (4.27) as (FD) variant of Maheshwari's one.

It should be mentioned that authors of paper [5] were made attempt to extend the King's and Kung-Traub's methods to these derivative-free variants. But the derived methods in [5] are different from our extensions (4.25) and (4.26).

Our family include five parameters $(\gamma, c, d, b, \omega)$. The coefficients in (4.24) may be depend on the number of iteration k. Our family of iteration (4.1) with parameter $\bar{\tau}_k$ given by (4.24) contains some well-known iterations as a special cases. In Table 4.1, we list some of them. Based on the conjecture of Kung and Traub, (4.1) method reached the optimality with efficiency index equals $EI = 4^{(1/3)} \approx 1.58$. Note that only $\bar{\tau}_k$ in method of Ren et al. [6] does not belongs to the family of $H(\theta_k)$ defined by (4.24). Thus, the proposed family (4.1) with parameter specified by (4.24) is a considerable generalization of the methods described in [6–22].

4.2 Derivative-Free Three-Point Iterations

Now, we consider the derivative free three-point methods:

$$y_k = x_k - \frac{f(x_k)}{\phi(x_k)},$$

$$z_k = y_k - \bar{\tau}_k\frac{f(y_k)}{\phi(x_k)},\tag{4.28}$$

$$x_{k+1} = z_k - \alpha_k\frac{f(z_k)}{\phi(x_k)},$$

Table 4.1 The Derivative-free iterative methods

Name of iterations	$\bar{\tau}_k$	Cases of H given by (4.24)
Methods [13, 18] Method [14] Petković's method [20]	$1 + \hat{d}_k \theta_k$	$c \neq 0, \quad d = b = \omega = 0$
Methods [22, 23] $h(t,s) = \frac{1}{1-t-s}$ [14] CTM in [9]	$\dfrac{1}{1 - \hat{d}_k \theta_k}$	$c = 1, \quad d = -\hat{d}_k, b = \omega = 0$
$h(t,s) = \frac{1+t}{1-s}$ [14, 24]	$\dfrac{1 + \gamma\phi_k + (\hat{d}_k(1+\gamma\phi_k) - 1)\theta_k}{1 + \gamma\phi_k - \theta_k}$	$c = 1 + \gamma\phi_k, d = -1,$ $b = \omega = 0$
Methods given in [8]	$\dfrac{1 + \phi_k + (\hat{d}_k(1+\phi_k) - (2-\phi_k))\theta_k}{1 + \phi_k - (2+\phi_k)\theta_k}$	$c = 1 + \phi_k, d = -(2 + \phi_k)$ $b = -1, \omega = 0$
Chebyshew-Halley [7]	$\dfrac{1 + \left(\hat{d}_k - \left(2\alpha + \frac{1}{1+\gamma\phi_k}\right)\right)\theta_k + \omega\theta_k^2}{1 - \left(2\alpha + \frac{1}{1+\gamma\phi_k}\right)\theta_k + \frac{2\alpha}{1+\gamma\phi_k}\theta_k^2}$	$c = 1, d = -\left(2\alpha + \frac{1}{1+\gamma\phi_k}\right)$ $b = \frac{2\alpha}{1+\gamma\phi_k}, \omega = 1$
Kung-Traub's and method in [21]	$\dfrac{1}{1 - d_k\theta_k + \frac{1}{1+\gamma\phi_k}\theta_k^2}$	$c = 1, d = -\hat{d}_k$ $b = \frac{1}{1+\gamma\phi_k}, \omega = 0$
Derivative-free form of Potra-Pták [19]	$1 + d_k\theta_k + \frac{d_k a}{2}\theta_k^2$	$c = 1, d = b = 0, \omega = \frac{d_k a}{2}$
King's type method [15]	$\dfrac{1 + (\gamma-1)\theta_k - \gamma\theta_k^2}{1 + \left(\gamma - 2 - \frac{1}{1-\beta\phi_k}\right)\theta_k + \frac{\gamma-2}{1-\beta\phi_k}\theta_k^2}$	$\omega = -\gamma$ $c = 1, d = \gamma - 2 - \frac{1}{1-\beta\phi_k}$ $b = (2 - \gamma)/(1 - \beta\phi_k)$
Methods given in [11]	$\dfrac{1 - \frac{\gamma\phi_k}{1+\gamma\phi_k}\theta_k}{1 - 2\theta_k + \theta_k^2}$	$c = b = 1, d = -2, \omega = 0$
Petković type ($P2$) methods [20] and method in [20]	$\dfrac{1 + \theta_k}{1 - \frac{\theta_k}{1+\gamma\phi_k}}$	$c = 1, d = -\frac{1}{1+\gamma\phi_k},$ $b = \omega = 0$
Methods [10]	$\dfrac{1 + \phi_k + 2\theta_k}{1 + \phi_k - \phi_k\theta_k}$	$c = 1 + \phi_k, d = -\phi_k$ $b = \omega = 0$
Methods [6]	$1 + \hat{d}_k\theta_k + \left(\alpha_1 + \frac{\alpha_2}{(1+\gamma\phi_k)^2}\right)$	$c = 1, d = b = 0$ $\omega = \alpha_1 + \frac{\alpha_2}{(1+\gamma\phi_k)^2}$

which is obtained from the three-point methods (3.26) replacing $f'(x_k)$ by $\phi(x_k)$. Note that the first two steps in (4.28) define an optimal fourth-order two-point method when $\bar{\tau}_k$ is given by formula (4.20). Our aim is to find α_k such that the order of (4.28) is eight.

4.2.1 Convergence Analysis

We use the Taylor expansion of $f(x_{k+1})$:

$$f(x_{k+1}) = f(z_k) - f'(z_k)\alpha_k \frac{f(z_k)}{\phi(x_k)} + O\left(f^2(z_k)\right),$$

$$= \left(1 - \alpha_k \frac{f'(z_k)}{\phi(x_k)}\right) f(z_k) + O\left(f^2(z_k)\right). \tag{4.29}$$

From last relation it clear that

$$f(x_{k+1}) = O\left(f^8(x_k)\right), \tag{4.30}$$

under condition

$$\alpha_k = \frac{\phi(x_k)}{f'(z_k)} + O\left(f^4(x_k)\right). \tag{4.31}$$

Now we approximate $f'(z_k)$ in (4.31) using available data $f(x_k)$, $f(y_k)$, $f(z_k)$ and $\phi(x_k)$ such that

$$f'(z_k) = a_k f(x_k) + b_k f(y_k) + c_k f(z_k) + d_k \phi(x_k) + O\left(f^4(x_k)\right). \tag{4.32}$$

Using Taylor expansion of $f(x)$ at point z_k we obtain the following system:

$$\begin{aligned}
&a_k + b_k + c_k = 0, \\
&a_k w_k + b_k \gamma_k + d_k = 1, \\
&a_k w_k^2 + b_k \gamma_k^2 + 2d_k\left(w_k + \frac{1}{2}\gamma f(x_k)\right) = 0, \\
&a_k w_k^3 + b_k \gamma_k^3 + 3d_k\left(w_k^2 + w_k \gamma f(x_k) + \frac{1}{3}\gamma^2 f^2(x_k)\right) = 0,
\end{aligned} \tag{4.33}$$

where

$$w_k = x_k - z_k, \quad \gamma_k = y_k - z_k. \tag{4.34}$$

The system (4.33) has a unique solution

$$\begin{aligned}
&c_k = -a_k - b_k, \\
&d_k = 1 - a_k w_k - b_k \gamma_k, \\
&b_k = \frac{w_k(w_k + \gamma f(x_k))}{\gamma_k(\gamma_k - w_k)(\gamma_k - w_k - \gamma f(x_k))}, \\
&a_k = \frac{\gamma_k}{w_k(\gamma_k - w_k)} \frac{(w_k - \gamma_k)(2w_k + \gamma f(x_k)) + (w_k + \gamma f(x_k))^2}{(w_k + \gamma f(x_k))(w_k - \gamma_k + \gamma f(x_k))}.
\end{aligned} \tag{4.35}$$

Substituting (4.35) into (4.32) we obtain

$$f'(z_k) = \phi_k$$
$$\times \left(1 + a_k w_k \left(\frac{f[z_k, x_k]}{\phi_k} - 1\right) + b_k \gamma_k \left(\frac{f[z_k, y_k]}{\phi_k} - 1\right)\right) + O\left(f^4(x_k)\right), \quad (4.36)$$

where

$$\phi_k = \phi(x_k) = f[x_k, \xi_k], \quad \xi_k = x_k + \gamma f(x_k).$$

According to (4.29) and (4.34) we have

$$w_k = \frac{f(x_k)}{\phi_k}\tau_k, \quad \gamma_k = (\tau_k - 1)\frac{f(x_k)}{\phi_k}, \quad \gamma_k - w_k = -\frac{f(x_k)}{\phi_k} = y_k - x_k, \quad (4.37)$$

$$\frac{\gamma_k}{w_k - \gamma_k} = \frac{y_k - z_k}{x_k - y_k} = \tau_k - 1 \rightarrow \tau_k = \frac{x_k - z_k}{x_k - y_k}, \quad (4.38)$$

$$w_k + \gamma f(x_k) = \frac{f(x_k)}{\phi_k}(\tau_k + \gamma\phi_k), \quad w_k - \gamma_k + \gamma f(x_k) = \frac{f(x_k)}{\phi_k}(1 + \gamma\phi_k). \quad (4.39)$$

Using (4.37)–(4.39) in (4.35) we obtain

$$b_k \gamma_k = \frac{\tau_k(\tau_k + \gamma\phi_k)}{1 + \gamma\phi_k}, \quad (4.40)$$

$$a_k w_k = (1 - \tau_k)\frac{2\tau_k + \gamma\phi_k + (\tau_k + \gamma\phi_k)^2}{(\tau_k + \gamma\phi_k)(1 + \gamma\phi_k)}.$$

Substituting (4.36) into (4.31) and neglecting small term $O\left(f^4(x_k)\right)$, we obtain

$$\alpha_k = \frac{1}{1 + a_k w_k \left(\frac{f[z_k, x_k]}{\phi_k} - 1\right) + b_k \gamma_k \left(\frac{f[z_k, y_k]}{\phi_k} - 1\right)}, \quad (4.41)$$

where $a_k w_k$ and $b_k \gamma_k$ are defined by formula (4.40). The expressions in parentheses in (4.41) can be rewritten in terms of second divided differences as:

$$\frac{f[z_k, x_k]}{\phi_k} - 1 = \frac{1}{\phi_k}f[z_k, x_k, \xi_k](z_k - \xi_k) = -\frac{f(x_k)}{\phi_k^2}f[z_k, x_k, \xi_k](\tau_k + \gamma\phi_k), \quad (4.42)$$

$$\frac{f[z_k, y_k]}{\phi_k} - 1 = -\frac{f(x_k)}{\phi_k^2}\left(f[y_k, z_k, x_k] + f[z_k, x_k, \xi_k](\tau_k + \gamma\phi_k)\right). \quad (4.43)$$

Substituting (4.40), (4.42) and (4.43) into (4.41) we obtain another representation of α_k:

$$\alpha_k = \frac{1}{1 - \frac{f(x_k)}{\phi_k^2(1+\gamma\phi_k)}\left(\tau_k(\tau_k + \gamma\phi_k)f[y_k, z_k, x_k] + C_k f[z_k, x_k, \xi_k]\right)},$$

where $C_k = (1 - \tau_k)(2\tau_k + \gamma\phi_k) + (\tau_k + \gamma\phi_k)^2$. Now we find the asymptotic of α_k given by formula (4.41). To this end we use the following formulae

$$\frac{f[z_k, x_k]}{\phi_k} - 1 = -\frac{(\bar{\tau}_k + \upsilon_k)\theta_k}{1 + \bar{\tau}_k\theta_k}, \tag{4.44}$$

$$\frac{f[z_k, y_k]}{\phi_k} - 1 = \frac{1 - \bar{\tau}_k - \upsilon_k}{\bar{\tau}_k}, \tag{4.45}$$

$$\tau_k = 1 + \bar{\tau}_k\theta_k, \quad \upsilon_k = \frac{f(z_k)}{f(y_k)}.$$

Analogously, the formula (4.40) can be also rewritten in term of $\bar{\tau}_k$, as (4.44) and (4.45). Taking into account this and (4.44) and (4.45) one can rewrite (4.41) as:

$$\alpha_k = \frac{1}{1 + \frac{A_1 + A_2 + A_3\upsilon_k}{(1+\gamma\phi_k)(1+\gamma\phi_k+\bar{\tau}_k\theta_k)(1+\bar{\tau}_k\theta_k)\bar{\tau}_k}}, \tag{4.46}$$

where

$$
\begin{aligned}
A_1 &= (1 + \theta_k\bar{\tau}_k)^2(1 + \gamma\phi_k + \bar{\tau}_k\theta_k)^2(1 - \bar{\tau}_k), \\
A_2 &= (2 + \gamma\phi_k + 2\bar{\tau}_k\theta_k + (1 + \gamma\phi_k + \bar{\tau}_k\theta_k)^2)\bar{\tau}_k^3\theta_k^2, \\
A_3 &= -(1 + \theta_k\bar{\tau}_k)^2(1 + \gamma\phi_k + \bar{\tau}_k\theta_k)^2 + (2 + \gamma\phi_k + 2\bar{\tau}_k\theta_k \\
&\quad + (1 + \gamma\phi_k + \bar{\tau}_k\theta_k)^2)\bar{\tau}_k^2\theta_k^2.
\end{aligned}
\tag{4.47}
$$

According to (4.44) we can write $\bar{\tau}_k$ as:

$$\bar{\tau}_k = 1 + \hat{d}_k\theta_k + \tilde{\beta}_k\theta_k^2 + \tilde{\gamma}_k\theta_k^3 + \cdots, \tag{4.48}$$

where $\tilde{\beta}_k$ and $\tilde{\gamma}_k$ are some constants. Then by Theorem 4.1 we have

$$f(y_k) = O\left(f^2(x_k)\right), \quad f(z_k) = O\left(f^4(x_k)\right), \quad \upsilon_k = O\left(f^2(x_k)\right). \tag{4.49}$$

Using (4.48) and (4.49) in (4.47) we get

$$
\begin{aligned}
A_1 &= -\theta_k(1 + \gamma\phi_k)^2(a_1 + a_2\theta_k + a_3\theta_k^2 + \cdots), \\
A_2 &= \theta_k^2(1 + \gamma\phi_k)^2(b_1 + b_2\theta_k + \cdots), \\
A_3 &= -(1 + \gamma\phi_k)^2(c_1 + c_2\theta_k + \cdots),
\end{aligned}
$$

where

$$a_1 = \hat{d}_k, \quad a_2 = \tilde{\beta} + 2\hat{d}_k^2, \quad a_3 = \tilde{\gamma} + 2\tilde{\beta}\hat{d}_k + \left(3\hat{d}_k^2 + \frac{2}{1 + \gamma\phi_k}\right)\hat{d}_k,$$

$$b_1 = 1 + \frac{\hat{d}_k}{1 + \gamma\phi_k}, \quad b_2 = \hat{d}_k - \hat{d}_k^2 + 3\hat{d}_k^3, \quad c_1 = 1, \quad c_2 = 2\hat{d}_k.$$

Analogously, we have

$$\frac{1}{\left(1 + \frac{\bar{\tau}_k\theta_k}{1+\gamma\phi_k}\right)(1 + \bar{\tau}_k\theta_k)\,\bar{\tau}_k} = 1 - 2\hat{d}_k\theta_k$$

$$+ \left(2\hat{d}_k^2 - \tilde{\beta} - \frac{1}{1 + \gamma\phi_k}\right)\theta_k^2 + O\left(f^3(x_k)\right).$$

Then we have

$$\frac{A_1 + A_2 + A_3 \upsilon_k}{(1 + \gamma\phi_k)(1 + \gamma\phi_k + \bar{\tau}_k\theta_k)(1 + \bar{\tau}_k\theta_k)\bar{\tau}_k}$$
$$= \left(1 - 2\hat{d}_k\theta_k + \left(2\hat{d}_k^2 - \tilde{\beta} - \frac{1}{1 + \gamma\phi_k}\right)\theta_k^2\right),$$
$$\times \left(-a_1\theta_k + (b_1 - a_2)\theta_k^2 + (b_2 - a_3)\theta_k^3 - (c_1 + c_2\theta_k)\upsilon_k\right),$$
$$= -a_1\theta_k + \left(b_1 - a_2 + 2a_1\hat{d}_k\right)\theta_k^2$$
$$+ \left(b_2 - a_3 - 2\hat{d}_k(b_1 - a_2) - a_1\left(2\hat{d}_k^2 - \tilde{\beta} - \frac{1}{1 + \gamma\phi_k}\right)\right)\theta_k^3$$
$$- (c_1 + (c_2 - 2c_1\hat{d}_k)\theta_k)\upsilon_k + O\left(f^4(x_k)\right).$$

Substituting, the last expression into (4.46) and using the well-known expansion (3.11) we obtain

$$\alpha_k = 1 + a_1\theta_k - (b_1 - a_2 + 2a_1\hat{d}_k)\theta_k^2 + \left(2\hat{d}_k(b_1 - a_2)\right.$$
$$- a_1\left(2\hat{d}_k^2 - \tilde{\beta} - \frac{1}{1 + \gamma\phi_k}\right) - (b_2 - a_3)\Big)\theta_k^3$$
$$+ (c_1 + (c_2 - 2c_1\hat{d}_k)\theta_k)\upsilon_k + a_1^2\theta_k^2 - 2a_1$$
$$\times \left(b_1 - a_2 + 2a_1\hat{d}_k\right)\theta_k^3 + 2a_1c_1\theta_k\upsilon_k + a_1^3\theta_k^3 + O\left(f^4(x_k)\right)$$
$$= 1 + a_1\theta_k + (a_1^2 - (b_1 - a_2) - 2a_1\hat{d}_k)\theta_k^2$$
$$+ (a_1^3 + 2\hat{d}_k(b_1 - a_2) + a_1\left(2\hat{d}_k^2 - \tilde{\beta} - \frac{1}{1 + \gamma\phi_k}\right)$$
$$- (b_2 - a_3) - 2a_1(b_1 - a_2 + 2a_1\hat{d}_k))\theta_k^3$$
$$+ (c_1 + (c_2 - 2c_1\hat{d}_k)\theta_k)\upsilon_k + O\left(f^4(x_k)\right),$$

or

$$
\alpha_k = 1 + \hat{d}_k\theta_k + \left(\tilde{\beta} + \frac{1}{1 + \gamma\phi_k}\right)\theta_k^2
$$
$$
+ \left(\tilde{\gamma} + \hat{d}_k\tilde{\beta} - \hat{d}_k - \frac{\hat{d}_k}{(1 + \gamma\phi_k)^2}\right)\theta_k^3
$$
$$
+ (1 + 2\hat{d}_k\theta_k)\upsilon_k + O\left(f^4(x_k)\right). \tag{4.50}
$$

When $\gamma = 0$ the formula (4.50) leads to (3.30) which is an asymptotic of α_k in three-point iteration methods (3.26). Now we give a precise analysis of convergence through the following theorem.

Theorem 4.2 *Assume that all assumption of Theorem 4.1 are fulfilled. Then the derivative-free three-point iteration methods (4.28) has a eight-order of convergence if and only if the iteration parameter $\bar{\tau}_k$ and α_k are given by formulae (4.48) and (4.50) respectively.*

Proof Let $\bar{\tau}_k$ and α_k be given by formulae (4.48) and (4.50) respectively. Then by Theorem 4.1, the first two steps in (4.28) define an optimal fourth-order method, i.e., $f(z_k) = O\left(f^4(x_k)\right)$. The α_k be given by (4.50) satisfies the condition (4.31). Hence, (4.30) is true. Conversely, assume that the order of convergence of iteration (4.28) is eight. Then from (4.28) and (4.30) it follows that $f(z_k) = O\left(f^4(x_k)\right)$ and formula (4.31). Therefore, by Theorem 4.1 the formula (4.48) is true with some constants $\tilde{\gamma}$ and $\tilde{\beta}$. Using approximation (4.36) in (4.31) we obtain (4.41) with accuracy $O\left(f^4(x_k)\right)$. Since (4.48) holds true then from (4.41) we arrive at the asymptotic (4.50). $\qquad\square$

If we take in (4.28) $\bar{\tau}_k = H(\theta_k)$ defined by (4.23) and (4.24)

$$
\bar{\tau}_k = H(\theta_k) = \frac{c + (\hat{d}_k c + d)\theta_k + \omega\theta_k^2}{c + d\theta_k + b\theta_k^2}, \quad c + d + b \neq 0, \quad c, d, b, \omega \in R,
$$

and

$$
\alpha_k = H(\theta_k) + \frac{1}{1 + \gamma\phi_k}\theta_k^2 + \hat{d}_k\left(\tilde{\beta} - \frac{2}{1 + \gamma\phi_k}\right)\theta_k^3 + (1 + 2\hat{d}_k\theta_k)\upsilon_k, \tag{4.51}
$$

then we obtain a family of derivative-free and optimal three-point iterations, because $\bar{\tau}_k$ and α_k be given by formulae (4.20) and (4.51) satisfy the condition (4.48) and (4.50) with constants

$$
\tilde{\beta} = \frac{\omega - b}{c} - \frac{d}{c}\left(\frac{d}{c} + \hat{d}_k\right),
$$
$$
\tilde{\gamma} = -\frac{(b + \omega)d}{c^2} + \frac{d^2 - bc}{c^2}\hat{d}_k,
$$

respectively. Thus, the generating function method allows us to design the family of derivative-free optimal eight-order iterations. Now we consider the following three-point derivative-free iterations

$$
\begin{aligned}
w_k &= x_k + \gamma f(x_k), \\
y_k &= x_k - \frac{f(x_k)}{f[x_k, w_k]}, \\
z_k &= \psi_4(x_k, y_k, w_k), \\
x_{k+1} &= z_k
\end{aligned}
\tag{4.52}
$$

$$
- \frac{f(z_k)}{f[z_k, y_k] + (z_k - y_k)f[z_k, y_k, x_k] + (z_k - y_k)(z_k - x_k)f[z_k, y_k, x_k, w_k]}.
$$

Here ψ_4 is any optimal fourth-order derivative-free method. From Theorem 4.2 immediately follows.

Theorem 4.3 *Let the all assumptions of Theorem 4.1 be fulfilled. Then the convergence order of the family of three-point derivative-free methods (4.52) is eight.*

Proof Since ψ_4 is any fourth-order iteration then z_k can be rewritten as

$$
z_k = y_k - \bar{\tau}_k \frac{f(y_k)}{\phi_k}, \quad \phi_k = f[x_k, w_k].
$$

By Theorem 4.1 we have $\bar{\tau}_k = 1 + \hat{d}_k \theta_k + O\left(f^2(x_k)\right)$. It means that the Taylor expansion of $\bar{\tau}_k$ (4.48) holds true. Using (4.48) the expansion α_k in (4.28) can be written as

$$
\alpha_k = \frac{\phi_k}{f[z_k, y_k] + (z_k - y_k)f[z_k, y_k, x_k] + (z_k - y_k)(z_k - x_k)f[z_k, y_k, x_k, w_k]}
$$

and satisfies the condition (4.50). Then by Theorem 4.2 the convergence order of iterations (4.52) is eight. $\square$

Remark 4.1 The convergence order of three-point iterations proposed in [12, 15, 22, 25] immediately follows from the Theorem 4.2. In this regard our Theorem 4.2 is extension of theorems given in [12, 15, 22, 25]. Note that all the existing optimal three-point derivative-free methods can be rewritten in the form (4.28) uniquely.

It is easy to verify that the parameters $\bar{\tau}_k$ and α_k of these methods have the same asymptotic behavior (4.48) and (4.50) with own constants $\tilde{\gamma}$ and $\tilde{\beta}$. Thus, the convergence of every existing optimal three-point methods can be proved using the sufficient convergence criteria (4.48) and (4.50) without using symbolic computation. Moreover, the application these sufficient convergence conditions allow us to design new optimal iterations in some cases [26]. Based on the conjecture of Kung and Traub (4.28) method reached the optimality with efficiency index equals $EI = 8^{1/4} \approx 1.68$.

4.2.2 Derivative-Free Three-Point Methods with Memory

As in preceding subsection, we can vary γ from step to step using information available from current and previous step and increase the order 8 without using any new function evaluations. Namely, we obtain three-point derivative-free iteration with memory: x_0, γ_0 are given, then $w_0 = x_0 + \gamma_0 f(x_0)$,

$$\gamma_k = -\frac{1}{N_4'(x_k)}, \quad w_k = x_k + \gamma_k f(x_k), \quad k = 1, 2, \ldots,$$

$$y_k = x_k - \frac{f(x_k)}{\phi(x_k, \gamma_k)},$$

$$z_k = \psi_4(x_k, y_k, w_k), \tag{4.53}$$

$$x_{k+1} = z_k$$

$$- \frac{f(z_k)}{f[z_k, y_k] + (z_k - y_k)f[z_k, y_k, x_k] + (z_k - y_k)(z_k - x_k)f[z_k, y_k, x_k, w_k]}.$$

Here, $N_4(t) = N_4(t, x_k, z_{k-1}, y_{k-1}, w_{k-1}, x_{k-1})$ is Newton interpolating polynomial of fourth degree, set through available nodal points $(t, x_k, z_{k-1}, y_{k-1}, w_{k-1}, x_{k-1})$. As in [7] it is easy to prove that the p-order of the methods with memory (4.53) is at least twelve.

4.3 Comparison of Some Optimal Derivative-Free Iterations

4.3.1 Some Optimal Derivative-Free Three-Point Iterations

At present there are many optimal derivative-free three-point iterations see, for example, [1, 7, 8, 12, 15, 22, 23, 25, 27, 28] and references therein. They mainly distinguished among themselves by approximations of $f'(z_k)$ at the last step. Let the values of $f(x)$ be known at points x_k, w_k, y_k and z_k. Usually often used three approaches for approximation $f'(z_k)$. First of them is used approximation [7, 12, 15, 22, 25]

$$f'(z_k) \approx N_3'(z_k), \tag{4.54}$$

where $N_3(z)$ is Newton's interpolation polynomial of degree three at the point x_k, w_k, y_k and z_k. The second approach is [23]

$$f'(z_k) \approx v_1 f(x_k) + v_2 f(w_k) + v_3 f(y_k) + v_4 f(z_k). \tag{4.55}$$

The real constants v_1, v_2, v_3 and v_4 are determined such that the relation (4.55) holds with equality for the four functions $f(x) = 1, x, x^2, x^3$. While in [1] was used the approximation

$$f'(z_k) \approx af(x_k) + bf(y_k) + cf(z_k) + d\phi(x_k),$$

$$\phi(x_k) = \frac{f(w_k) - f(x_k)}{w_k - x_k} = f[x_k, w_k]. \tag{4.56}$$

The real constants a, b, c and d in (4.56) are determined such that the equality (4.56) holds with accuracy $O\left(f^4(x_k)\right)$. Note that in last years have been appeared papers, in which were used another approximations such as Pade approximant [8] and rational approximations [27] and so on. As we seen from (4.54), (4.55) and (4.56) more suitable and guaranteed approximation is (4.56). In general, all these three approaches turn out to be identical. This is well-known long ago fact [29]. This idea motivated us to make detail comparison of methods based on (4.54), (4.55) and (4.56). The well-known Zheng et al. methods (Z8) [22] based on (4.54) and has a form

$$y_k = x_k - \frac{f(x_k)}{f[x_k, w_k]}, \quad w_k = x_k + \gamma f(x_k), \quad \gamma \in R \setminus \{0\},$$

$$z_k = y_k - \frac{f(y_k)}{f[x_k, y_k] + f[y_k, w_k] - f[x_k, w_k]}, \tag{4.57}$$

$$x_{k+1} = z_k$$

$$- \frac{f(z_k)}{f[z_k, y_k] + (z_k - y_k)f[z_k, y_k, x_k] + (z_k - y_k)(z_k - x_k)f[z_k, y_k, x_k, w_k]}.$$

The well-known Khattri et al. methods (KS8) [23] based on (4.55) and has a form

$$y_k = x_k - \frac{f(x_k)}{f[x_k, w_k]},$$

$$z_k = y_k - \frac{f(y_k)}{\frac{x_k - y_k + \gamma f(x_k)}{(x_k - y_k)\gamma} - \frac{(x_k - y_k)f(w_k)}{(w_k - y_k)\gamma f(x_k)} - \frac{(2x_k - 2y_k + \gamma f(x_k))f(y_k)}{(x_k - y_k)(w_k - y_k)}}, \tag{4.58}$$

$$x_{k+1} = z_k - \frac{f(z_k)}{H_1 + H_2 + H_3 - H_4}.$$

Here

$$H_1 = -\frac{(y_k - z_k)(w_k - z_k)}{(x_k - z_k)\gamma(x_k - y_k)},$$

$$H_2 = \frac{(y_k - z_k)(x_k - z_k)f(w_k)}{(w_k - z_k)(w_k - y_k)\gamma f(x_k)}, \tag{4.59}$$

$$H_3 = \frac{(x_k - z_k)(w_k - z_k)f(y_k)}{(y_k - z_k)(w_k - y_k)(x_k - y_k)},$$

$$H_4 = \frac{\gamma(x_k - 2z_k + y_k)f(x_k) + x_k^2 + (-4z_k + 2y_k)x_k + 3z_k^2 - 2y_k z_k}{(y_k - z_k)(x_k - z_k)(w_k - z_k)}f(z_k).$$

In [23], the authors pointed out that these methods given by (4.58), (4.59) is similar to the already known methods proposed in [4, 12, 15, 25], in particular to method in [22], however, they are not the same methods. From (4.57) and (4.58) we see that the second and third subteps in (4.58) are much complicated as compared with (4.57). The formula, requiring many mathematical operations absolutely unfitted for numerical and stability points of view. Hence, the formula (4.58) needed further simplifications. The families of derivative-free optimal methods proposed in [1] are based on (4.56) and have a form [30]

$$
\begin{aligned}
y_k &= x_k - \frac{f(x_k)}{\phi(x_k)}, \\
z_k &= y_k - \bar{\tau}_k \frac{f(y_k)}{\phi(x_k)}, \\
x_{k+1} &= z_k - \alpha_k \frac{f(z_k)}{\phi(x_k)},
\end{aligned}
\tag{4.60}
$$

where

$$
\bar{\tau}_k = \frac{c + (\hat{d}_k c + d)\theta_k + \omega\theta_k^2}{c + d\theta_k + b\theta_k^2}, \quad c + d + b \neq 0, \quad c, d, b, \omega \in R,
\tag{4.61}
$$

and

$$
\alpha_k = \frac{1}{\left(1 + a_k w_k \left(\frac{f[z_k, x_k]}{f[x_k, w_k]} - 1\right) + b_k \gamma_k \left(\frac{f[z_k, y_k]}{f[x_k, w_k]} - 1\right)\right)}
\tag{4.62}
$$

with

$$
\begin{aligned}
a_k w_k &= (1 - \tau_k)\frac{2\tau_k + \gamma\phi_k + (\tau_k + \gamma\phi_k)^2}{(\tau_k + \gamma\phi_k)(1 + \gamma\phi_k)}, \\
b_k \gamma_k &= \frac{\tau_k(\tau_k + \gamma\phi_k)}{1 + \gamma\phi_k}, \quad \phi_k = f[x_k, w_k], \\
\tau_k &= 1 + \bar{\tau}_k\theta_k, \quad \theta_k = \frac{f(y_k)}{f(x_k)}.
\end{aligned}
$$

We call the representation (4.60) of three-point methods as canonical form. Each derivative-free three-point methods, in particular the methods (4.57) and (4.58) can be written in canonical form uniquely. Note that all the considered methods (4.57), (4.58) and (4.60) are optimal eighth-order methods. So they has an efficiency index $8^{1/4} \approx 1.682$. The methods (4.57) and (4.58) contain one free parameter γ, whereas the methods (4.60) contain, in addition γ, yet four parameter c, d, b and w. Hence, in our opinion, the families (4.60) represent a wide class of optimal three-point methods. Our aim is to compare the above mentioned methods in detail. First, we will show that the optimal derivative-free methods (4.57) and (4.58) are identical.

Theorem 4.4 *The optimal derivative-free methods (4.57) and (4.58) are equivalent.*
Proof Using easily verifying relations

$$f[x_k, y_k] = \phi_k(1 - \theta_k), \quad f[y_k, w_k] = \phi_k\left(1 - \frac{\theta_k}{1 + \gamma\phi_k}\right),$$

the second-step in (4.57) and (4.58) can be easily rewritten as

$$z_k = y_k - \bar{\tau}_k \frac{f(y_k)}{\phi(x_k)}, \tag{4.63}$$

where

$$\bar{\tau}_k = \frac{1}{1 - \hat{d}_k\theta_k}, \quad \hat{d}_k = \frac{2 + \gamma\phi_k}{1 + \gamma\phi_k}. \tag{4.64}$$

Thus, the first two sub-steps of (4.57) and (4.58) are the same. In passing, we obtain very simple calculation formula (4.63) for iteration method (4.58). It remains to compare the third sub-steps in (4.57) and (4.58). The third sub-steps in (4.57) and (4.58) can be rewritten as

$$x_{k+1} = z_k - \alpha_k \frac{f(z_k)}{\phi(x_k)},$$

where

$$\alpha_k = \frac{\phi_k}{f[z_k, y_k] + (z_k - y_k)f[z_k, y_k, x_k] + (z_k - y_k)(z_k - x_k)f[z_k, y_k, x_k, w_k]}, \tag{4.65}$$

for iteration (4.57) and

$$\alpha_k = \frac{\phi_k}{H_1 + H_2 + H_3 - H_4}, \tag{4.66}$$

for iteration (4.58). Using the following relations

$$f[z_k, y_k] = \frac{\phi_k}{\bar{\tau}_k}(1 - \upsilon_k), \quad \upsilon_k = \frac{f(z_k)}{f(y_k)}, \quad f[z_k, x_k] = \frac{\phi_k}{\tau_k}(1 - \theta_k\upsilon_k),$$

$$f[z_k, y_k, x_k] = \frac{\phi_k^2}{\tau_k f(x_k)} \frac{\bar{\tau}_k(1 - \theta_k) - (1 - \upsilon_k)}{\bar{\tau}_k}, \tag{4.67}$$

$$f[z_k, y_k, x_k, w_k] = \frac{\phi_k^3}{f^2(x_k)(\tau_k + \gamma\phi_k)}\left(\frac{\theta_k}{1 + \gamma\phi_k} - \frac{\bar{\tau}_k - \tau_k + \upsilon_k}{\bar{\tau}_k\tau_k}\right),$$

one can write (4.65) as

$$\alpha_k = \frac{1}{\frac{\tau_k(\tau_k + \gamma\phi_k)}{(\tau_k - 1)(1 + \gamma\phi_k)}\theta_k + (1 - \tau_k)\frac{2\tau_k + \gamma\phi_k}{\tau_k(\tau_k + \gamma\phi_k)} - \frac{\tau_k(3\tau_k - 2) + \gamma\phi_k(2\tau_k - 1)}{\tau_k(\tau_k - 1)(\tau_k + \gamma\phi_k)}\theta_k\upsilon_k}. \tag{4.68}$$

In a similar way, using (4.67), the expression (4.66) can be easily rewritten as (4.68).
Thus, the third-step of (4.57) and (4.58) also coincides with each other. Therefore,
the iterations (4.57) and (4.58) are identical. $\qquad\square$

So the methods (4.58) can be considered as rediscovered variant of Zheng et al.
ones.

Now, we use the relations (4.67) in (4.62). After some algebraic manipulations
we again arrive at (4.68). It means that the third sub-step of iterations (4.57) and
(4.60) are the same. Therefore, the iterations (4.57), (4.58) and (4.60) can be written
in more convenient and unified form as

$$
y_k = x_k - \frac{f(x_k)}{f[x_k, w_k]},
$$
$$
z_k = y_k - \bar{\tau}_k \frac{f(y_k)}{f[x_k, w_k]},
$$
$$
x_{k+1} = z_k
$$
$$
- \frac{f(z_k)}{f[z_k, y_k] + (z_k - y_k)f[z_k, y_k, x_k] + (z_k - y_k)(z_k - x_k)f[z_k, y_k, x_k, w_k]},
$$
(4.69)

where $\bar{\tau}_k$ is given by (4.61) for (4.60) and is given by (4.64) for (4.57) and (4.58).
When $c = 1$, $d = -\hat{d}_k$ and $\omega = b = 0$ in (4.61), $\bar{\tau}_k$ coincides with (4.63). In this
case the iterations (4.57) and (4.58) and (4.69) are identical. So our iterations (4.60)
contain the methods (4.57) and (4.58) as particular cases. In addition, the iterations
(4.69) contain some well-known iterations as particular cases. For example, when
$c = 1$, $d = -1/(1 + \gamma\phi_k)$, $b = 0$ and $\omega = \hat{d}_k a/2$ then the iteration (4.69) produces
method given by Lotfi et al. in [12]. When $c = 1$, $d = \beta - 1 - \hat{d}_k$, $b = (2 - \beta)/(1 +
\gamma\phi_k)$, $\omega = -\beta$ in (4.61) then the iteration (4.69) produces the King's type method
[15]. When $c = 1$, $d = -1/(1 + \gamma\phi_k)$, $b = \omega = 0$ in (4.61) then the iteration (4.69)
produces method given by Sharma et al. in [25]. When $c = 1$, $d = -2\alpha - 1/(1 +
\gamma\phi_k)$, $b = 2\alpha/(1 + \gamma\phi_k)$, $\omega = H(\theta_k)$ in (4.61) then the iteration (4.69) produces the
Chebyshev-Halley type methods [7]. So, our iterations (4.60) or (4.69) represent a
wide class of optimal derivative-free iterations.

Now, we denote the method (4.69) with $c = 1$, $d = -\hat{d}_k$, $b = -1/(1 + \gamma\phi_k)$ and
$\omega = 0$ by M1. These parameters are chosen to have a large region of convergence
and a big basin of attraction for family (4.17). Moreover, the iteration (4.69) can be
rewritten as

$$
y_k = x_k - \frac{f(x_k)}{f[x_k, w_k]}, \quad w_k = x_k + \gamma f(x_k), \quad \gamma \in R \setminus \{0\},
$$
$$
z_k = \psi_4(x_k, y_k, z_k),
$$
$$
x_{k+1} = z_k
$$
(4.70)
$$
- \frac{f(z_k)}{f[z_k, y_k] + (z_k - y_k)f[z_k, y_k, x_k] + (z_k - y_k)(z_k - x_k)f[z_k, y_k, x_k, w_k]},
$$

where ψ_4 is any optimal fourth-order derivative-free method. From (4.70) we see that the each iteration of the family of derivative-free optimal three-point iterations (4.70) essentially depends on the choice ψ_4 or the choice of iteration parameter $\bar{\tau}_k$ in (4.69).

4.3.2 Convergence Analysis

Generally, the convergence properties of family of iterations (4.69) essentially depend on the convergence of iterations consisting of the first two sub-steps in (4.69), i.e.,

$$y_k = x_k - \frac{f(x_k)}{\phi(x_k)},$$

$$z_k = y_k - \bar{\tau}_k \frac{f(y_k)}{\phi(x_k)}, \qquad (4.71)$$

where $\bar{\tau}_k$ is given by (4.61). It is easy to show that if the iterations (4.71) converge then its convergence order is four. Moreover, if the iterations (4.71) converge, so does (4.69) with convergence order eight. From this clear that in order to establish the convergence of (4.69) it suffice to establish the convergence of iterations (4.71). To this end we use Taylor expansion of function $f \in C^2(D)$ and another form of second-step in (4.71) as

$$z_k = x_k - \tau_k \frac{f(x_k)}{\phi_k}, \qquad \tau_k = 1 + \bar{\tau}_k \theta_k.$$

As a result, we have

$$f(z_k) = \left(1 - \frac{f'(x_k)}{\phi_k}\tau_k + \frac{\zeta_k}{2}\frac{(f'(x_k))^2}{\phi_k^2}\tau_k^2\right) f(x_k), \qquad (4.72)$$

where

$$\zeta_k = \frac{f''(x_k)f(x_k)}{(f'(x_k))^2}.$$

From (4.72) it follows

$$|f(z_k)| \le \bar{q}|f(x_k)|, \qquad (4.73)$$

where

$$\bar{q} = \left|1 - \eta_k + \frac{\zeta_k}{2}\eta_k^2\right|, \qquad \eta_k = \frac{f'(x_k)}{\phi_k}\tau_k. \qquad (4.74)$$

From (4.73) we see that the convergence of iterations (4.71) is expected only when

$$\bar{q} \le 1. \tag{4.75}$$

Thus, it suffices to find conditions for which (4.75) holds true. It is easy to prove that.

Lemma 4.1 *Let the $\zeta_k \in (-2, 1)$. Then the inequality (4.75) holds true under conditions*

$$\begin{aligned}
0 < \eta_k < 2 \quad &when \quad 0 < \zeta_k < 1, \\
0 < \eta_k < 1 \quad &when \quad -2 < \zeta_k < 0.
\end{aligned} \tag{4.76}$$

Theorem 4.5 *Let $1 + \gamma \phi_k > 0$ and $\zeta_k \in (-2, 1)$. Then the two-point iterative methods (4.71) converge under condition*

$$|\theta_k| < 1 + \gamma \phi_k. \tag{4.77}$$

Proof Using the following relations

$$\frac{f'(x_k)}{\phi_k} = 1 - \frac{\gamma \phi_k}{1 + \gamma \phi_k} \theta_k + O\left(f^2(x_k)\right),$$

and

$$\tau_k = 1 + \theta_k + \hat{d}_k \theta_k^2 + \cdots,$$

in (4.74) we obtain

$$\eta_k = 1 + \frac{1}{1 + \gamma \phi_k} \theta_k + O\left(f^2(x_k)\right). \tag{4.78}$$

If we use (4.78) then the condition (4.76) can be written in term of θ_k as (4.77) within the accuracy $O\left(f^2(x_k)\right)$. In other words, (4.75) holds true under condition (4.77). $\qquad\square$

From (4.61) we obtain

$$\bar{\tau}_k - 1 = \frac{\theta_k(\hat{d}_k c + (\omega - b)\theta_k)}{c + d\theta_k + b\theta_k^2} = \frac{\theta_k \varphi(\theta_k)}{c + d\theta_k + b\theta_k^2}, \tag{4.79}$$

where

$$\varphi(\theta_k) = \hat{d}_k c + (\omega - b)\theta_k.$$

Let $|\omega - b| < \hat{d}_k c$. Then $\varphi(\theta_k) > 0$ on $\theta_k \in [-1, 1]$. Then from (4.79) we deduce that the following relations

$$\bar{\tau}_k \to 1, \quad \theta_k \to 0,$$

are equivalent and the convergence of sequences $f(z_k)$ and θ_k to zero as $k \to \infty$ expected simultaneously with equal order four. On the other hand, the iteration (4.71) can be considered as damped Newton's method

$$y_k = x_k - \frac{f(x_k)}{f'(x_k)} \eta_k,$$

with damping parameter η_k given by (4.78). As is known that, the damped Newton's method converges [31] if

$$0 < \eta_k < 2. \tag{4.80}$$

In term of θ_k the condition (4.80) gives the same result (4.78).

4.3.3 Dynamical Behavior of Derivative-Free Methods

We will make comparison of method M1 and other methods based on the dynamical behaviour. Further, we will use the abbreviated names for methods (see last column of Table 4.2). Now, we consider method (CH8) using the weight function $H(\theta_k) =$

Table 4.2 Choices of parameters

c	d	b	w	$\bar{\tau}_k$	Methods
1	$-\hat{d}_k$	0	0	$\dfrac{1}{1-\hat{d}_k\theta_k}$	(KS8), (Z8) [22, 23]
1	$-\dfrac{1}{1+\gamma\phi_k}$	0	$\dfrac{a\hat{d}_k}{2}$	$\dfrac{1+\theta_k+a\hat{d}_k\frac{\theta_k^2}{2}}{1-\frac{\theta_k}{1+\gamma\phi_k}}$	Lotfi (L8) [12]
1	$\beta-1-\hat{d}_k$	$\dfrac{2-\beta}{1+\gamma\phi_k}$	β	$\dfrac{1+(\beta-1)\theta_k+\beta\theta_k^2}{1+\left(\beta-2-\frac{1}{1+\gamma\phi_k}\right)\theta_k+\frac{\beta-2}{1+\gamma\phi_k}\theta_k^2}$	King's type (K8) [15]
1	$-\dfrac{1}{1+\gamma\phi_k}$	0	0	$\dfrac{1+\theta_k}{1-\frac{\theta_k}{1+\gamma\phi_k}}$	Sharma (S8) [25]
1	$-2\alpha-\dfrac{1}{1+\gamma\phi_k}$	$\dfrac{2\alpha}{1+\gamma\phi_k}$	$H(\theta_k)$	$\dfrac{1}{1-2\alpha\theta_k}\dfrac{H(\theta_k)}{1-\frac{\theta_k}{1+\gamma\phi_k}}$	Chebyshev-Halley (CH8) [7]
1	$-\hat{d}_k$	$\dfrac{\hat{d}_k^2}{4}$	0	$\dfrac{1}{(1-\frac{\hat{d}_k}{2}\theta_k)^2}$	[32]
1	$-\hat{d}_k$	$\dfrac{1}{1+\gamma\phi_k}$	0	$\dfrac{1}{1-\hat{d}_k\theta_k+\frac{1}{1+\gamma\phi_k}\theta_k^2}$	Thukral (T8) [21]
					Kung-Traub (KT8) [3]
1	$-\hat{d}_k$	$\dfrac{1}{1-\phi_k}$	0	$\dfrac{1}{\left(1-\frac{\theta_k}{1-\phi_k}\right)(1-\theta_k)}$	Soleymani (SS8) [33]
1	-1	0	$\dfrac{1}{(1+\gamma\phi_k)^2}$	$\left(1+\dfrac{\theta_k}{(1+\gamma\phi_k)}+\dfrac{\theta_k^2}{(1+\gamma\phi_k)^2}\right)\dfrac{1}{1-\theta_k}$	Soleymani (SV8) [5]
1	$-\hat{d}_k$	$-\dfrac{1}{1+\gamma\phi_k}$	0	$\dfrac{1}{1-\hat{d}_k\theta_k-\frac{\theta_k^2}{1+\gamma\phi_k}}$	M1

$1 + (1 - 2\alpha)\theta_k$ with values of the parameter $\alpha = 0, \pm 1$ (see [7]), method (L8) for $(a = 0, \pm 1)$ and method (K8) for $(\beta = 0, \pm 1)$. In addition to compare family (4.69) with other methods we also consider some optimal methods, which third substeps are different from method (4.69). Namely, we used the following substeps.

Derivative-free Thukral's three-step method (T8) [21] has the following substep:

$$x_{k+1} = z_k - \left(1 - \frac{f(z_k)}{f(w_k)}\right)^{-1}$$
$$\times \left(1 + \frac{2f^3(y_k)}{f^2(w_k)f(x_k)}\right)^{-1} \left(\frac{f(z_k)}{f[z_k, y_k] - f[x_k, y_k] + f[z_k, x_k]}\right).$$

Derivative-free Soleymani et al. three-step method (SS8) [33] has the following substep:

$$x_{k+1} = z_k - \frac{f(z_k)f(w_k)}{(f(w_k) - f(y_k))f[x_k, y_k]}$$
$$\times \left(1 + \frac{f(z_k)}{f(y_k)}\right)\left(1 + (2 - f[x_k, w_k])\frac{f(z_k)}{f(w_k)}\right)$$
$$\times \left(1 + \left(\frac{f(z_k)}{f(x_k)}\right)^2\right)\left(1 + (1 - f[x_k, w_k])\left(\frac{f(y_k)}{f(w_k)}\right)^2\right).$$

Derivative-free Soleymani et al. three-step method (SV8) [5] has the following substep:

$$x_{k+1} = z_k - \frac{f(z_k)}{f[z_k, y_k]}\left(1 - \frac{1}{f[x_k, w_k] - 1}\left(\frac{f(y_k)}{f(x_k)}\right)^2 + (2 - f[z_k, y_k])\frac{f(z_k)}{f(w_k)}\right).$$

Derivative-free Kung-Traub's three-step method (KT8) [3] has the following substep:

$$x_{k+1} = z_k - \frac{f(y_k)f(w_k)\left(y_k - x_k + \frac{f(x_k)}{f[x_k, z_k]}\right)}{(f(y_k) - f(z_k))(f(w_k) - f(z_k))}.$$

Generally, higher-order convergence methods consist of multi-steps which may use more evaluations of functions than the original one. In this case, multi-point methods may have the extraneous fixed points (black points). In order to find the extraneous fixed points, we rewrite any three-point method as [32]:

$$x_{k+1} = x_k - \frac{f(x_k)}{f[x_k, w_k]}H_f(x_k),$$

where $H_f = 1 + \theta_k(\bar{\tau}_k + \alpha_k \upsilon_k)$. Clearly, the root x^* of $f(x)$ is a fixed point of the method. The points $\xi \neq x^*$ for which $H_f(\xi) = 0$ are also fixed points of the method.

These fixed points are called extraneous fixed points. As we all know, a fixed point ξ is called:

- attractive if $|R'(\xi)| < 1$,
- repulsive if $|R'(\xi)| > 1$,
- parabolic if $|R'(\xi)| = 1$,

where $R(z) = z - f(z)/f[z, w]H_f(z)$ is the iteration function. In addition, if $|R'(\xi)| = 0$, the fixed point is superattracting. Now, we will discuss the extraneous fixed points of each method for comparison. To make it easier, we have taken the simple quadratic polynomial $p(z) = z^2 - 1$, whose roots are $z = \pm 1$.

In Table 4.3, we have collected the extraneous fixed points of the methods Z8, KS8, M1, L8, K8, S8, CH8. Next four methods T8, SS8, KT8 and SV8 are analyzed and found that they unable to compare with other methods. Because T8, SS8, KT8 and SV8 have 48, 94, 39, 52 extraneous fixed points respectively. Therefore, we have not include those results in Table 4.3.

For methods Z8 and KS8, we found that the methods have same ten extraneous fixed points. All fixed points are repulsive. For method L8, we have presented only the case $a = 0$. The reason is that other values of a lead to not good results. All fixed points are repulsive for this value of a. It is clear that method S8 is special case of method L8 when $a = 0$. Therefore, the results of finding are not appended to in Table 4.3. For methods CH8, we have presented the case $\alpha = 1$. The reason is that other values of α lead to not good results. All fixed points are repulsive. For K8 we have presented only the case $\beta = 1$. The reason is that other values of β lead to not good results. It is hence omitted. All fixed points are repulsive well as.

The basins of attraction of iterative methods is another tool for comparing them. Thus, we compare our methods (4.17) with other methods by using the basins of attraction for polynomials $p(z) = z^k - 1$, $k = 3, 4$.

In order to study dynamical behavior of the iterative methods, we consider as in [32] a square $[-3, 3] \times [-3, 3]$ of complex plane with a mesh of 600×600 points, which contains all roots of concerned nonlinear equation. The basins of attraction for 12 methods are displayed in Fig. 4.1. In this figure, the red, green and blue colors are assigned for the attraction basins of the three zeros and the roots of function are marked with white points. Black color is shown lack of convergence to any of the roots. In this cases, the stopping criterion $\varepsilon = 10^{-4}$ and maximum of 25 iterations are used. These dynamical planes have been generated by using the Mathematica 11. From Fig. 4.1 and Table 4.3, we can also see that methods M1 and Z8 is much more stable than the others. It can be observed from the figures that the methods M1 along with the existing methods Z8 have wide attraction basins to corresponding zeros than other methods. Z8 also has the least number of black points.

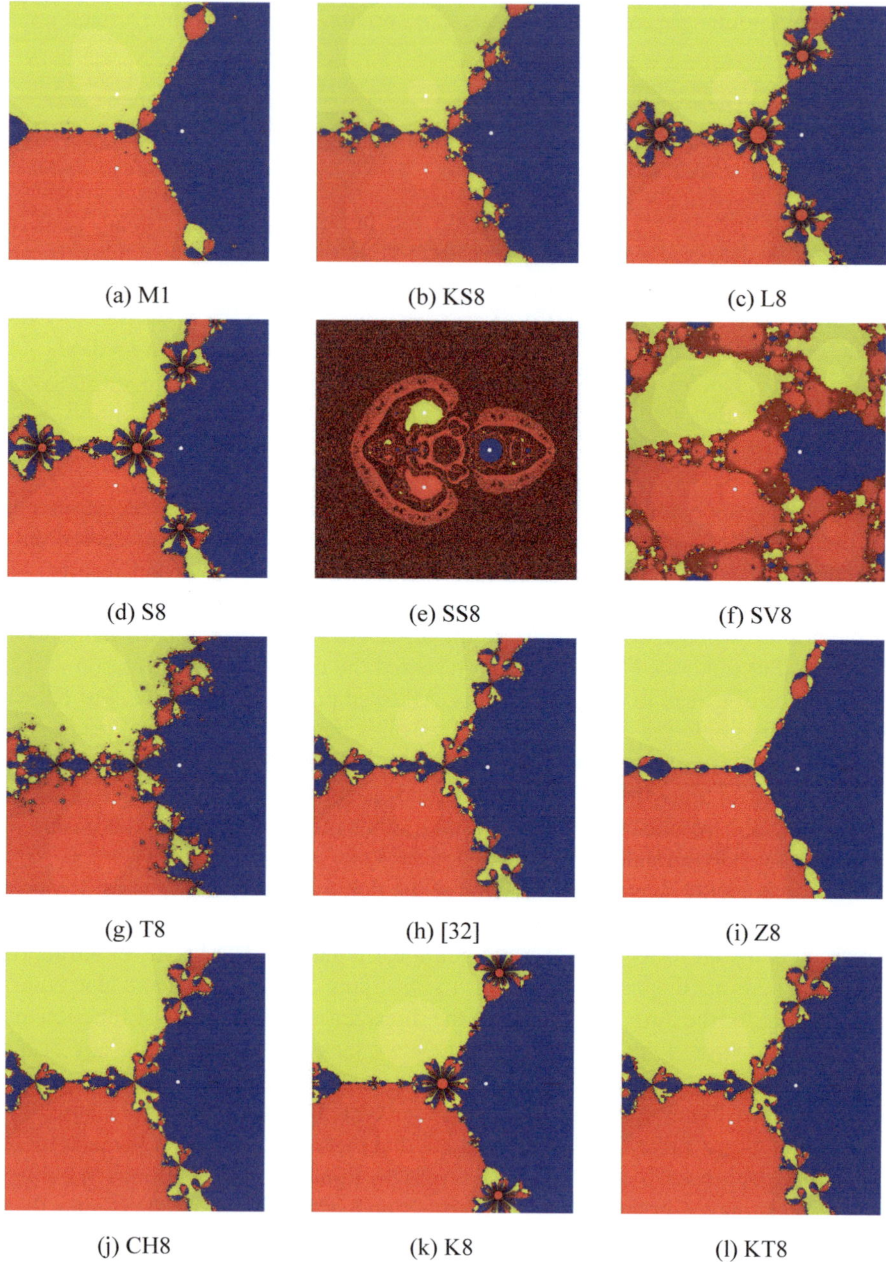

(a) M1　　　　　　(b) KS8　　　　　　(c) L8

(d) S8　　　　　　(e) SS8　　　　　　(f) SV8

(g) T8　　　　　　(h) [32]　　　　　　(i) Z8

(j) CH8　　　　　　(k) K8　　　　　　(l) KT8

Fig. 4.1 Basins of attraction of different derivative-free three-point iterations on $z^3 - 1$

Table 4.3 The extraneous fixed points

Methods	The extraneous fixed points ξ	Numbers of ξ
Z8	$-0.555220397255420 \pm 1.15928646739103\iota$	10
	$-0.460115602837211 \pm 0.456390703516719\iota$	
	$-0.450000501793328 \pm 0.129063966758804\iota$	
	$1.89303155290658 \pm 0.233570409469479\iota$	
	$1.79931236664623, 2.67863086464586$	
KS8	$-0.555220397255420 \pm 1.15928646739103\iota$	10
	$-0.460115602837211 \pm 0.456390703516719\iota$	
	$-0.450000501793328 \pm 0.129063966758804\iota$	
	$1.89303155290658 \pm 0.233570409469479\iota$	
	$1.79931236664623, 2.67863086464586$	
L8 $a = 0$	$-0.667945591214872 \pm 0.100425568541848\iota$	20
	$-0.664810542206285 \pm 0.236676931399034i$	
	$-0.256812572074558 \pm 0.169283171474748\iota$	
	$-0.235816064609120 \pm 0.101845050899904\iota$	
	$1.785182636 \pm .18650901311\iota$	
	$2.10884950227391 \pm 1.165904110639611\iota$	
	$2.176232698 \pm .5457413141\iota$	
	$2.15196391166621, 6.87682348522886$	
	$2.10230777448310, 1.71672219502143$	
M1	$-0.676558832763406 \pm 1.36018262584118\iota$	16
	$-0.624888463964184 \pm 0.20890104128772\iota$	
	$-0.493766364512498 \pm 0.607060501953625\iota$	
	$-0.461962845726289 \pm 0.221119195986523\iota$	
	$-0.204327487662501 \pm 0.86651046669376\iota$	
	$1.932083323 \pm 0.1163156841\iota$	
	$2.004864313 \pm 0.7365790432\iota$	
	$2.083325978 \pm 0.4554281653\iota$	

4.4 On Optimal Choices of Parameters in Two-Point Iterative Methods

4.4.1 Optimal Two-Point Iterations

The main goal is to find the optimal choices of parameters τ_k and γ, λ in the two point iterative methods. We consider the two-point iterations of the form (3.1). The sufficient fourth-order convergence condition (3.12) is often used not only to check the order of convergence of iterations (3.1), but also to derive new optimal methods. As a example, we consider the well-known cubically convergent Laguerre's family of iterations (or Hansen-Patrick's iterations) (3.1) with parameter τ_k given by

$$\tau_k = \frac{\alpha + 1}{\alpha + \text{sign}(\alpha)\sqrt{1 - (\alpha + 1)\frac{f''(x_k)f(x_k)}{(f'(x_k))^2}}}, \quad \alpha \neq -1. \tag{4.81}$$

Using Taylor expansion of $f(y_k)$ at point x_k, one can easily shown that

$$\theta_k = \frac{f''(x_k)f(x_k)}{2(f'(x_k))^2} + O\left(f^2(x_k)\right).$$

Then (4.81) leads to

$$\tau_k = \frac{\alpha + 1}{\alpha \pm \sqrt{1 - 2(\alpha + 1)\theta_k + O\left(f^2(x_k)\right)}}. \tag{4.82}$$

Neglecting the small term $O\left(f^2(x_k)\right)$ in (4.82), we get

$$\tau_k = \frac{\alpha + 1}{\alpha + \sqrt{1 - 2(\alpha + 1)\theta_k}}. \tag{4.83}$$

Kansel et al. in [34] was considered the iterations (3.1) with parameter defined by (4.83). Using the well-known relation

$$(1 - x)^\alpha = 1 - \alpha x + \frac{\alpha(\alpha - 1)}{2}x^2 + \frac{\alpha(\alpha - 1)(\alpha - 2)}{6}x^3 + \cdots, \quad |x| < 1, \tag{4.84}$$

one can easily shown that (4.83) has the following asymptotic

$$\tau_k = 1 + \theta_k + \frac{\alpha + 3}{2}\theta_k^2 + O\left(\theta_k^3\right). \tag{4.85}$$

Comparison of (4.85) with (3.12) implies that the iterations (3.1) with parameter τ_k given by (4.83) is not optimal. The exception is only for $\alpha = 1$ i.e.,

$$\tau_k = \frac{2}{1 + \sqrt{1 - 4\theta_k}}, \tag{4.86}$$

which satisfies the condition (3.12). It should be mentioned that, via accelerating procedure for τ_k was obtained in Sect. 1.3.3 fourth-order iterations (3.1) with τ_k given by (4.86). By this reason the value defined by (4.86) called the optimal one. Note also that in [35] was made an attempt to search optimal parameter α of Lagguerre's family from the convergence point of view.

Generally, the iteration (3.1) with parameter τ_k given by (4.83) is only third-order of convergence. Using the condition (3.12) one can find optimal fourth-order modification of Hansen-Patrick's family. To this end, we seek for

$$\tau_k = \frac{\alpha + 1}{\alpha + \sqrt{1 - 2(\alpha + 1)\theta_k}} H(\theta_k), \tag{4.87}$$

where H is real valued function satisfying the conditions

$$H(0) = 1, \quad H'(0) = a, \quad H''(0) = 2b. \tag{4.88}$$

We find a and b in (4.88) such that (4.87) satisfies the condition (3.12). Using Taylor expansion of $H(\theta_k)$ and (4.85) in (4.87) we obtain

$$\tau_k = 1 + (a + 1)\theta_k + \left(a + b + \frac{\alpha + 3}{2}\right)\theta_k^2 + \cdots . \tag{4.89}$$

The comparison (4.89) with (3.12) gives

$$a = 0, \quad b = \frac{1 - \alpha}{2}.$$

Thus, we obtain optimal fourth-order variant of Hansen-Patrick's family (3.1) with parameter given by

$$\tau_k = \frac{\alpha + 1}{\alpha + \sqrt{1 - 2(\alpha + 1)\theta_k}}\left(1 + \frac{1 - \alpha}{2}\theta_k^2\right), \quad \alpha \neq -1. \tag{4.90}$$

When $\alpha = 1$, (4.90) leads to (4.86). In general, one can pass from any third-order iterations (3.1) to optimal two-point iterations, using the condition (3.12). Analogously, one can easily shown that, the iterative methods (3.1) has a optimal fourth-order of convergence if

$$\tau_k = \frac{\alpha + 1}{\alpha + \sqrt{1 - 2(\alpha + 1)\theta_k}} + \frac{1 - \alpha}{2}\theta_k^2.$$

Note that in [9], the authors proposed a new optimal fourth-order modification of Hansen-Patrick's family (3.1) with parameter given by

$$\tau_k = \frac{\alpha + 1}{\alpha + \sqrt{\frac{1 - (\alpha + 3)\theta_k - (\alpha^2 - 1)\theta_k^2}{1 + (\alpha - 1)\theta_k}}}. \tag{4.91}$$

In spite of being optimal with efficiency index $EI = 4^{1/3} \approx 1.587$ the iterations (3.1) require the evaluation of first-order derivative at each iteration step and hence cannot be applied to non-smooth functions equations. In the preceding subsection was given a rule for transition of iterations to their optimal derivative-free variant and vise versa. According to this rule one can easily obtain the derivative-free variant of (3.1) with (4.91). This has a form

$$y_k = x_k - \frac{f(x_k)}{\phi_k},$$

$$x_{k+1} = x_k - \tau_k \frac{f(x_k)}{\phi_k} \quad \text{or} \quad x_{k+1} = y_k - \bar{\tau}_k \frac{f(y_k)}{\phi_k}, \qquad (4.92)$$

where

$$w_k = x_k + \gamma f(x_k), \quad \phi_k = f[x_k, w_k],$$

and

$$\tau_k = 1 + \bar{\tau}_k \theta_k,$$

$$\bar{\tau}_k = \frac{1}{\theta_k}\left(\frac{\alpha + 1}{\alpha + \sqrt{\frac{1-(\alpha+3)\theta_k-(\alpha^2-1)\theta_k^2}{1+(\alpha-1)\theta_k}}} - 1 \right) + (\hat{d}_k - 2)\theta_k, \qquad (4.93)$$

$$\hat{d}_k = \frac{2 + \gamma\phi_k}{1 + \gamma\phi_k}.$$

On the other hand, using the sufficient fourth-order convergence condition [1]

$$\bar{\tau}_k = 1 + \hat{d}_k\theta_k + O\left(f^2(x_k)\right),$$

for (4.92) one can easily construct the derivative-free variant of iterations (3.1), (4.90). It can be written as (4.92) with parameter

$$\bar{\tau}_k = \frac{1}{\theta_k}\left(\frac{\alpha + 1}{\alpha + \sqrt{1 - 2(\alpha + 1)\theta_k}}\left(1 + \left(\hat{d}_k - 2 - \frac{\alpha - 1}{2}\right)\theta_k^2\right) - 1 \right). \quad (4.94)$$

Thus, we have the families of derivative-free Hansen-Patrick's type iterations (4.92) with parameter given by two choices (4.93) and (4.94).

Remark 4.2 In general, we can consider the following weight function

$$W(\theta_k, \alpha, m) = \frac{\alpha + 1}{\alpha + \sqrt[m]{1 - m(\alpha + 1)\theta_k}} = 1 + \theta_k + \left(1 - \frac{1 - m}{2}(\alpha + 1)\right)\theta_k^2 + \cdots$$

in the iteration (3.1). We call $W(\theta_k, \alpha, m)$ the generalized Hansen-Patrick's type weight function. $W(\theta_k, \alpha, 2)$ leads to (4.83). It easy to show that the iterative methods (3.1) has a optimal fourth-order of convergence when τ_k satisfies one of the following conditions

$$\tau_k = W(\theta_k, \alpha, m) + \left(1 + \frac{1 - m}{2}(\alpha + 1)\right)\theta_k^2$$

and

$$\tau_k = W(\theta_k, \alpha, m)H(\theta_k),$$

where H is real valued function satisfying the following conditions

$$H(0) = 1, \quad H'(0) = 0, \quad H''(0) = 2\left(1 + \frac{1-m}{2}(\alpha + 1)\right).$$

As for H one can choose, for examples, the following functions

$$H_1 = 1 + \left(1 + \frac{1-m}{2}(\alpha + 1)\right)\theta_k^2,$$

$$H_2 = \frac{1}{1 - \left(1 + \frac{1-m}{2}(\alpha + 1)\right)\theta_k^2},$$

$$H_3 = \sqrt{1 + (2 + (1 - m)(\alpha + 1))\,\theta_k^2}$$

so on.

4.4.2   Two Parametric Optimal Fourth-Order Iterations

Now we consider two parametric derivative-free iterations

$$y_k = x_k - \frac{f(x_k)}{\phi_k + \lambda f(w_k)}, \quad \lambda \in R,$$

$$x_{k+1} = y_k - \bar{\tau}_k \frac{f(y_k)}{\phi_k + \lambda f(w_k)}, \tag{4.95}$$

where $w_k = x_k + \gamma f(x_k)$, $\gamma \in R$, $\phi_k = f[x_k, w_k] = (f(w_k) - f(x_k))/(\gamma f(x_k))$. Our aim is to find $\bar{\tau}_k$ in (4.95) such that the iterations (4.95) has optimal fourth-order convergence. To do this we first use the Taylor expansion of function $f(w_k) = f(x_k)(1 + \gamma \phi_k)$ at point x_k. Then we have

$$\phi_k = f'(x_k)\left(1 + \frac{a_k}{2}f'(x_k)\gamma\right) + O\left(f^2(x_k)\right), \tag{4.96}$$

where

$$a_k = \frac{f''(x_k)f(x_k)}{(f'(x_k))^2}. \tag{4.97}$$

Let $\eta_k = f'(x_k)/\phi_k$. Then using (4.96), we get

$$\eta_k = \frac{1}{1 + \frac{a_k}{2}f'(x_k)\gamma + O\left(f^2(x_k)\right)} = 1 - \frac{a_k}{2}f'(x_k)\gamma + O\left(f^2(x_k)\right). \tag{4.98}$$

The Taylor expansion of $f(y_k)$ at point x_k gives

$$
\begin{aligned}
f(y_k) &= f(x_k)\left(1 - \eta_k\left(1 - \frac{\lambda f(w_k)}{\phi_k}\right)\right) + O\left(f^2(x_k)\right) \\
&= f(x_k)\left(1 - \left(1 - \frac{a_k}{2}f'(x_k)\gamma\right)\left(1 - \frac{\lambda f(w_k)}{\phi_k}\right)\right) + O\left(f^2(x_k)\right).
\end{aligned}
\tag{4.99}
$$

According to (4.97) we have $f(y_k) = O\left(f^2(x_k)\right)$. Analogously, from the second step in (4.95) we obtain

$$
f(x_{k+1}) = \left(1 - \bar{\tau}_k\frac{f'(y_k)}{\phi_k + \lambda f(w_k)}\right)f(y_k) + O\left(f^2(y_k)\right).
\tag{4.100}
$$

From (4.99) and (4.100) it clear that

$$
f(x_{k+1}) = O\left(f^4(x_k)\right),
$$

if we choose $\bar{\tau}_k$ such that

$$
1 - \bar{\tau}_k\frac{f'(y_k)}{\phi_k + \lambda f(w_k)} = O\left(f^2(x_k)\right),
$$

or

$$
\bar{\tau}_k = \frac{\phi_k + \lambda f(w_k)}{f'(y_k)} + O\left(f^2(x_k)\right).
\tag{4.101}
$$

Taylor expansion of $f'(y_k)$ gives

$$
f'(y_k) = f'(x_k)\left(1 - \frac{f''(x_k)f(x_k)}{f'(x_k)\phi_k(1 + \frac{\lambda f(w_k)}{\phi_k})}\right) + O\left(f^2(x_k)\right).
$$

Using (4.97) and (4.98) in the last relation we obtain

$$
f'(y_k) = f'(x_k)(1 - a_k) + O\left(f^2(x_k)\right).
\tag{4.102}
$$

Substituting (4.96) and (4.102) into (4.101) we get

$$\bar{\tau}_k = \frac{1 + \frac{a_k}{2} f'(x_k)\gamma + \lambda \frac{(1+\gamma\phi_k)f(x_k)}{f'(x_k)} + O\left(f^2(x_k)\right)}{1 - a_k + O\left(f^2(x_k)\right)}$$

$$= \left(1 + \frac{a_k}{2} f'(x_k)\gamma + \frac{\lambda(1+\gamma\phi_k)f(x_k)}{f'(x_k)}\right)(1 + a_k) + O\left(f^2(x_k)\right) \quad (4.103)$$

$$= 1 + \left(1 + \frac{f'(x_k)\gamma}{2}\right)a_k + \frac{\lambda(1+\gamma\phi_k)f(x_k)}{f'(x_k)} + O\left(f^2(x_k)\right).$$

According to (4.97) and (4.98) we have $f'(x_k) = \phi_k + O\left(f(x_k)\right)$. Therefore, we can replace $f'(x_k)$ by ϕ_k in (4.103) without loss of accuracy and as a result, we get

$$\bar{\tau}_k = 1 + \frac{2 + \gamma\phi_k}{2} a_k + \frac{\lambda(1+\gamma\phi_k)f(x_k)}{\phi_k} + O\left(f^2(x_k)\right).$$

Further using Taylor expansion of $f(y_k)$ and (4.98) one can easily obtain

$$\theta_k = \frac{a_k}{2}(1 + \gamma f'(x_k)) + \frac{\lambda(1+\gamma\phi_k)}{\phi_k} f(x_k) + O\left(f^2(x_k)\right)$$

$$= \frac{a_k}{2}(1 + \gamma\phi_k) + \frac{\lambda(1+\gamma\phi_k)}{\phi_k} f(x_k) + O\left(f^2(x_k)\right) \quad (4.104)$$

$$= (1 + \gamma\phi_k)\left(\frac{a_k}{2} + \lambda\frac{f(x_k)}{\phi_k}\right) + O\left(f^2(x_k)\right).$$

From this we find

$$\frac{a_k}{2} = \frac{\theta_k}{1 + \gamma\phi_k} - \frac{\lambda f(x_k)}{\phi_k} + O\left(f^2(x_k)\right). \quad (4.105)$$

Substituting (4.105) into (4.104) we obtain

$$\bar{\tau}_k = 1 + \hat{d}_k\theta_k - \frac{\lambda f(x_k)}{\phi_k} + O\left(f^2(x_k)\right). \quad (4.106)$$

Thus, we can formulate the obtained results as follows.

Theorem 4.6 *Assume that function $f : D \subset R \to R$ is sufficiently differentiable and has a simple zero $x^* \in D$. The initial guess x_0 is sufficiently close to x^* and the parameter $\bar{\tau}_k$ satisfies the condition (4.106). Then the iterative methods defined by (4.95) has a optimal fourth-order convergence.*

Kansal et al. in [34] proposed a new tri-parametric optimal derivative-free family of Hansen-Patrick's type iterations

$$y_k = x_k - \frac{f(x_k)}{\phi_k + \lambda f(w_k)},$$

$$x_{k+1} = y_k - \bar{\tau}_k \frac{f(y_k)}{f[y, w_k] + \lambda f(w_k)}, \qquad (4.107)$$

where

$$\bar{\tau}_k = \frac{1}{\theta_k}\left(\frac{\alpha + 1}{\alpha + \sqrt{1 - 2(\alpha + 1)\theta_k}} - 1\right) H(\theta_k).$$

Here H is a real variable weight function satisfying condition

$$H(0) = 1, \quad H'(0) = -\frac{\alpha + 1}{2}, \quad |H''(0)| < \infty. \qquad (4.108)$$

Note that the iterations (4.107) has a dissimilarity in the denominator of the second step as compared with (4.95). Using the easily verified relation

$$f[y, w_k] + \lambda f(w_k) = (\phi_k + \lambda f(w_k))$$
$$\times \left(1 - \frac{\phi_k \theta_k - \lambda f(w_k)}{(1 + \gamma \phi_k)(\phi + \lambda f(w_k))}\right) + O\left(f^2(x_k)\right), \qquad (4.109)$$

one can rewrite the second step in (4.107) as:

$$x_{k+1} = y_k - \bar{\tau}_k \frac{f(y_k)}{\phi_k + \lambda f(w_k)},$$

where

$$\bar{\tau}_k = \frac{1}{\theta_k}\left(1 + \frac{\phi_k \theta_k - \lambda f(w_k)}{(1 + \gamma \phi_k)(\phi + \lambda f(w_k))}\right)\left(\frac{\alpha + 1}{\alpha + \sqrt{1 - 2(\alpha + 1)\theta_k}} - 1\right) H(\theta_k). \qquad (4.110)$$

It is easy to show that $\bar{\tau}_k$ defined by (4.110) satisfies the condition (4.106). Thus we prove that the iterations (4.107), (4.108) has fourth-order convergence without using symbolic computation used in [34]. Thus, the two parametric iteration (4.95) with $\bar{\tau}_k$ given by (4.110) represent new variant of derivative-free family of Hansen-Patrick's type iterations. Analogously, using formula (4.109) one can easily shown that the two-parametric fourth-order derivative-free methods given in [7, 16, 23, 36] satisfies the condition (4.106).

4.5 Optimal Sixth and Seventh-Order Iterations

Let $\gamma = 0$ in (4.95). Then (4.95) leads to one parameter iterations

$$
\begin{aligned}
y_k &= x_k - \frac{f(x_k)}{f'(x_k) + \lambda f(x_k)}, \\
x_{k+1} &= y_k - \bar{\tau}_k \frac{f(y_k)}{f'(x_k) + \lambda f(x_k)}.
\end{aligned}
\tag{4.111}
$$

By Theorem 4.6 the iterations (4.111) has an optimal fourth-order convergence, when $\bar{\tau}_k$ defined by

$$
\bar{\tau}_k = 1 + 2\theta_k - \frac{\lambda f(x_k)}{f'(x_k)} + O\left(f^2(x_k)\right).
$$

The iterations (4.111) require three functional evaluations $f(x_k)$, $f(y_k)$ and $f'(x_k)$. The efficiency index of the iterations is $EI = \sqrt[3]{4} \approx 1.587$. Now, we shall try to find the optimal value of free parameter λ in (4.111). To this end, first we use Taylor expansion of $f(y_k)$ at point x_k and the relation (4.84). As a result, we get

$$
\begin{aligned}
f(y_k) = f^2(x_k) &\left(\frac{\lambda}{f'(x_k)} + \frac{f''(x_k)}{2(f'(x_k))^2} - \frac{\lambda^2 f(x_k)}{(f'(x_k))^2} \right. \\
&\left. - \frac{\lambda f(x_k) f''(x_k)}{(f'(x_k))^3} - \frac{f'''(x_k) f(x_k)}{6(f'(x_k))^3} \right) + O\left(f^4(x_k)\right).
\end{aligned}
\tag{4.112}
$$

It means that $f(y_k) = O\left(f^2(x_k)\right)$ for any λ. If we choose

$$
\lambda = \lambda_k = -\frac{f''(x_k)}{2f'(x_k)},
\tag{4.113}
$$

in (4.112), then we get $f(y_k) = O\left(f^3(x_k)\right)$. We call the value λ_k defined by (4.113) optimal one in the sense that it increases the convergence order of sequence y_k from 2 to 3. Under choice (4.113) the expression (4.112) becomes

$$
f(y_k) = f(x_k) \left(\frac{a_k^2}{4} - \frac{f'''(x_k) f^2(x_k)}{6(f'(x_k))^3} \right) + O\left(f^4(x_k)\right).
$$

From this it follows that

$$
\theta_k = \frac{a_k^2}{4} - \frac{f'''(x_k) f^2(x_k)}{6(f'(x_k))^3} + O\left(f^3(x_k)\right),
$$

which implies

$$\frac{f'''(x_k)f^2(x_k)}{(f'(x_k))^3} = \frac{3}{2}a_k^2 + O\left(f^3(x_k)\right) - 6\theta_k. \tag{4.114}$$

Now, we consider the Taylor expansion of $f(x_{k+1})$ at point y_k

$$f(x_{k+1}) = f(y_k)\left(1 - \frac{f'(y_k)}{f'(x_k)}\bar{\tau}_k\left(1 - \frac{\lambda f(x_k)}{f'(x_k)}\right)\right) + O\left(f^2(y_k)\right). \tag{4.115}$$

Above, we show that $f(y_k) = O\left(f^3(x_k)\right)$ under choice (4.113). Hence, from (4.115) it clear that

$$f(x_{k+1}) = O\left(f^6(x_k)\right), \tag{4.116}$$

if we choose $\bar{\tau}_k$ such that

$$1 - \frac{f'(y_k)}{f'(x_k)}\bar{\tau}_k\left(1 - \frac{\lambda f(x_k)}{f'(x_k)}\right) = O\left(f^3(x_k)\right)$$

or

$$\bar{\tau}_k = \frac{1}{1 - \frac{\lambda f(x_k)}{f'(x_k)}}\frac{f'(x_k)}{f'(y_k)} + O\left(f^3(x_k)\right). \tag{4.117}$$

Using the Taylor expansion of $f'(y_k)$ at point x_k, one can easily shown that

$$f'(y_k) = f'(x_k)\left(1 - a_k - \frac{a_k^2}{2} + \frac{f'''(x_k)f^2(x_k)}{2(f'(x_k))^3}\right) + O\left(f^3(x_k)\right).$$

From this and taking into account (4.114), we obtain

$$\frac{f'(x_k)}{f'(y_k)} = \frac{1}{1 - a_k + a_k^2/4 - 3\theta_k + O\left(f^3(x_k)\right)}$$

$$= 1 + a_k + \frac{3a_k^2}{4} + 3\theta_k + O\left(f^3(x_k)\right). \tag{4.118}$$

Substituting (4.113) and (4.118) into (4.117), we get

$$\bar{\tau}_k = 1 + \frac{a_k}{2} + \frac{a_k^2}{4} + 3\theta_k + O\left(f^3(x_k)\right). \tag{4.119}$$

This means that the relation (4.116) holds when the elections (4.119) and (4.113). So we have.

Theorem 4.7 *Suppose that the function $f : D \subset R \to R$ is sufficiently differentiable and has a simple zero $x^* \in D$. Let the initial approximation x_0 be close enough*

*to x^*, and the parameters λ_k and $\bar{\tau}_k$ satisfy the conditions (4.113) and (4.119). Then
the iterative methods (4.111) have sixth-order convergence.*

4.5.1 New Two Parametric Optimal Sixth-Order Methods

Based on Theorem 4.7, the following two parametric sixth-order method is formulated:

$$
\begin{aligned}
&x_0, \lambda_0 \quad \text{are given. Then,} \\[2mm]
&\lambda_k = -\frac{\Delta_k}{2 f'(x_k)}, \qquad k = 1, 2, \cdots, \\[2mm]
&\bar{\tau}_k = 1 - \frac{\lambda_k f(x_k)}{f'(x_k)} + \left(\frac{\lambda_k f(x_k)}{f'(x_k)} \right)^2 + 3\theta_k, \\[2mm]
&y_k = x_k - \frac{f(x_k)}{f'(x_k) + \lambda_k f(x_k)}, \\[2mm]
&x_{k+1} = y_k - \bar{\tau}_k \frac{f(y_k)}{f'(x_k) + \lambda_k f(x_k)}, \quad k = 0, 1, \cdots,
\end{aligned}
\tag{4.120}
$$

where

$$
\Delta_k = \frac{f(x_k + \gamma f(x_k)) - 2 f(x_k) + f(x_k - \gamma f(x_k))}{(\gamma f^2(x_k))}, \quad \gamma \in R \setminus \{0\}.
$$

Remark 4.3 The error equation obtained by symbolic computation plays an important role to construct new derivative-free methods with memory see [4, 8, 9, 16, 23,
36–38] and references therein. For instance with choice

$$
\lambda = -c_2 = -\frac{f''(x^*)}{2 f'(x^*)}, \quad \lim_{k \to \infty} \lambda_k = \lambda,
$$

the order of convergence of methods increased, while we have exact analytical formula (4.113) at each iteration step.

Analogously, it is easy to show that

$$
f(x_{k+1}) = O\left(f^5(x_k) \right) \quad \text{if} \quad \bar{\tau}_k = 1 + \frac{a_k}{2} + O\left(f^2(x_k) \right).
\tag{4.121}
$$

It should be pointed out that similar to (4.111) methods were considered by Wang et
al. in [39]. They have a form

$$y_k = x_k - \frac{f(x_k)}{f'(x_k) + \lambda f(x_k)},$$

$$x_{k+1} = y_k - \frac{f(y_k)}{f'(x_k) + \gamma f(x_k)} G(\theta_k), \quad \lambda, \gamma \in R, \tag{4.122}$$

and shown that it is of fourth-order convergence when

$$\gamma = 2\lambda, \quad G(0) = 1, \quad G'(0) = 2 \quad \text{and} \quad |G''(0)| < \infty. \tag{4.123}$$

They also used self-accelerating parameter

$$\lambda_k = -\frac{H_m''(x_k)}{2 H_m'(x_k)}, \quad m = 2, 3, 4 \tag{4.124}$$

in (4.122) and proven that the p-order of convergence of iterative methods (4.122) with parameter given by (4.124) with memory is at least $(5 + \sqrt{17})/2 \approx 4.5616$, $(5 + \sqrt{21})/2 \approx 4.7913$ and 5, respectively. Here $H_m(x_k)$ is the Hermite interpolating polynomial with degree $m = 2, 3, 4$ satisfying the condition $H_m'(x_k) = f'(x_k)$. The iterations (4.122) and (4.123) can be rewritten as (4.111) with

$$\bar{\tau}_k = \left(1 - \frac{\lambda f(x_k)}{f'(x_k) + \lambda f(x_k)} + \cdots\right)(1 + 2\theta_k + \cdots) = 1 + \frac{a_k}{2} + O\left(f^2(x_k)\right),$$

i.e., $\bar{\tau}_k$ satisfies the condition (4.121). If we choose λ_k as the parameter in method (4.120), then the iteration with memory:

$$x_0, \lambda_0 \quad \text{are} \quad \text{given,} \quad \text{then,}$$

$$\lambda_k = -\frac{\Delta_k}{2 f'(x_k)}, \quad a_k = -\frac{2\lambda_k f(x_k)}{f'(x_k)}, \quad \bar{\tau}_k = 1 + \frac{a_k}{2},$$

$$y_k = x_k - \frac{f(x_k)}{f'(x_k) + \lambda_k f(x_k)},$$

$$x_{k+1} = y_k - \bar{\tau}_k \frac{f(y_k)}{f'(x_k) + \lambda_k f(x_k)},$$

has a fifth-order of convergence.

Remark 4.4 As above mentioned that when considered the Hansen-Patrick's type of iterations the $\bar{\tau}_k$ is given by (4.110).

4.5.2 *Seventh-Order Iterative Method with Memory*

Now we proceed to construct new iterative methods with memory from (4.95) using two self-accelerating parameters γ and λ. It is easy to show that

$$w_k = x_k - \frac{f(x_k)}{f'(x_k)},$$

and

$$f(w_k) = \frac{f''(x_k)f^2(x_k)}{2(f'(x_k))^2} + O\left(f^3(x_k)\right), \tag{4.125}$$

under choice

$$\gamma = \gamma_k = -\frac{1}{f'(x_k)}. \tag{4.126}$$

Let $f(x_k) \in C^4(D)$. Using the Taylor expansion for $f(w_k)$ and (4.126) we have

$$\phi_k = f'(x_k)\left(1 - \frac{a_k}{2} + \frac{f'''(x_k)f^2(x_k)}{6(f'(x_k))^3}\right) + O\left(f^3(x_k)\right).$$

Hence

$$\eta_k = \frac{f'(x_k)}{\phi_k} = 1 + \frac{a_k}{2} + \frac{a_k^2}{4} - \frac{f'''(x_k)f^2(x_k)}{6(f'(x_k))^3} + O\left(f^3(x_k)\right). \tag{4.127}$$

The Taylor expansion of $f(y_k)$ around point x_k gives

$$f(y_k) = f(x_k)\left(1 - \frac{\eta_k}{1 + \frac{\lambda f(w_k)}{\phi_k}} + \frac{a_k}{2}\left(\frac{\eta_k}{1 + \frac{\lambda f(w_k)}{\phi_k}}\right)^2\right.$$
$$\left. - \frac{f'''(x_k)f^2(x_k)}{6(f'(x_k))^3}\left(\frac{\eta_k}{1 + \frac{\lambda f(w_k)}{\phi_k}}\right)^3\right) + O\left(f^4(x_k)\right). \tag{4.128}$$

By virtue of (4.125) and (4.127) we have

$$\frac{\eta_k}{1 + \frac{\lambda f(w_k)}{\phi_k}} = \left(1 + \frac{a_k}{2} + \frac{a_k^2}{4} - \frac{f'''(x_k)f^2(x_k)}{6(f'(x_k))^3} + \cdots\right)\left(1 - \frac{\lambda f(w_k)}{\phi_k} + \cdots\right)$$
$$= \left(1 + \frac{a_k}{2} + \frac{a_k^2}{4} - \frac{f'''(x_k)f^2(x_k)}{6(f'(x_k))^3} - \frac{\lambda f(w_k)}{\phi_k} + (f^3(x_k))\right). \tag{4.129}$$

Using (4.129) in (4.128) we obtain

$$f(y_k) = f(x_k) \left(-\frac{a_k}{2} - \frac{a_k^2}{4} + \frac{f'''(x_k)f^2(x_k)}{6(f'(x_k))^3} + \frac{\lambda f(w_k)}{\phi_k} \right.$$
$$\left. + \frac{a_k}{2}(1 + a_k) - \frac{f'''(x_k)f^2(x_k)}{6(f'(x_k))^3} \right) + O\left(f^4(x_k)\right)$$
$$= f(x_k) \left(\frac{a_k^2}{4} + \frac{\lambda f(w_k)}{\phi_k} \right) + O\left(f^4(x_k)\right). \tag{4.130}$$

From (4.130) it is clear that

$$f(y_k) = O\left(f^4(x_k)\right), \tag{4.131}$$

if

$$\frac{a_k^2}{4} + \frac{\lambda f(w_k)}{\phi_k} = 0,$$

or

$$\lambda_k = -\frac{a_k^2 \phi_k}{4 f(w_k)}. \tag{4.132}$$

Using (4.125) and (4.127) in (4.132) we get

$$\lambda_k = -\frac{f''(x_k)}{2 f'(x_k)}, \tag{4.133}$$

i.e., the relation (4.131) holds under choice (4.133). Further from (4.100) and (4.131) it clear that

$$f(x_{k+1}) = O\left(f^7(x_k)\right),$$

if

$$\bar{\tau}_k = -\frac{\phi_k + \lambda f(w_k)}{f'(y_k)} + O\left(f^3(x_k)\right),$$

or

$$\bar{\tau}_k = \frac{\phi_k}{f'(y_k)}(1 - \frac{a_k^2}{4}) + O\left(f^3(x_k)\right). \tag{4.134}$$

The Taylor expansion of $f'(y_k)$ at point x_k gives

$$f'(y_k) = f'(x_k)$$
$$\times \left(1 - a_k \left(\frac{\eta_k}{1 - \frac{a_k^2}{4}} \right) + \frac{f'''(x_k)f^2(x_k)}{2(f'(x_k))^3} \left(\frac{\eta_k}{1 - \frac{a_k^2}{4}} \right)^2 \right) + O\left(f^3(x_k)\right). \tag{4.135}$$

Since

$$\frac{\eta_k}{1 - \frac{a_k^2}{4}} = \left(1 + \frac{a_k}{2} + \frac{a_k^2}{4} - \frac{f'''(x_k)f^2(x_k)}{6(f'(x_k))^3} + \cdots\right)\left(1 + \frac{a_k^2}{4} + \cdots\right)$$

$$= 1 + \frac{a_k}{2} + \frac{a_k^2}{2} - \frac{f'''(x_k)f^2(x_k)}{6(f'(x_k))^3} + O\left(f^3(x_k)\right).$$

From (4.135) we get

$$f'(y_k) = f'(x_k)\left(1 - a_k - \frac{a_k^2}{2} + \frac{f'''(x_k)f^2(x_k)}{2(f'(x_k))^3}\right) + O\left(f^3(x_k)\right).$$

Using last expression and (4.127) in (4.134) we get

$$\bar{\tau}_k = 1 - \frac{a_k}{2} + \frac{3}{4}a_k^2 + 2(1 + \gamma_k\phi_k) + O\left(f^3(x_k)\right),$$

in which we have used the formula

$$1 + \gamma_k\phi_k = \frac{a_k}{2} - \frac{f'''(x_k)f^2(x_k)}{6(f'(x_k))^3} + O\left(f^3(x_k)\right). \tag{4.136}$$

Thus, we can formulate the obtained results as follows.

Theorem 4.8 *Assume that function $f : D \subset R \to R$ is sufficiently differentiable and has a simple zero $x^* \in D$. The initial guess x_0 is sufficiently close to x^* and the parameters γ and λ in (4.95) are chosen as (4.126) and (4.133) and $\bar{\tau}_k$ is defined by formula (4.106), (4.136). Then the iterative methods (4.95) has a convergence order seven.*

Thus the optimal choices of parameters allow to increase the convergence order from 4 to 7. However, the values of $f'(x_k)$ and $f''(x_k)$ are not available in practice and such acceleration of convergence can not be realized. But we would approximate the parameters γ_k and λ_k. They can be computed by using information available from the current and previous iterations.

Based on the choices (4.126) and (4.133) one can construct new seventh-order derivative-free two-point iterations with memory:

$$x_0, \lambda_0, \gamma_0 \quad \text{are} \quad \text{given}, \quad \text{then}$$

$$w_0 = x_0 + \gamma_0 f(x_0), \quad \gamma_k = -\frac{1}{N_3'(x_k)},$$

$$w_k = x_k + \gamma_k f(x_k), \quad \lambda_k = -\frac{N_4''(x_k)}{2N_4'(x_k)}, \quad k = 1, 2, \ldots,$$

$$y_k = x_k - \frac{f(x_k)}{\phi_k + \lambda_k f(w_k)},$$

$$x_{k+1} = y_k - \bar{\tau}_k \frac{f(y_k)}{\phi_k + \lambda_k f(w_k)}, \quad k = 0, 1, \ldots,$$

where $\bar{\tau}_k$ satisfies the condition (4.106). Here, $N_3(t, x_k, y_{k-1}, x_{k-1}, w_{k-1})$ and $N_4(t, w_k, x_k, w_{k-1}, y_{k-1}, x_{k-1})$ are Newton interpolation polynomial of third and fourth degree, set through available nodal points $(x_k, x_{k-1}, y_{k-1}, w_{k-1})$ and $(x_k, w_k, x_{k-1}, y_{k-1}, w_{k-1})$, respectively. Note that in [40] was shown the above mentioned property holds when

$$\lambda_k = -\frac{N_4''(w_k)}{2N_4'(w_k)}.$$

References

1. T. Zhanlav, K. Otgondorj, O. Chuluunbaatar, A derivative-free families of optimal two-and three-point iterative methods for solving nonlinear equations. Comput. Math. Math. Phys. **59**, 864–880 (2019)
2. T. Zhanlav, K. Otgondorj, R. Mijiddorj, Constructive theory of designing optimal eighth-order derivative-free methods for solving nonlinear equations. Am. J. Comput. Math. **10**, 100–117 (2020)
3. H.T. Kung, J.F. Traub, Optimal order of one-point and multi-point iteration. J. Assoc. Comput. Math. **21**, 643–651 (1974)
4. T. Zhanlav, O. Chuluunbaatar, V. Ulziibayar, Generating function method for constructing new iterations. Appl. Math. Comput. **315**, 414–423 (2017)
5. F. Soleymani, S.K. Vanani, Optimal Steffensen-type methods with eighth order of convergence. Comput. Math. Appl. **62**, 4619–4626 (2011)
6. H. Ren, Q. Wu, W. Bi, A class of two-step Steffensen type methods with fourth-order convergence. Appl. Math. Comput. **209**, 206–210 (2009)
7. I.K. Argyros, M. Kansal, V. Kanwar, S. Bajaj, Higher-order derivative-free families of Chebyshev-Halley type methods with or without memory for solving nonlinear equations. Appl. Math. Comput. **315**, 224–245 (2017)
8. A. Cordero, J.L. Hueso, E. Martínez, J.R. Torregrosa, A new technique to obtain derivative-free optimal iterative methods for solving nonlinear equations. J. Comput. Appl. Math. **252**, 95–102 (2013)
9. A. Cordero, J.R. Torregrosa, A class of Steffensen type methods with optimal order of convergence. Appl. Math. Comput. **217**, 7653–7659 (2011)
10. S.K. Khattri, R.P. Agarwal, Derivative-free optimal iterative methods. Comput. Meth. Appl. Math. **10**, 368–375 (2010)
11. Z. Liu, Q. Zheng, P. Zhao, A variant of Steffensen's method of fourth-order convergence and its applications. Appl. Math. Comput. **216**, 1978–1983 (2010)
12. T. Lotfi, F. Soleymani, M. Ghorbanzadeh, P. Assari, On the construction of some tri-parametric iterative methods with memory. Numer. Algorithms **70**, 835–845 (2015)
13. Y. Peng, H. Feng, Q. Li, X. Zhang, A fourth-order derivative-free algorithm for nonlinear equations. J. Comput. Appl. Math. **235**, 2551–2559 (2011)
14. M.S. Petković, B. Neta, L.D. Petković, J. Džunić, Multipoint methods for solving nonlinear equations. Appl. Math. Comput. **226**, 635–660 (2014)
15. S. Sharifi, S. Siegmund, M. Salimi, Solving nonlinear equations by a derivative-free form of the King's family with memory. Calcolo **53**, 201–215 (2016)
16. J.R. Sharma, R.K. Goyal, Fourth-order derivative-free methods for solving non-linear equations. Int. J. Comput. Math. **83**, 101–106 (2006)
17. F. Soleymani, Optimal fourth-order iterative methods free from derivative, Miskolc. Math. Notes **12**, 255–264 (2011)
18. F. Soleymani, S.K. Khattri, Finding simple roots by seventh-and eighth-order derivative-free methods. Int. J. Math. Model. Methods Appl. Sci. **6**, 45–52 (2012)

19. F. Soleymani, R. Sharma, X. Li, E. Tohidi, An optimized derivative-free form of the Potra-Pták method. Math. Comput. Model. **56**, 97–104 (2012)
20. R. Thukral, A family of three-point derivative-free methods of eighth-order for solving nonlinear equations. J. Mod. Methods Numer. Math. **3**, 11–21 (2012)
21. R. Thukral, Eighth-order iterative methods without derivatives for solving nonlinear equations. ISRN Appl. Math. *2011*(1), 12 (2011). Article ID: 693787
22. Q. Zheng, J. Li, F. Huang, An optimal Steffensen-type family for solving nonlinear equations. Appl. Math. Comput. **217**, 9592–9597 (2011)
23. S.K. Khattri, T. Steihaug, Algorithm for forming derivative-free optimal methods. Numer. Algorithms **65**, 809–842 (2014)
24. M.S. Petković, S. Ilić, J. Džunić, Derivative-free two-point methods with and without memory for solving nonlinear equations. Appl. Math. Comput. **217**, 1887–1895 (2010)
25. J.R. Sharma, R.K. Guha, P. Gupta, Some efficient derivative free methods with memory for solving nonlinear equations. Appl. Math. Comput. **219**, 699–707 (2012)
26. F. Zafar, N. Yasmin, M.A. Kutbi, M. Zeshan, Construction of tri-parametric derivative free fourth order with and without memory iterative method. J. Nonlinear Sci. Appl. **9**, 1410–1423 (2016)
27. R. Behl, D. Gonzalez, P. Maroju, S.S. Motsa, An optimal and efficient general eighth-order derivative-free scheme for simple roots. J. Comput. Appl. Math. **330**, 666–675 (2018)
28. M.S. Petković, B. Neta, L.D. Petković, J. Džunić, *Multipoint Methods for Solving Nonlinear Equations* (Elsevier, 2013)
29. R.W. Hamming, *Numerical Methods for Scientists and Engineers* (Dover Publications, New-York, 1962)
30. T. Zhanlav, K. Otgondorj, Comparison of some optimal derivative-free three-point iterations. J. Numer. Anal. Approx. Theory **49**, 76–90 (2020)
31. T. Zhanlav, O. Chuluunbaatar, G. Ankhbayar, Relationship between the inexact Newton method and the continuous analogy of Newton's method. Rev. Anal. Numer. Theor. Approx. **40**, 182–189 (2011)
32. C. Chun, B. Neta, Comparative study of eighth-order methods for finding simple roots of nonlinear equations. Numer. Algorithms **74**, 1169–1201 (2017)
33. F. Soleymani, S. Shateyi, Two optimal eighth-order derivative–free classes of iterative methods. Abstr. Appl. Anal. **14**(1), 14 (2012). Article ID: 318165
34. M. Kansal, V. Kanwar, S. Bhatia, Efficient derivative-free variants of Hansen-Patrick's family with memory for solving nonlinear equations. Numer. Algorithms **73**, 1017–1036 (2016)
35. L.D. Petković, M.S. Petković, B. Neta, On optimal parameter of Laguerre's family of zero-finding methods. Int. J. Comput. Math. **95**, 692–707 (2018)
36. X. Wang, L. Liu, New eighth-order iterative methods for solving nonlinear equations. J. Comput. Appl. Math. **234**, 1611–1620 (2010)
37. R. Thukral, M.S. Petković, A family of three-point methods of optimal order for solving nonlinear equations. J. Comput. Appl. Math. **233**, 2278–2284 (2010)
38. T. Zhanlav, V. Ulziibayar, O. Chuluunbaatar, Necessary and sufficient conditions for the convergence of two- and three-point Newton-type iterations. Comput. Math. Math. Phys. **57**, 1090–1100 (2017)
39. X. Wang, T. Zhang, A new family of Newton-type iterative methods with and without memory for solving nonlinear equations. Calcolo **51**, 1–15 (2014)
40. J. Džunić, On efficient two-parameter methods for solving nonlinear equations. Numer. Algorithms **63**, 549–569 (2013)

Part II
Higher Order Iterations for Systems of Nonlinear Equations

Chapter 5
Higher Order Newton-Type Iterations

Abstract We established necessary and sufficient conditions for two- and three-step iterations to be of order p ($p = 4, 5, \ldots, 10$). They are written in terms of parameter matrices. Based on these conditions, we developed new multi-parametric higher-order Jarratt and Newton-type iterations with different selections of parameters and implementation algorithms. The proposed family methods include some well-known methods as special cases. Some of the proposed iterative methods are extended to multi-step iterations. We also discuss the total cost comparison between the methods and show that our methods are competitive in computational cost, as well as in the order of convergence of the existing ones.

5.1 Newton's Method for Solving Systems of Nonlinear Equations

For a given nonlinear system $F(x) : D \subset R^n \to R^n$, the problem is to find a vector x^* such that $F(x) = 0$. This problem is important and interesting and often appears in numerical analysis and engineering. The most widely used method for solving this problem is the quadratically convergent Newton's method given by

$$x_{k+1} = x_k - \left(F'(x_k)\right)^{-1} F(x_k), \quad k = 0, 1, \ldots,$$

where x_0 is the initial approximation and $(F'(x))^{-1}$ is the inverse of Fréchet derivative $F'(x)$ of the function $F(x)$. Following [1] we use:

Definition 5.1 Let $\{x_k\}_{k\geq 0}$ be a sequence in R^n which converges to x^*. Then, the convergence is called of order p, $p > 1$, if

$$\frac{\|F(x_{k+1})\|}{\|F(x_k)\|^p} = O(1)$$

or

$$F(x_{k+1}) = O\left(h^p\right), \quad h = F(x_k), \quad h^p = \overbrace{(h, h, \cdots, h)}^{p}.$$

T. Zhanlav and O. Chuluunbaatar, *New Developments of Newton-Type Iterations for Solving Nonlinear Problems*, Mathematical Engineering, https://doi.org/10.1007/978-3-031-63361-4_5

Usually, in order to find the convergence order of iteration methods used the lemma for Taylor's expansion on vector functions (see [2]).

Lemma 5.1 *Letting $F : D \subseteq R^n \to R^n$ be p-times Fréchet differentiable in a convex set $D \subseteq R^n$, then for any $x, \hat{h} \in D$ the following expression holds:*

$$F(x + \hat{h}) = F(x) + F'(x)\hat{h} + \frac{1}{2!}F''(x)\hat{h}^2 + \frac{1}{3!}F'''(x)\hat{h}^3 + \cdots + \frac{1}{p!}F^{(p)}(x)\hat{h}^p + R_p,$$

where

$$\|R_p\| \le \frac{1}{p!} \sup_{0 < t < 1} \|F^{(p)}(x + t\hat{h})\| \|\hat{h}\|^p,$$

and $\| \cdot \|$ denotes any norm in R^n, or a corresponding operator norm.

The following notation was introduced in [3], but we present it for completeness. The q-th derivative of F at $u \in R^n$, $q \ge 1$, is the q-linear function $F^{(q)}(u) : R^n \times \ldots \times R^n \to R^n$ such that $F^{(q)}(u)\left(v_1, \ldots, v_q\right) \in R^n$,

   (i) $F^{(q)}(u)\left(v_1, \ldots, v_{q-1}, \cdot\right) \in L\left(R^n\right)$,

   (ii) $F^{(q)}(u)\left(v_{\sigma(1)}, \ldots, v_{\sigma(q)}\right) = F^{(q)}(u)\left(v_1, \ldots, v_q\right)$, for all permutations σ of $\{1, 2, \ldots, q\}$.

From the above properties, we can use the following notations:

   (a) $F^{(q)}(u)\left(v_1, \ldots, v_q\right) = F^{(q)}(u)v_1 \ldots v_q$,

   (b) $F^{(q)}(u)v^{q-1}F^{(p)}(u)v^p = F^{(q)}(u)F^{(p)}(u)v^{q+p-1}$.

5.1.1   Two-Step Iterative Methods

We consider the two-step iterative method

$$y_k = x_k - \left(F'(x_k)\right)^{-1} F(x_k), \tag{5.1a}$$

$$x_{k+1} = y_k - \bar{\tau}_k \left(F'(x_k)\right)^{-1} F(y_k), \tag{5.1b}$$

where $\bar{\tau}_k$ is some $n \times n$ matrix to be determined properly. Using the Taylor expansion of $F(x_{k+1})$ around y_k, we have

$$F(x_{k+1}) = \left(I - F'(y_k)\bar{\tau}_k \left(F'(x_k)\right)^{-1}\right) F(y_k) + O\left(F^2(y_k)\right), \tag{5.2}$$

where I is identity matrix. From (5.2) it clear that $F(x_{k+1}) = O\left(h^4\right)$ if we choose $\bar{\tau}_k$, such that

$$I - F'(y_k)\bar{\tau}_k \left(F'(x_k)\right)^{-1} = 0 \quad \text{or} \quad \bar{\tau}_k = F'(y_k))^{-1}F'(x_k). \tag{5.3}$$

Substituting (5.3) into (5.1b) we obtain

$$x_{k+1} = y_k - (F'(y_k))^{-1} F(y_k). \tag{5.4}$$

Our task is to approximate $(F'(y_k))^{-1}$ with accuracy $O\left(h^2\right)$. We use the expansion of $F'(y_k)$ at point x_k

$$\begin{aligned} F'(y_k) &= F'(x_k) - F''(x_k)\left(F'(x_k)\right)^{-1} F(x_k) + O\left(h^2\right) \\ &= F'(x_k)\left(I - P_k\right) + O\left(h^2\right), \end{aligned} \tag{5.5}$$

where

$$P_k = \left(F'(x_k)\right)^{-1} F''(x_k)\left(F'(x_k)\right)^{-1} F(x_k).$$

We can assume that

$$\|P_k\| \le 1,$$

for x_k sufficiently close to x^*. By virtue of Banach lemma of inverse, from (5.5) we get

$$(F'(y_k))^{-1} = (I - P_k)^{-1}\left(F'(x_k)\right)^{-1} + O\left(h^2\right) = (I + P_k)\left(F'(x_k)\right)^{-1} + O\left(h^2\right), \tag{5.6}$$

because

$$(I - P_k)^{-1} = \sum_{j=0}^{\infty} P_k^j = I + P_k + O\left(h^2\right), \quad \|P\| < 1. \tag{5.7}$$

Using the approximate formula obtained from (5.6)

$$(F'(y_k))^{-1} \approx (I + P_k)\left(F'(x_k)\right)^{-1}, \tag{5.8}$$

in (5.4) gives

$$x_{k+1} = y_k - (I + P_k)\left(F'(x_k)\right)^{-1} F(y_k). \tag{5.9}$$

Using (5.8) in (5.3) we have

$$\bar{\tau}_k = I + P_k + O\left(h^2\right) = I + 2\Theta_k + O\left(h^2\right). \tag{5.10}$$

Here

$$\Theta_k = \frac{P_k}{2} = \frac{1}{2}\left(F'(x_k)\right)^{-1} F''(x_k)\left(F'(x_k)\right)^{-1} F(x_k). \tag{5.11}$$

Since

$$F(y_k) = \frac{F''(x_k)}{2}\left(\left(F'(x_k)\right)^{-1} F(x_k)\right)^2 + O\left(h^3\right), \tag{5.12}$$

$$\left(F'(x_k)\right)^{-1} F(y_k) = \frac{\left(F'(x_k)\right)^{-1} F''(x_k)}{2}\left(\left(F'(x_k)\right)^{-1} F(x_k)\right)^2 + O\left(h^3\right). \tag{5.13}$$

Substituting (5.11) into (5.13) we obtain

$$\left(F'(x_k)\right)^{-1} F(y_k) = \Theta_k \left(F'(x_k)\right)^{-1} F(x_k) + O\left(h^3\right). \tag{5.14}$$

On the other hand, using (5.1a) and (5.14) one can rewrite (5.9) as

$$x_{k+1} = x_k - \left(I + (I + P_k)\frac{P_k}{2}\right)\left(F'(x_k)\right)^{-1} F(x_k),$$

or

$$x_{k+1} = x_k - \tau_k \left(F'(x_k)\right)^{-1} F(x_k), \tag{5.15}$$

where

$$\tau_k = I + (I + P_k)\frac{P_k}{2} = I + \Theta_k + 2\Theta_k^2 + O\left(h^3\right). \tag{5.16}$$

From (5.10) and (5.16) one can find the connection of τ_k and $\bar{\tau}_k$ as:

$$\bar{\tau}_k = (\tau_k - I)\Theta_k^{-1} = I + 2\Theta_k + O\left(h^2\right). \tag{5.17}$$

Thus Theorem 1 in [4] is extended to system of nonlinear equations.

Theorem 5.1 *Let $F(x) : D \subset R^n \to R^n$ be a sufficiently Fréchet differentiable function in a convex set $D \subset R^n$ containing a zero x^* of $F(x)$. Suppose that $F'(x)$ is continuous and nonsingular in x^*. Then, for an initial approximation sufficiently close to x^*, the sequence $\{x_k\}_{k\geq 0}$, $x_0 \in D$ obtained by (5.15) has order four if and only if the parameter matrix τ_k is given by (5.16).*

Proof The Taylor expansion of function $F(x_{k+1})$ at point y_k and use of (5.17) and (5.14) gives

$$F(x_{k+1}) = F(y_k) + F'(y_k)(\tau_k - I)(y_k - x_k) + O\left(((\tau_k - I)(y_k - x_k))^2\right). \tag{5.18}$$

Since $\tau_k - I = O(h)$ and $y_k - x_k = O(h)$, using (5.5) and (5.14) into (5.18) we have

$$F(x_{k+1}) = F(y_k) - F'(x_k)(I - P_k)(\tau_k - I)\Theta_k^{-1}\left(F'(x_k)\right)^{-1} F(y_k) + O\left(h^4\right)$$
$$= F'(x_k)\left(I - (I - 2\Theta_k)(\tau_k - I)\Theta_k^{-1}\right)\left(F'(x_k)\right)^{-1} F(y_k) + O\left(h^4\right). \tag{5.19}$$

By choice (5.16) we have $F(x_{k+1}) = O\left(h^4\right)$. The converse follows from (5.19). $\quad\square$

From Theorem 5.1 immediately follows that the two-step iterative method (5.1) has a order four if and only if the parameter matrix $\bar{\tau}_k$ is given by (5.17). The main practical difficulty related to parameter matrix Θ_k is the evaluation of the second-order Fréchet derivative. For a nonlinear system of n equations and n unknowns, the first Fréchet derivative is a matrix with n^2 values, while the second Fréchet derivative for continuous functions has $(n^3 + n^2)/2$ values. This implies a huge amount of operations in order to evaluate every iteration. To overcome these difficulties we will find some approximate formulas for Θ_k. In general, we assume that y_k is defined by

$$y_k = x_k - a \left(F'(x_k) \right)^{-1} F(x_k), \quad a \neq 0. \tag{5.20}$$

In order to compute Θ_k with some accuracy we will use the first order divided difference of $F(x)$ [5]

$$[F; x + h_1, x] := \int_0^1 F'(x + th_1)dt$$

$$= F'(x) + \frac{1}{2}F''(x)h_1 + \frac{1}{6}F'''(x)h_1^2 + O\left(h_1^3\right), \tag{5.21}$$

where $h_1^i = (h_1, h_1, \overset{i}{\ldots}, h_1)$, $h_1 \in R^n$. Using (5.20) and (5.21) we have

$$\left(F'(x_k)\right)^{-1} [F; y_k, x_k] = I - a\Theta_k$$

$$+ \frac{a^2}{6} \left(F'(x_k)\right)^{-1} F'''(x_k) \left(\left(F'(x_k)\right)^{-1} F(x_k) \right)^2 + O\left(h^3\right). \tag{5.22}$$

Analogously, using Taylor expansion of $F'(y_k)$ at point x_k, we obtain

$$\left(F'(x_k)\right)^{-1} F'(y_k) = I - 2a\Theta_k$$

$$+ \frac{a^2}{2} \left(F'(x_k)\right)^{-1} F'''(x_k) \left(\left(F'(x_k)\right)^{-1} F(x_k) \right)^2 + O\left(h^3\right). \tag{5.23}$$

The inverse of $\left(F'(x_k)\right)^{-1} F'(y_k)$ exists and by virtue of (5.23) we have

$$(F'(y_k))^{-1} F'(x_k) = I + 2a\Theta_k + O\left(h^2\right). \tag{5.24}$$

Then from (5.23) and (5.24) it follows that

$$\left(F'(x_k)\right)^{-1} F'(y_k) + (F'(y_k))^{-1} F'(x_k) = 2I + O\left(h^2\right). \tag{5.25}$$

From (5.22) we find

$$\Theta_k = \frac{1}{a} \left(I - \left(F'(x_k) \right)^{-1} [F; y_k, x_k] \right.$$
$$\left. + \frac{a^2}{6} \left(F'(x_k) \right)^{-1} F'''(x_k) \left(\left(F'(x_k) \right)^{-1} F(x_k) \right)^2 \right) + O\left(h^3 \right). \quad (5.26)$$

It follows from (5.23) and (5.26) that

$$\Theta_k = \frac{1}{a} \left(F'(x_k) \right)^{-1} \left([F; y_k, x_k] - F'(y_k) \right) + O\left(h^3 \right). \qquad (5.27)$$

Elimination of term with factor a^2 from (5.22) and (5.23) gives

$$\Theta_k = \frac{1}{a} \left(F'(x_k) \right)^{-1} \left(2F'(x_k) + F'(y_k) - 3[F; y_k, x_k] \right) + O\left(h^3 \right). \qquad (5.28)$$

As a consequence of (5.28) and (5.27) we get

$$\Theta_k = \frac{1}{a} \left(F'(x_k) \right)^{-1} \left(F'(x_k) - [F; y_k, x_k] \right) + O\left(h^3 \right), \qquad (5.29)$$

and
$$\Theta_k = \frac{1}{2a} \left(I - \left(F'(x_k) \right)^{-1} F'(y_k) \right) + O\left(h^3 \right). \qquad (5.30)$$

Thus, we have four approximate formulas for Θ_k that will be used for determining the parameter matrix $\bar{\tau}_k$ in (5.17). In some cases it may be useful to have less accurate formula for Θ_k. Using (5.25) in (5.30) we get

$$\Theta_k = \frac{1}{2a} \left(-I + (F'(y_k))^{-1} F'(x_k) \right) + O\left(h^2 \right). \qquad (5.31)$$

In a similar way, (5.25) can be used in (5.27), (5.28) to obtain less accurate formulas. The divided difference $[F; y, x]$ of F is an $n \times n$ matrix with elements (see [5])

$$[F; y, x]_{i,j} = \frac{F_i(y_{(1)}, \ldots, y_{(j)}, x_{(j+1)}, \ldots, x_{(n)}) - F_i(y_{(1)}, \ldots, y_{(j-1)}, x_{(j)}, \ldots, x_{(n)})}{y_{(j)} - x_{(j)}},$$

where $1 \leq i, j \leq n$.

Remark 5.1 Note that the fourth-order convergence of iteration (5.1) holds true if Θ_k in (5.17) is replaced by one of approximate formulae (5.27)–(5.31) with $a = 1$.

Analogously, we can prove that

Theorem 5.2 *Assume that all the assumptions of Theorem 5.1 are fulfilled. Then the two-step iteration (5.15) has a third-order of convergence if and only if the iteration matrix τ_k is chosen such that*

$$\tau_k = I + \Theta_k + O\left(\Theta_k^2 \right). \qquad (5.32)$$

In what follows we will use two step iterative method (5.1a), (5.1b) as

$$y_k = x_k - \left(F'(x_k)\right)^{-1} F(x_k),$$
$$x_{k+1} = x_k - \tau_k \left(F'(x_k)\right)^{-1} F(x_k), \tag{5.33}$$

which can be considered as a continuous analogue of Newton's method (*CANM for short*) or damped Newton's method [6]. Following the idea of generating functions method in [7] we can state the following.

Theorem 5.3 *Suppose that all assumptions of Theorem 5.1 are fulfilled. Then the two-step iteration (5.33) has a fourth-order of convergence if and only if the iteration matrix τ_k is chosen such that*

$$\tau_k = (I - \alpha\Theta_k)^{-1}\left(I + (1-\alpha)\Theta_k + (2-\alpha)\Theta_k^2 + \omega\Theta_k^3\right), \quad \alpha, \omega \in R. \tag{5.34}$$

Proof It is easy to show that τ_k given by (5.34) can be rewritten as:

$$\tau_k = \left(I + \alpha\Theta_k + \alpha^2\Theta_k^2 + \alpha^3\Theta_k^3 + \cdots\right)\left(I + (1-\alpha)\Theta_k + (2-\alpha)\Theta_k^2 + \omega\Theta_k^3\right)$$
$$= I + \Theta_k + 2\Theta_k^2 + O\left(h^3\right).$$

Thus by Theorem 5.1 the convergence order of iteration (5.33) is equal to four. $\square$

From the approximate formulas (5.27)–(5.30) we see that the Theorem 5.3 is valid when Θ_k is defined by one of the formulas (5.27)–(5.30) with $a = 1$. When $\alpha = 0$ the formula (5.34) leads to (5.16).

Now we consider another two-point iterative method

$$y_k = x_k - a\left(F'(x_k)\right)^{-1} F(x_k), \quad x_{k+1} = x_k - \tau_k\left(F'(x_k)\right)^{-1} F(x_k). \tag{5.35}$$

It is easy to show that

$$F(y_k) = O\left(h^\sigma\right), \quad \sigma = \begin{cases} 2, & a = 1, \\ 1, & a \neq 1. \end{cases}$$

We consider the Taylor expansion of $F(x_{k+1})$ at point x_k. We have

$$F(x_{k+1}) = F(x_k) - F'(x_k)\tau_k\left(F'(x_k)\right)^{-1} F(x_k) + \frac{F''(x_k)}{2}\left(\tau_k\left(F'(x_k)\right)^{-1} F(x_k)\right)^2$$
$$- \frac{F'''(x_k)}{6}\left(\tau_k\left(F'(x_k)\right)^{-1} F(x_k)\right)^3 + O\left(h^4\right). \tag{5.36}$$

As the preceding case, we seek for τ_k in the form

$$\tau_k = I + \Theta_k + c_k\Theta_k^2 + d_k + O\left(h^3\right). \tag{5.37}$$

Substituting (5.37) into (5.36) and taking into account the following relations

$$
\left(\tau_k \left(F'(x_k) \right)^{-1} F(x_k) \right)^2 = (I + \Theta_k) \left(\left(F'(x_k) \right)^{-1} F(x_k) \right)^2
$$
$$
+ \left(F'(x_k) \right)^{-1} F(x_k) \Theta_k \left(F'(x_k) \right)^{-1} F(x_k) + O\left(h^4 \right),
$$
$$
\left(\tau_k \left(F'(x_k) \right)^{-1} F(x_k) \right)^3 = \left(\left(F'(x_k) \right)^{-1} F(x_k) \right)^3 + O\left(h^4 \right), \tag{5.38}
$$

we get

$$
F(x_{k+1}) = -F'(x_k) \left(\Theta_k + c_k \Theta_k^2 + d_k \right) \left(F'(x_k) \right)^{-1} F(x_k)
$$
$$
+ \frac{F''(x_k)}{2} \left((I + \Theta_k)((F'(x_k))^{-1} F(x_k))^2 \right.
$$
$$
\left. + \left(F'(x_k) \right)^{-1} F(x_k) \Theta_k \left(F'(x_k) \right)^{-1} F(x_k) \right)
$$
$$
- \frac{F'''(x_k)}{6} \left(\left(F'(x_k) \right)^{-1} F(x_k) \right)^3 + O\left(h^4 \right). \tag{5.39}
$$

Multiplying by $\left(F'(x_k) \right)^{-1}$ both sides of (5.39) and taking (5.11) into account we have

$$
\left(F'(x_k) \right)^{-1} F(x_{k+1}) = \left((1 - c_k) \Theta_k^2 + \frac{1}{2} \left(F'(x_k) \right)^{-1} F''(x_k) \Theta_k \left(F'(x_k) \right)^{-1} F(x_k) \right.
$$
$$
\left. - \left(d_k + \frac{1}{6} \left(F'(x_k) \right)^{-1} F'''(x_k)((F'(x_k))^{-1} F(x_k))^2 \right) \right)
$$
$$
\times \left(F'(x_k) \right)^{-1} F(x_k) + O\left(h^4 \right). \tag{5.40}
$$

From (5.40) we see that
$$
F(x_{k+1}) = O\left(h^4 \right),
$$

if we choose c_k and d_k such that

$$
c_k = I + \frac{1}{2} \left(F'(x_k) \right)^{-1} F''(x_k) \Theta_k \left(F'(x_k) \right)^{-1} F(x_k) \Theta_k^{-2} + O(h), \tag{5.41}
$$
$$
d_k = -\frac{1}{6} \left(F'(x_k) \right)^{-1} F'''(x_k) \left(\left(F'(x_k) \right)^{-1} F(x_k) \right)^2 + O\left(h^3 \right). \tag{5.42}
$$

From (5.23) we find d_k given by (5.42) as

$$
d_k = -\frac{1}{3a^2} \left(\left(F'(x_k) \right)^{-1} F'(y_k) - (I - 2a\Theta_k) \right) + O\left(h^3 \right). \tag{5.43}
$$

Substituting (5.41) and (5.43) into (5.37), we get

$$\tau_k = I + \Theta_k + \Theta_k^2 + \frac{1}{2} \left(F'(x_k)\right)^{-1} F''(x_k)\Theta_k \left(F'(x_k)\right)^{-1} F(x_k)$$

$$-\frac{\left(F'(x_k)\right)^{-1} F'(y_k) - (I - 2a\Theta_k)}{3a^2} + O\left(h^3\right). \tag{5.44}$$

If we use the following equality (which holds due to (ii) property)

$$\frac{1}{2} \left(F'(x_k)\right)^{-1} F''(x_k)\Theta_k \left(F'(x_k)\right)^{-1} F(x_k) = \Theta_k^2 + O\left(h^3\right), \tag{5.45}$$

then (5.44) is written as

$$\tau_k = I + \Theta_k + 2\Theta_k^2 + d_k + O\left(h^3\right). \tag{5.46}$$

Note that the assumption (5.45) fulfilled for scalar equation case and τ_k given by (5.46) coincides with that of scalar equation case [7]. Furthermore, the assumption (5.45) holds true for some method (5.33). As an example, we consider the generalized Jarratt's method given in [8], as (5.35) with $a = 2/3$ and

$$\tau_k = \frac{1}{2} \left(3F'(y_k) - F'(x_k)\right)^{-1} \left(3F'(y_k) + F'(x_k)\right).$$

The method M43 [9] is a special case of M4, which is when $a = 1$ and (5.29) is used for Θ_k. Using (5.24) it is easy to show that τ_k satisfies (5.46). It means that the assumption (5.45) fulfilled in this case. We summarize the above result in the following theorem:

Theorem 5.4 *The two-step method (5.35) has a fourth-order convergence if τ_k is given by (5.46).*

Analogously, using (5.40) one can prove the following:

Theorem 5.5 *The two-step iteration (5.35) has a p-order ($p = 2, 3$) of convergence if τ_k is given by*

$$\tau_k = I + \Theta_k + O\left(h^2\right) \quad and \quad \tau_k = I + O(h),$$

respectively.

The fifth and sixth-order scheme is composed of three steps; the first two steps are same as that of fourth-order scheme, whereas the third step contains another parameter matrix α_k. Generally, The fifth and sixth-order methods, which are based on above the fourth-order methods, are given by

$$y_k = x_k - \left(F'(x_k)\right)^{-1} F(x_k),$$
$$z_k = \phi_p(x_k, y_k),$$
$$x_{k+1} = z_k - \alpha_k \left(F'(x_k)\right)^{-1} F(z_k),$$

where $z_k = \phi_p(x_k, y_k)$ is any iteration function of order $p \geq 2$.

In this study, we shall follow the basic principle of numerical analysis that a genuine ranking of numerical algorithms can be attained using computational efficiency that is, always proportional to quality of an algorithm and inversely proportional to its computational cost. Quality of an algorithm concerns with the convergence speed of algorithm along with its structure.

5.1.2 Three-Step Iterative Methods

Now we consider three-step iterative method

$$y_k = x_k - \left(F'(x_k)\right)^{-1} F(x_k), \tag{5.47a}$$

$$z_k = y_k - \bar{\tau}_k \left(F'(x_k)\right)^{-1} F(y_k), \tag{5.47b}$$

$$x_{k+1} = z_k - \alpha_k \left(F'(x_k)\right)^{-1} F(z_k). \tag{5.47c}$$

The order of convergence of iteration (5.47) is given by the following theorem:

Theorem 5.6 *Suppose that all the assumptions of Theorem 5.1 are fulfilled. Then the iteration (5.47) has a order p if and only if the iteration matrices $\bar{\tau}_k$ and α_k are given by formulas in Table 5.1.*

Proof Let τ_k be defined by formula (5.16). Then according to Theorem 5.1 we have $F(z_k) = O\left(h^4\right)$. Then using expansion of $F(x_{k+1})$ around z_k we have

$$F(x_{k+1}) = \left(I - F'(z_k)\alpha_k \left(F'(x_k)\right)^{-1}\right) F(z_k) + O\left(F(z_k)^2\right). \tag{5.48}$$

From (5.48) it follows that

$$F(x_{k+1}) = O\left(F(z_k)^2\right),$$

Table 5.1 Choices of parameters

p	α_k	$\bar{\tau}_k$
5	$I + O(h)$	$I + 2\Theta_k + \beta\Theta_k^2$
	$I + 2\Theta_k + O\left(h^2\right)$	$I + O(h)$
6	$I + 2\Theta_k + O\left(h^2\right)$	$I + 2\Theta_k + O\left(h^2\right)$
7	$I + 2\Theta_k + 6\Theta_k^2 + 3d_k + O\left(h^3\right)$	$I + 2\Theta_k + O\left(h^2\right)$
	$I + 2\Theta_k + O\left(h^2\right)$	$I + 2\Theta_k + 5\Theta_k^2 + 3d_k + O\left(h^3\right)$
8	$I + 2\Theta_k + 6\Theta_k^2 + 3d_k + O\left(h^3\right)$	$I + 2\Theta_k + 5\Theta_k^2 + 3d_k + O\left(h^3\right)$

For completeness, we present the results of Theorems 5.7 and 5.22 in the last two rows of the Table 5.1

under condition

$$\alpha_k = (F'(z_k))^{-1} F'(x_k). \tag{5.49}$$

Using Taylor expansion of $F'(z_k)$ around x_k we have

$$F'(z_k) = F'(x_k) \left(I - (F'(x_k))^{-1} F''(x_k)\tau_k (F'(x_k))^{-1} F(x_k) \right) + O\left(h^2\right). \tag{5.50}$$

From (5.50) it follows that

$$(F'(z_k))^{-1} = \left(I - (F'(x_k))^{-1} F''(x_k)\tau_k (F'(x_k))^{-1} F(x_k) \right)^{-1}$$
$$\times (F'(x_k))^{-1} + O\left(h^2\right).$$

Using equality type (5.7) we have

$$(F'(z_k))^{-1} = \left(I + (F'(x_k))^{-1} F''(x_k)\tau_k (F'(x_k))^{-1} F(x_k) \right)$$
$$\times (F'(x_k))^{-1} + O\left(h^2\right). \tag{5.51}$$

If we take into account (5.16) and $\Theta_k = O\left(F(x_k)\right)$, then substituting (5.51) into (5.49) we find α_k with accuracy $O\left(h^2\right)$ as

$$\alpha_k = I + 2\Theta_k + O\left(h^2\right). \tag{5.52}$$

As a result, from (5.48) we get

$$F(x_{k+1}) = O\left(F^6(x_k)\right). \tag{5.53}$$

Conversely, let (5.53) hold. From (5.49) and (5.51) easily seen that

$$\alpha_k = (F'(z_k))^{-1} F'(x_k) = I + (F'(x_k))^{-1} F''(x_k)\tau_k (F'(x_k))^{-1} F(x_k) + O\left(h^2\right).$$

Substituting it into (5.48) we get

$$F(x_{k+1}) = O\left(h^2\right) F(z_k) + O\left(F(z_k)^2\right).$$

From this we conclude that

$$F(z_k) = O\left(h^4\right),$$

because of (5.53). Hence, by Theorem 5.1 we obtain (5.16). Substituting (5.50) into (5.48) we have

$$F(x_{k+1}) = \left(I - F'(x_k)\left(I - \left(F'(x_k)\right)^{-1} F''(x_k)\tau_k \left(F'(x_k)\right)^{-1} F(x_k)\right)\right.$$
$$\left.\times \alpha_k \left(F'(x_k)\right)^{-1}\right) F(z_k) + O\left(h^6\right). \tag{5.54}$$

Then from (5.53) and (5.54) it follows that

$$I - F'(x_k)(I - \left(F'(x_k)\right)^{-1} F''(x_k)\tau_k \left(F'(x_k)\right)^{-1} F(x_k))\alpha_k \left(F'(x_k)\right)^{-1} = O\left(h^2\right),$$

or

$$\left(I - \left(F'(x_k)\right)^{-1} F''(x_k) \left(F'(x_k)\right)^{-1} F(x_k)\right)\alpha_k = I + O\left(h^2\right),$$

in which we have used (5.16). From this we obtain

$$\alpha_k = (I - 2\Theta_k)^{-1}(I + O\left(h^2\right)) = I + 2\Theta_k + O\left(h^2\right).$$

The proof for $p = 6$ is completed.

Let the parameter matrix τ_k be defined by (5.32) and α_k is given by (5.52). Then by Theorem 5.2 we have $F(z_k) = O\left(F^3(x_k)\right)$. Using (5.32) in (5.50) we get

$$F'(z_k) = F'(x_k)(I - 2\Theta_k) + O\left(h^2\right).$$

Hence

$$I - F'(z_k)\alpha_k \left(F'(x_k)\right)^{-1}$$
$$= I - F'(x_k)(I - 2\Theta_k)(I + 2\Theta_k) \left(F'(x_k)\right)^{-1} + O\left(h^2\right) = O\left(h^2\right) \tag{5.55}$$

because of $\Theta_k = O\left(F(x_k)\right)$. Using (5.55) in (5.47c) we get

$$F(x_{k+1}) = O\left(F^5(x_k)\right), \tag{5.56}$$

i.e., $p = 5$ when τ_k and α_k are defined by (5.32) and (5.52), respectively. Conversely, let (5.56) hold. From (5.48) it is clear that

$$F(z_k) = O\left(F^3(x_k)\right),$$

and

$$I - F'(z_k)\alpha_k \left(F'(x_k)\right)^{-1} = O\left(h^2\right). \tag{5.57}$$

Hence, by Theorem 5.2 we get $\tau_k = 1 + \Theta_k + O\left(h^2\right)$. Then from (5.50) we obtain

$$F'(z_k) = F'(x_k)(I - 2\Theta_k) + O\left(h^2\right). \tag{5.58}$$

Substituting (5.58) into (5.57) we obtain

$$I - F'(x_k)(I - 2\Theta_k)\alpha_k \left(F'(x_k)\right)^{-1} = O\left(h^2\right).$$

From the last expression immediately follows (5.52), i.e., we have proved that $p = 5$ when τ_k and α_k are defined by (5.32) and (5.52), respectively. In a similar way, one can show that $p = 5$ for τ_k and α_k defined by the formulas in the first row of Table 5.1. It remains to prove for $p = 7$. To do this we use the first divided differences

$$[F; y_k, x_k] = F'(x_k) - \frac{1}{2}F''(x_k)\left(F'(x_k)\right)^{-1}F(x_k)$$
$$+ \frac{1}{6}F'''(x_k)\left(\left(F'(x_k)\right)^{-1}F(x_k)\right)^2 + O\left(h^3\right),$$
$$[F; z_k, x_k] = F'(x_k) - \frac{1}{2}F''(x_k)\tau_k\left(F'(x_k)\right)^{-1}F(x_k)$$
$$+ \frac{1}{6}F'''(x_k)\left(\tau_k\left(F'(x_k)\right)^{-1}F(x_k)\right)^2 + O\left(h^3\right). \tag{5.59}$$

Not that in deriving (5.59) we used the relations (5.11), (5.12) and (5.17). According to (5.16) and Theorem 5.1, we have $F(z_k) = O\left(F^4(x_k)\right)$. Using (5.16), (5.17) and (5.38) we obtain

$$\left(F'(x_k)\right)^{-1}[F; y_k, x_k] = I - \Theta_k - D_k + O\left(h^3\right), \tag{5.60}$$
$$\left(F'(x_k)\right)^{-1}[F; z_k, x_k] = I - \Theta_k - \frac{1}{2}C_k - D_k + O\left(h^3\right), \tag{5.61}$$

where

$$C_k = \left(F'(x_k)\right)^{-1}F''(x_k)\Theta_k\left(F'(x_k)\right)^{-1}F(x_k),$$
$$D_k = -\frac{1}{6}\left(F'(x_k)\right)^{-1}F'''(x_k)\left(\left(F'(x_k)\right)^{-1}F(x_k)\right)^2. \tag{5.62}$$

From (5.60), (5.61) we find

$$C_k = 2\left(F'(x_k)\right)^{-1}([F; y_k, x_k] - [F; z_k, x_k]) + O\left(h^3\right), \tag{5.63}$$
$$D_k = I - \Theta_k - \left(F'(x_k)\right)^{-1}[F; y_k, x_k] + O\left(h^3\right). \tag{5.64}$$

Expanding $F'(z_k)$ around x_k and using (5.16) we have

$$F'(z_k) = F'(x_k)\left(I - (2\Theta_k + C_k + 3D_k) + O\left(h^3\right)\right). \tag{5.65}$$

Since $C_k = O\left(F^2(x_k)\right)$ and $D_k = O\left(F^2(x_k)\right)$ then there exists the inverse of matrix in brackets and by Banach lemma we have

$$\left(I - (2\Theta_k + C_k + 3D_k) + O\left(h^3\right)\right)^{-1} = I + 2\Theta_k + C_k + 3D_k + 4\Theta_k^2 + O\left(h^3\right).$$

Hence, from (5.65) we have

$$\left(F'(z_k)\right)^{-1} = \left(I + 2\Theta_k + C_k + 3D_k + 4\Theta_k^2 + O\left(h^3\right)\right)\left(F'(x_k)\right)^{-1}. \tag{5.66}$$

The Taylor expansion of $F(x_{k+1})$ around z_k gives

$$F(x_{k+1}) = \left(I - F'(z_k)\alpha_k\left(F'(x_k)\right)^{-1}\right)F(z_k) + O\left(F(z_k)^2\right). \tag{5.67}$$

From (5.67) it is clear that

$$F(x_{k+1}) = O\left(F^7(x_k)\right), \tag{5.68}$$

provided that

$$I - F'(z_k)\alpha_k\left(F'(x_k)\right)^{-1} = O\left(h^3\right)$$

or

$$F'(x_k) - F'(z_k)\alpha_k = O\left(h^3\right). \tag{5.69}$$

From (5.69) we get

$$\alpha_k = (F'(z_k))^{-1}F'(x_k) + O\left(h^3\right). \tag{5.70}$$

Substituting (5.66) into (5.70) we obtain

$$\alpha_k = I + 2\Theta_k + C_k + 3D_k + 4\Theta_k^2 + O\left(h^3\right). \tag{5.71}$$

Using (5.63) and (5.64) in (5.71) we have

$$\begin{aligned}\alpha_k = {} & I + 2\Theta_k + 4\Theta_k^2 + 3(I - \Theta_k)\\ & - \left(F'(x_k)\right)^{-1}\left([F; y_k, x_k] + 2[F; z_k, x_k]\right) + O\left(h^3\right).\end{aligned}$$

If we use assumption (5.45) then, by formula (5.62) we get $C_k = 2\Theta_k^2$ and

$$\Theta_k^2 = \left(F'(x_k)\right)^{-1}\left([F; y_k, x_k] - [F; z_k, x_k]\right) + O\left(h^3\right).$$

Finally, we get

$$\alpha_k = I + 2\Theta_k + 6\left(F'(x_k)\right)^{-1}\left([F; y_k, x_k] - [F; z_k, x_k]\right) + O\left(h^3\right). \tag{5.72}$$

Conversely, let (5.68) hold. Then from (5.67) it follows that $F(z_k) = O\left(F^4(x_k)\right)$ (indeed, if $F(z_k) = O\left(h^3\right)$ then $F(x_{k+1}) = O\left(F^6(x_k)\right)$ for any choice of α_k. This contradicts (5.68)). By Theorem 5.1 we get (5.16). As a consequence, the above obtained relations (5.66), (5.67) and (5.72) are valid too. This completes the proof of Theorem 5.6. □

Moreover, based on the generating functions method one can propose p-order ($p = 5, 6$) iterative methods (5.47) with parameter matrices τ_k given by (5.34) and α_k given by formula

$$\alpha_k = (I - \beta\Theta_k)^{-1}(I + (2 - \beta)\Theta_k), \quad \beta \in R.$$

Remark 5.2 Quite recently Sharma et al. [10] proposed the fifth-order three-step iterative method

$$y_k = x_k - \left(F'(x_k)\right)^{-1} F(x_k),$$
$$z_k = y_k - 5\left(F'(x_k)\right)^{-1} F(y_k), \tag{5.73}$$
$$x_{k+1} = y_k - \frac{9}{5}\left(F'(x_k)\right)^{-1} F(y_k) - \frac{1}{5}\left(F'(x_k)\right)^{-1} F(z_k).$$

If we compare (5.73) with (5.47), then $\bar{\tau}_k = 5I$. Hence by virtue of (5.17) we have $\tau_k = I + 5\Theta_k = I + O\left(F(x_k)\right)$. Then according to Theorem 1 in [4] we have $F(z_k) = O\left(F^2(x_k)\right)$. On the other hand, using relation (5.14) one can write the third-step in (5.73) as

$$x_{k+1} = z_k - \frac{1}{5}(I - 16\Theta_k)\left(F'(x_k)\right)^{-1} F(z_k) + O\left(F^3(x_k)\right),$$

which does not have such a form as in (5.47). Hence our theory (Theorem 5.6) does not work in this case.

In Table 5.2 we cite some existing p-order methods. It is easy to show that the matrices $\bar{\tau}_k$ and α_k for the methods in Table 5.2 meet p-order sufficient conditions in Theorem 5.1 and Theorem 5.6. The proposed p-order methods with different choices of parameter matrix Θ_k include some existing methods as special cases. For example, if we choose Θ_k by formula (5.29) with $a = 1$ then we obtain fifth-order methods given in [11, 12] and [13]. The choice (5.28) with $a = 1$ gives the fourth and sixth-order methods obtained by Sharma et al. [9]. The fifth-order method presented by M. Grau-Sánchez et al. in [14] is obtained for the choice given by (5.28) with $a = 1$.
Now we consider the following method:

$$y_k = x_k - a\left(F'(x_k)\right)^{-1} F(x_k),$$
$$z_k = \phi_p(x_k, y_k), \tag{5.74}$$
$$x_{k+1} = z_k - \alpha_k\left(F'(x_k)\right)^{-1} F(z_k).$$

Note that $z_k = \phi_p(x_k, y_k)$ is the iteration function of order $p \geq 2$.

Table 5.2 Some existing p-order methods

N	Methods	Order	$\bar{\tau}_k$	α_k
1	Cordero et al. [11]	4	$2I - \left(F'(x_k)\right)^{-1} F(y_k)$	
2	Sharma [9]		$3I - 2\left(F'(x_k)\right)^{-1} [F; y_k, x_k]$	
3	Grau-Sánchez et al. [15]		$(2[F; y_k, x_k] - \left(F'(x_k)\right)^{-1})F'(x_k)$	
			$2[F; y_k, x_k]^{-1} F'(x_k) - I$	
4	Madhu et al. [16]		$\tau_k = I - \frac{3}{4}(s_k - I) + \frac{9}{8}(s_k - I)^2$ $s_k = \left(F'(x_k)\right)^{-1} F'(y_k)$	
	Sharma et al. [17]		$\tau_k = \frac{1}{2}\Big(-I + \frac{9}{4}(F'(y_k))^{-1}F'(x_k) + \frac{3}{4}\left(F'(x_k)\right)^{-1} F'(y_k)\Big)$	
5	Grau-Sánchez et al. [14]	5		
	Xiao et al. [13]		$\tau_k = \frac{1}{2}(I + (F'(y_k))^{-1}F'(x_k))$	$(F'(y_k))^{-1}F'(x_k)$
6	Cordero et al. [18]		$2(I - \left(F'(x_k)\right)^{-1} F'(y_k))^{-1}$	$(F'(y_k))^{-1}F'(x_k)$
7	Xiao et al. [13]		$y_k = x_k - a\left(F'(x_k)\right)^{-1} F'(x_k)$ $\bar{\tau}_k = $ $\left((1 - \frac{1}{2a})I + \frac{1}{2a}\left(F'(x_k)\right)^{-1} F'(y_k)\right)^{-1}$	$-I + 2(\frac{1}{2a}F'(y_k) + (1 - \frac{1}{2a})F'(x_k))^{-1}F'(x_k)$
	Sharma et al. [12]		$a = \frac{1}{2},\ ((F'(y_k))^{-1})F'(x_k) - I)\Theta_k^{-1}$	$2(F'(y_k))^{-1}F'(x_k) - I$
8	Sharma et al. [9]	6	$3I - 2\left(F'(x_k)\right)^{-1} [F; y_k, x_k]$	$3I - 2\left(F'(x_k)\right)^{-1} [F; y_k, x_k]$
9	Xiao et al. [13]		$y_k = x_k - a\left(F'(x_k)\right)^{-1} F'(x_k)$ $\tau_k = \frac{1}{2}(-I + \frac{9}{4}(F'(y_k))^{-1})F'(x_k) + \frac{3}{4}\left(F'(x_k)\right)^{-1})F'(y_k))$	$(1 - \frac{1}{a})I + \frac{1}{a}(F'(y_k))^{-1}F'(x_k)$
10	Grau-Sánchez et al. [15]		$(2[F; y_k, x_k] - F'(x_k))^{-1}F'(x_k)$ $(2[F; y_k, x_k]^{-1} - \left(F'(x_k)\right)^{-1})F'(x_k)$	$(2[F; y_k, x_k] - F'(x_k))^{-1}F'(x_k)$ $2[F; y_k, x_k]^{-1}F'(x_k) - I$
11	Cordero et al. [19]		$a = \frac{1}{2},\ (F'(x_k) - 2F'(y_k))^{-1}(3F'(x_k)\Theta_k^{-1} - 4F(x_k))$	$(F'(x_k) - 2F'(y_k))^{-1}F(x_k)$
12	Madhu et al. [16]		$a = \frac{2}{3},\ \tau_k = H_1$	$I - \frac{3}{2}(s_k - I) + \frac{1}{2}(s_k - I)^2$

Theorem 5.7 *Let $F : D \subseteq R^n \to R^n$ be a sufficiently Fréchet differentiable function in a neighborhood $D \subseteq R^n$ containing a zero x^* of $F(x)$. Suppose that $F'(x)$ is continuous and nonsingular in x^*. Then, for an initial approximation sufficiently close to x^*, the sequence $\{x_k\}$, $x_0 \in D$ obtained by (5.74) has order of convergence $p + 2$ if and only if the parameter matrix α_k is given by*

$$\alpha_k = I + 2\Theta_k + O\left(h^2\right). \tag{5.75}$$

Proof Since $F(z_k) = O\left(h^p\right)$, then from (5.74) we get

$$F(x_{k+1}) = \left(I - F'(z_k)\alpha_k \left(F'(x_k)\right)^{-1}\right) F(z_k) + O\left(h^{2p}\right). \tag{5.76}$$

If we choose α_k such that

$$I - F'(z_k)\alpha_k \left(F'(x_k)\right)^{-1} = O\left(h^2\right),$$

or

$$\alpha_k = \left(F'(z_k)\right)^{-1} F'(x_k) + O\left(h^2\right), \tag{5.77}$$

then from (5.76) we get

$$F(x_{k+1}) = O\left(h^{p+2}\right), \quad p \geq 2.$$

On other hand, the second step in (5.74) can be rewritten as

$$z_k = \phi_p(x_k, y_k) = x_k - \tau_k \left(F'(x_k)\right)^{-1} F(x_k).$$

Then, we have an expansion

$$F'(z_k) = F'(x_k) \left(I - \left(F'(x_k)\right)^{-1} F''(x_k)\tau_k \left(F'(x_k)\right)^{-1} F(x_k)\right) + O\left(h^2\right). \tag{5.78}$$

For small $O\left(F(x_k)\right)$ the matrix $I - \left(F'(x_k)\right)^{-1} F''(x_k)\tau_k \left(F'(x_k)\right)^{-1} F(x_k)$ is invertible and hence from (5.78) we have

$$\begin{aligned}
\left(F'(z_k)\right)^{-1} &= \left(I - \left(F'(x_k)\right)^{-1} F''(x_k)\tau_k F(x_k)^{-1} F(x_k)\right)^{-1} \\
&\quad \times \left(F'(x_k)\right)^{-1} + O\left(h^2\right).
\end{aligned} \tag{5.79}$$

Using (5.7) in (5.79) we have

$$\left(F'(z_k)\right)^{-1} = \left(I + \left(F'(x_k)\right)^{-1} F''(x_k)\tau_k \left(F'(x_k)\right)^{-1} F(x_k)\right)\left(F'(x_k)\right)^{-1} + O\left(h^2\right).$$

By Theorems 5.4 and 5.5 we can replace τ_k by I in the last expansion without loss of accuracy. Hence by virtue of (5.11), the (5.77) has a form

$$\alpha_k = \left(I + \left(F'(x_k)\right)^{-1} F''(x_k)\left(F'(x_k)\right)^{-1} F(x_k)\right) + O\left(h^2\right) = I + 2\Theta_k + O\left(h^2\right).$$

The converse is obvious from (5.76) to (5.77). $\square$

There are many possibilities to obtain α_k satisfying (5.75). We consider some choices for α_k. First we use approximate formula (5.30). Substituting (5.30) into (5.75) we have

$$\alpha_k = \left(1 + \frac{1}{a}\right) I - \frac{1}{a}\left(F'(x_k)\right)^{-1} F'(y_k). \tag{5.80}$$

For instance, if $a = 2/3$, then (5.80) leads to [20]

$$\alpha_k = \frac{1}{2}\left(5I - 3\left(F'(x_k)\right)^{-1} F'(y_k)\right).$$

For $a = 1/2, -1$ and 1 we obtain

$$\alpha_k = 3I - 2\left(F'(x_k)\right)^{-1} F'(y_k),$$
$$\alpha_k = \left(F'(x_k)\right)^{-1} F'(y_k),$$

and

$$\alpha_k = 2I - \left(F'(x_k)\right)^{-1} F'(y_k) = (F'(y_k))^{-1} F'(x_k),$$

respectively. The last case is sixth-order method presented in [18]. Another possible case is to use less accurate formula (5.31). Using (5.31) in (5.75) we get [13]

$$\alpha_k = \left(1 - \frac{1}{a}\right) I + \frac{1}{a}(F'(y_k))^{-1} F'(x_k). \tag{5.81}$$

The simplest generating functions method [7] for (5.75) is

$$\alpha_k = (I - 2\Theta_k)^{-1}. \tag{5.82}$$

Table 5.3 The cost of each iteration for fourth-order methods

Method	Evaluation of F and Jacobian	Scalar vector multiply	Matrix vector multiply	Linear solve	Total
Jarratt [8]	$n\mu_0 + 2n^2\mu_1$	$4n$	0	$3\alpha + 3\beta$ $+ \left(\frac{3\beta}{2} + 3n\right)\ell$	n^3 $+ \left(2\mu_1 + \frac{3+3\ell}{2}\right)n^2$ $+ (\mu_0 + \frac{3+3\ell}{2})n$
Jarratt-like [20]	$n\mu_0 + 2n^2\mu_1$	$4n$	$2n^2$	$\alpha + 3\beta$ $+ \left(\frac{\beta}{2} + 3n\right)\ell$	$n^3/3$ $+ \left(2\mu_1 + \frac{9+\ell}{2}\right)n^2$ $+ (\mu_0 + \frac{7+15\ell}{6})n$
Cordero et al. [11]	$2n\mu_0 + 2n^2\mu_1$	$3n$	n^2	$\alpha + 3\beta$ $+ \left(\frac{\beta}{2} + 3n\right)\ell$	$n^3/3$ $+ \left(2\mu_1 + \frac{7+\ell}{2}\right)n^2$ $+ (2\mu_0 + \frac{1+15\ell}{6})n$
M43 [9]	$2n\mu_0 + n^2\mu_1$	$3n$	n^2	$\alpha + 3\beta$ $+ \left(\frac{\beta}{2} + 3n\right)\ell$	$n^3/3$ $+ \left(\mu_1 + \frac{7+\ell}{2}\right)n^2$ $+ (2\mu_0 + \frac{1+15\ell}{6})n$
M4	$2n\mu_0 + 2n^2\mu_1$	$3n$	n^2	$\alpha + 3\beta$ $+ \left(\frac{\beta}{2} + 3n\right)\ell$	$n^3/3$ $+ \left(2\mu_1 + \frac{7+\ell}{2}\right)n^2$ $+ (2\mu_0 + \frac{1+15\ell}{6})n$

Using (5.30) and (5.31) in (5.82) we get

$$\alpha_k = \left(\left(1 - \frac{1}{a}\right)F'(x_k) + \frac{1}{a}F'(y_k)\right)^{-1}F'(x_k), \qquad (5.83)$$

and

$$\alpha_k = \left(\left(1 + \frac{1}{a}\right)F'(y_k) - \frac{1}{a}F'(x_k)\right)^{-1}F'(y_k). \qquad (5.84)$$

Thus, we propose four-type choices (5.80), (5.81), (5.83) and (5.84) for α_k. When $a = 2/3$ the formula (5.83) leads to that of in [3], while when $a = 1/2$ the formula (5.83) leads to that of in [19]. Thus, the Theorem 5.7 is more general than the particular Theorems presented in [3, 13, 18–20].

Remark 5.3 The choices (5.80), (5.81), (5.83) and (5.84) for α_k and $\bar{\tau}_k$ also valid for three-step method (5.47) with sixth-order of convergence.

5.1.3 Total Cost Comparison Between Methods

We see Tables 5.3, 5.4 and 5.5 to check the total cost of each iteration for each method, where n is the system dimension, $\alpha = n(n-1)(2n-1)/6$, $\beta = n(n-1)$, μ_0 and μ_1 are relative costs of evaluation of F and Jacobian, respectively, in terms

Table 5.4 The cost of each iteration for fifth-order methods

Method	Evaluation of F and Jacobian	Scalar vector multiply	Matrix vector multiply	Linear solve	Total
Sharma and Arora [12]	$2n\mu_0 + 2n^2\mu_1$	$4n$	0	$2\alpha + 4\beta + (\beta + 4n)\ell$	$2n^3/3 + (2\mu_1 + 3 + \ell)n^2 + (2\mu_0 + \frac{1+9\ell}{3})n$
Cordero et al. [18]	$2n\mu_0 + 2n^2\mu_1$	$3n$	0	$3\alpha + 3\beta + \left(\frac{3\beta}{2} + 3n\right)\ell$	$n^3 + \left(2\mu_1 + \frac{3+3\ell}{2}\right)n^2 + (2\mu_0 + \frac{1+3\ell}{2})n$
Xiao et al. [21]	$2n\mu_0 + 2n^2\mu_1$	$4n$	0	$2\alpha + 4\beta + (\beta + 4n)\ell$	$2n^3/3 + (2\mu_1 + 3 + \ell)n^2 + (2\mu_0 + \frac{1+9\ell}{3})n$
M5	$3n\mu_0 + n^2\mu_1$	$5n$	$2n^2$	$\alpha + 5\beta + \left(\frac{\beta}{2} + 5n\right)\ell$	$n^3/3 + \left(\mu_1 + \frac{13+\ell}{2}\right)n^2 + (3\mu_0 + \frac{1+27\ell}{6})n$

of multiplications and ℓ is the relative cost of division in terms of multiplications. The total cost ranges from $n^3/3$ to n^3, not including lower powers of the dimension n of the system. Here we denote our fourth-order method (5.1) with Θ_k given by (5.30) by M4, fifth-order method (5.47) with Θ_k at (5.29) by M5, and sixth-order method (5.47) with Θ_k defined by (5.29) by M6. We chose $a = 1$ for M4, M5, M6. The most expensive method is Cordero et al. [18] closely followed by Jarratt [8] for which the total cost is n^3. Four methods Sharma et al. [12], Xiao et al. [21, 22], M61 [9] and M62 [9] cost $2n^3/3$. All other methods cost $n^3/3$. We note that our methods M4, M5 and M6 all cost $n^3/3$. Hence they are competitive under both of computational cost and order of convergence to existing methods. The efficiency index of methods is given by $E = \rho^{1/C}$, where ρ is the order of convergence and C is computational cost per iteration. For large system C tends to infinity as $n \to \infty$. Hence $E \to 1$ for all methods, i.e., E is not distinguished each other, as the scalar equation case. To compare the efficiency of different methods at first needed to be estimated computational cost per iteration (see more details [9, 13, 21, 22]). It mainly consists of evaluations of functions and derivatives (divided difference) and of estimation of products and quotients for matrix inversion, multiplications of a matrix by a matrix and a vector. Therefore, to estimate the computational cost we first keep in mind the number of above mentioned operations per iteration. Obviously, it is

Table 5.5 The cost of each iteration for sixth-order methods

Method	Evaluation of F and Jacobian	Scalar vector multiply	Matrix vector multiply	Linear solve	Total
Jarratt-like [20]	$2n\mu_0 + 2n^2\mu_1$	$6n$	$3n^2$	$\alpha + 5\beta + \left(\frac{\beta}{2} + 5n\right)\ell$	$n^3/3 + \left(2\mu_1 + \frac{15+\ell}{2}\right)n^2 + \left(2\mu_0 + \frac{7+27\ell}{6}\right)n$
Sharma and Arora [9]	$3n\mu_0 + n^2\mu_1$	$5n$	$2n^2$	$\alpha + 5\beta + \left(\frac{\beta}{2} + 5n\right)\ell$	$n^3/3 + \left(\mu_1 + \frac{13+\ell}{2}\right)n^2 + \left(3\mu_0 + \frac{1+27\ell}{6}\right)n$
M61 [9]	$3n\mu_0 + n^2\mu_1$	$3n$	0	$2\alpha + 3\beta + (\beta + 3n)\ell$	$2n^3/3 + (\mu_1 + 2 + \ell)n^2 + \left(3\mu_0 + \frac{1+6\ell}{3}\right)n$
M62 [9]	$3n\mu_0 + n^2\mu_1$	$5n$	0	$2\alpha + 5\beta + (\beta + 5n)\ell$	$2n^3/3 + (\mu_1 + 4 + \ell)n^2 + \left(3\mu_0 + \frac{1+12\ell}{3}\right)n$
M63 [9]	$3n\mu_0 + n^2\mu_1$	$5n$	$2n^2$	$\alpha + 5\beta + \left(\frac{\beta}{2} + 5n\right)\ell$	$n^3/3 + \left(\mu_1 + \frac{13+\ell}{2}\right)n^2 + \left(3\mu_0 + \frac{1+27\ell}{6}\right)n$
M6	$3n\mu_0 + n^2\mu_1$	$6n$	$3n^2 + \left(\frac{\beta}{2} + 6n\right)\ell$	$\alpha + 6\beta$	$n^3/3 + \left(\mu_1 + \frac{17+\ell}{2}\right)n^2 + \left(3\mu_0 + \frac{1+33\ell}{6}\right)n$

necessary to compare methods by CPU time. In Tables 5.3, 5.4 and 5.5 we present the number of evaluations per iteration for different fourth-, fifth- and sixth-order methods.

5.2 Higher Order Jarratt-Type Iterations for Solving Systems of Nonlinear Equations

In this section, we propose a new family of methods, such as Jarratt, with the fifth and sixth-order. This includes some popular methods as special cases. We propose four different selection for parameter matrix T_k. The main advantage of the proposed methods is that they work well for any value of parameter "a" in the first stage of iterations, while the existing methods work only for some a (2/3 or 1/2). Thus, we extend essentially the domain of applicability of the original ones. Based on the computational efficiency analysis, we also made a selection of some high-efficiency ones among the families.

5.2.1   Iterative Methods with $(p + 2)$-Order of Convergence

Three-step method: we considered is given below

$$
\begin{aligned}
y_k &= x_k - a \left(F'(x_k)\right)^{-1} F(x_k), \quad a \in R \setminus \{0\}, \\
z_k &= \Psi_p(x_k, y_k), \\
x_{k+1} &= z_k - \tilde{\alpha}_k \left(F'(x_k)\right)^{-1} F(z_k).
\end{aligned}
\tag{5.85}
$$

Here $\Psi_p(x_k, y_k)$ is an iteration function of a method of order $p \geq 2$. We call the iterations (5.85) with $a \neq 1$ Jarratt-like methods, although Jarratt previously considered iterations (5.85) with $a = 2/3$ [8]. In Sect. 5.1 we proved that (5.85) has $(p + 2)$-order (see Theorem 5.7) under condition

$$
\tilde{\alpha}_k = I + 2\Theta_k + O\left(h^2\right),
\tag{5.86}
$$

where

$$
\Theta_k = \frac{1}{2} \left(F'(x_k)\right)^{-1} F''(x_k) \left(F'(x_k)\right)^{-1} F(x_k).
\tag{5.87}
$$

Note if $\tilde{\alpha}_k = I$ in (5.85) then its convergence order reduces from $p + 2$ to $p + 1$. To find $\tilde{\alpha}_k$ satisfying (5.86), we need some approximate formulae for Θ_k. Applying Taylor's expansion of $F'(y_k)$ around x_k and the first equation in (5.85) we can get the below

$$
F'(y_k) = F'(x_k) \left(I - 2a\Theta_k - 3a^2 d_k\right) + O\left(h^3\right),
\tag{5.88}
$$

where

$$
d_k = -\frac{1}{6} \left(F'(x_k)\right)^{-1} F'''(x_k) \left(\left(F'(x_k)\right)^{-1} F(x_k)\right)^2.
\tag{5.89}
$$

We use the notations

$$
t_k = \left(F'(x_k)\right)^{-1} F'(y_k), \quad s_k = \left(F'(y_k)\right)^{-1} F'(x_k).
$$

Then from (5.88) it follows that

$$
t_k = I - 2a\Theta_k - 3a^2 d_k + O\left(h^3\right).
\tag{5.90}
$$

Further, from (5.88) we find

$$
\left(F'(y_k)\right)^{-1} = -\left(-I + 2a\Theta_k + 3a^2 d_k\right)^{-1} \left(F'(x_k)\right)^{-1} + O\left(h^3\right).
\tag{5.91}
$$

Since $\Theta_k = O(h)$ and $d_k = O\left(h^2\right)$ then using the well known expansion (5.7) in (5.91) we have

$$\left(F'(y_k)\right)^{-1} = \left(I + 2a\Theta_k + 3a^2 d_k + 4a^2 \Theta_k^2\right)\left(F'(x_k)\right)^{-1} + O\left(h^3\right). \tag{5.92}$$

From (5.92) it follows that

$$s_k = I + 2a\Theta_k + 3a^2 d_k + 4a^2 \Theta_k^2 + O\left(h^3\right). \tag{5.93}$$

Adding (5.90) and (5.93) we get

$$t_k + s_k = 2I + 4a^2 \Theta_k^2 + O\left(h^3\right). \tag{5.94}$$

We find Θ_k from (5.90) and it has a form

$$\Theta_k = \frac{I - t_k}{2a} - \frac{3a}{2}d_k + O\left(h^3\right). \tag{5.95}$$

Then

$$\Theta_k + d_k = \frac{I - t_k}{2a} + \left(1 - \frac{3a}{2}\right)d_k + O\left(h^3\right). \tag{5.96}$$

Since $d_k = O\left(h^2\right)$ then from (5.94), (5.95) it clear that

$$\Theta_k = \frac{I - t_k}{2a} + O\left(h^2\right), \tag{5.97}$$

$$\Theta_k = \frac{1}{2a}(-I + s_k) + O\left(h^2\right).$$

Based on the idea of method [7], one can choose $\tilde{\alpha}_k$ as:

$$\tilde{\alpha}_k = (I + \delta\Theta_k)^{-1}(I + (2 + \delta)\Theta_k), \quad \delta \in R. \tag{5.98}$$

Using (5.97) in (5.98) we get

$$\tilde{\alpha}_k = \left(\left(1 + \frac{\delta}{2a}\right)F'(x_k) - \frac{\delta}{2a}F'(y_k)\right)^{-1}$$
$$\times \left(\left(1 + \frac{2+\delta}{2a}\right)F'(x_k) - \frac{2+\delta}{2a}F'(y_k)\right), \quad \delta \in R, \tag{5.99}$$

which satisfy the condition (5.86). Thus, we obtain the iteration

$$y_k = x_k - a\left(F'(x_k)\right)^{-1}F(x_k),$$
$$z_k = \Psi_p(x_k, y_k), \tag{5.100}$$
$$x_{k+1} = z_k - \tilde{\alpha}_k\left(F'(x_k)\right)^{-1}F(z_k),$$

where $\tilde{\alpha}_k$ in (5.100) is defined by (5.99). According to Theorem 3.2 in [23] the iteration (5.100) has $(p+2)$-order of convergence. Note that the iteration (5.100), (5.98) with $\delta = -1$ and $\delta = -2a$ converted to those obtained by Xiao et al. in [21] and [13], respectively. The most efficient methods are obtained when $\delta = 0$ and $\delta = -2$ in (5.99). As a second step in (5.100) we consider

$$z_k = x_k - T_k \left(F'(x_k) \right)^{-1} F(x_k),$$

where T_k has the following form

$$T_k = I + \Theta_k + O\left(h^2\right) \tag{5.101}$$

and

$$T_k = I + \Theta_k + 2\Theta_k^2 + d_k + O\left(h^3\right). \tag{5.102}$$

If T_k satisfies

$$T_k = I + \Theta_k + (p-3)(2\Theta_k^2 + d_k), \tag{5.103}$$

then according to Theorems 2.4 and 2.5 in [23] we can easily obtain

$$F(z_k) = O\left(h^p\right), \quad p = 3, 4.$$

Instead of (5.100) we have the following three-step iteration

$$
\begin{aligned}
y_k &= x_k - a \left(F'(x_k) \right)^{-1} F(x_k), \\
z_k &= x_k - T_k \left(F'(x_k) \right)^{-1} F(x_k), \\
x_{k+1} &= z_k - \tilde{\alpha}_k \left(F'(x_k) \right)^{-1} F(z_k),
\end{aligned}
\tag{5.104}
$$

where $\tilde{\alpha}_k$ and T_k are given by (5.99) and (5.103), respectively. Note that the condition (5.102) was obtained in [23] under the condition (5.45), i.e.,

$$\frac{1}{2} \left(F'(x_k) \right)^{-1} F''(x_k) \Theta_k \left(F'(x_k) \right)^{-1} F(x_k) = \Theta_k^2 + O\left(h^3\right). \tag{5.105}$$

The above obtained results can be formulated as:

Theorem 5.8 *The three-step iteration (5.104) has order of convergence $p+2$, where $p = 3, 4$.*

Table 5.6 Choices of parameters a, T_k and $\tilde{\alpha}_k$

Methods	a	T_k	$\tilde{\alpha}_k$
Sharma et al. [24]	a	(5.107), $\beta = 0$	(5.99), $\delta = 0$
Cordero et al. [8]	1	(5.107), $\beta = 1$	(5.99), $\delta = -2$
Grau-Sánchez et al. [14]	1	(5.107), $\beta = 2$	(5.99), $\delta = -2$
Sharma et al. [12]	1/2	(5.107), $\beta = 1$	(5.99), $\delta = -1$
Xiao et al. [13]	1	(5.107), $\beta = 2$	(5.99), $\delta = -2$
Xiao et al. [21]	a	(5.107), $\beta = 1$	(5.99), $\delta = -1$

5.2.2 Development of Fifth-Order Methods

Now we proceed to explain how to choose T_k satisfying the condition (5.101). The most easy way to obtain T_k satisfying the condition (5.101) is to replace Θ_k in (5.101) by formula (5.97). As a result, we have

$$T_k = I + \frac{1}{2a}(I - t_k).$$

But we prefer, as in (5.98), more general choice:

$$T_k = (I - \beta\Theta_k)^{-1}(I + (1 - \beta)\Theta_k), \quad \beta \in R, \tag{5.106}$$

satisfying the condition (5.101) or (5.103) with $p = 3$. Using (5.97) in (5.106) we obtain

$$T_k = \left(\left(1 - \frac{\beta}{2a}\right)F'(x_k) + \frac{\beta}{2a}F'(y_k)\right)^{-1}$$
$$\times \left(\left(1 + \frac{1-\beta}{2a}\right)F'(x_k) - \frac{1-\beta}{2a}F'(y_k)\right), \quad \beta \in R, \tag{5.107}$$

which used for constructing fifth-order methods. By virtue of Theorem 5.8, the iterations (5.104) with T_k and $\tilde{\alpha}_k$ given by (5.107) and (5.99) has a fifth-order of convergence. Our families of two parametric iterations (5.104) contain many of known methods as particular cases, see Table 5.6.

5.2.3 Development of Sixth-Order Methods

Here we give four different choices for parameter matrix T_k, which guarantee sixth-order convergence of iteration (5.104). In what following we assume $a \neq 0, 1$ in (5.104). Now we shall try to find T_k satisfying (5.102). We first use formula (5.96) in (5.102). Then we get

$$T_k = I + \frac{I - t_k}{2a} + \frac{(I - t_k)^2}{2a^2} + \left(1 - \frac{3}{2}a\right) d_k + O\left(h^3\right). \tag{5.108}$$

When $a = 2/3$ the formula (5.108) leads to

$$T_k = I + \frac{3}{4}(I - t_k) + \frac{9}{8}(I - t_k)^2,$$

which is the most easy and perfect choice. The problem how to find T_k satisfying (5.102) for $a \neq 2/3$ is still open. It turns out to be preferable to use additional information about $F(x)$. Let w_k be given by

$$w_k = x_k + a\left(F'(x_k)\right)^{-1} F(x_k). \tag{5.109}$$

From (5.109) and from the first equation in (5.104) we get

$$w_k = 2x_k - y_k.$$

Taylor's expansions of $F(y_k)$ and $F(w_k)$ around x_k give

$$F(y_k) = F'(x_k)\left((1 - a)I + a^2\Theta_k + a^3 d_k\right)\left(F'(x_k)\right)^{-1} F(x_k) + O\left(h^3\right), \tag{5.110}$$

and

$$F(w_k) = F'(x_k)\left((1 + a)I + a^2\Theta_k - a^3 d_k\right)\left(F'(x_k)\right)^{-1} F(x_k) + O\left(h^3\right), \tag{5.111}$$

in which (5.87) and (5.89) are used. Analogously, we can get

$$F'(w_k) = F'(x_k)\left(I + 2a\Theta_k - 3a^2 d_k\right) + O\left(h^3\right). \tag{5.112}$$

Now we are ready to prove the following theorem that gives the rule of choice d_k in (5.108).

Theorem 5.9 *The quantity d_k in (5.108) is defined as a solution of the linear system*

$$F'(x_k)d_k = \frac{2F'(x_k) - F'(y_k) - F'(w_k)}{6a^2}. \tag{5.113}$$

Proof From (5.88) and (5.112) we attain

$$\left(F'(x_k)\right)^{-1}\left(F'(y_k) + F'(w_k)\right) = 2I - 6a^2 d_k + O\left(h^3\right).$$

From this we find that

$$d_k = -\frac{\left(F'(x_k)\right)^{-1}\left(-2F'(x_k) + F'(y_k) + F'(w_k)\right)}{6a^2} + O\left(h^3\right),$$

i.e., we arrive at (5.113). $\qquad\square$

The sixth-order convergence of iteration (5.104), (5.108), (5.99) was proven by Sharma et al. in [20] only when $a = 2/3$. Analogously, one can prove the following:

Theorem 5.10 *The choice*

$$T_k = \left(1 + \frac{a-2}{2a^2}\right)I + \frac{1}{2a^2}s_k + \frac{1-a}{2a^2}t_k + \frac{2-3a}{2}d_k, \qquad (5.114)$$

with d_k given by (5.113) satisfies (5.102).

Proof From (5.90) and (5.93) it follows that

$$s_k - t_k = 4a(\Theta_k + d_k) + 4a^2\Theta_k^2 + 2a(3a - 2)d_k.$$

Hence, from last expression we obtain

$$\Theta_k + d_k = \frac{s_k - t_k}{4a} - a\Theta_k^2 - \frac{3a-2}{2}d_k. \qquad (5.115)$$

Taking into account (5.94) and (5.115) the condition (5.102) can be written as

$$T_k = I + \Theta_k + 2\Theta_k^2 + d_k = I + (2-a)\Theta_k^2 + \frac{s_k - t_k}{4a} + \frac{2-3a}{2}d_k$$

$$= \left(1 + \frac{a-2}{2a^2}\right)I + \frac{1}{2a^2}s_k + \frac{1-a}{2a^2}t_k + \frac{2-3a}{2}d_k,$$

i.e., we obtain (5.114). $\qquad\square$

When $a = 2/3$ the formula (5.114) converted to

$$T_k = \frac{1}{2}\left(-I + \frac{9}{4}s_k + \frac{3}{4}t_k\right),$$

which ensures $F(z_k) = O\left(h^4\right)$ [17]. Besides that, the formula (5.99) with $\delta = -2a$ leads to

$$\tilde{\alpha}_k = \left(1 - \frac{1}{a}\right)I + \frac{1}{a}s_k. \qquad (5.116)$$

The formula (5.116), when $a = 2/3$, leads to

$$\tilde{\alpha}_k = \frac{1}{2}(3s_k - 1).$$

Thus, the iteration (5.104) with T_k and $\tilde{\alpha}_k$ given by (5.114) and (5.116) for $a = 2/3$ leads to one proposed by Xiao et al. in [13]. Hence, the iteration (5.104), (5.114), (5.116) can be considered as a generalization of method given by Xiao et al. [13].

The natural way to construct T_k satisfying the condition (5.102) is the use its representation like (5.106). To this end, we consider the expansion of $(I + \gamma(I - t_k)/(2a))^{-1}$, i.e.,

$$\left(I + \gamma \frac{I - t_k}{2a}\right)^{-1} = I - \gamma \frac{I - t_k}{2a} + \gamma^2 \left(\frac{I - t_k}{2a}\right)^2 + O\left(h^3\right). \tag{5.117}$$

Using (5.95) in (5.117) we have

$$\left(I + \gamma \frac{I - t_k}{2a}\right)^{-1} = I - \gamma \Theta_k + \gamma^2 \Theta_k^2 - \frac{3a}{2} \gamma d_k + O\left(h^3\right). \tag{5.118}$$

Then, one can easily shown that

$$T_k = \left(I + \gamma \frac{I - t_k}{2a}\right)^{-1} \left(I + (1 + \gamma)\Theta_k + (\gamma + 2)\Theta_k^2 + \left(1 + \frac{3a}{2}\gamma\right) d_k\right), \tag{5.119}$$

satisfies the sufficient convergence condition (5.102) for $F(z_k)$ to be of order $O\left(h^4\right)$. The direct application of (5.119) is the following result

Theorem 5.11 *The choice*

$$T_k = \left(\left(1 + \frac{\gamma}{2a}\right) F'(x_k) - \frac{\gamma}{2a} F'(y_k)\right)^{-1} \left(\left(1 + \frac{2 + 3a\gamma}{6a^2}\right) F'(x_k)\right.$$
$$- \frac{3a + 2 + 6a\gamma}{12a^2} F'(y_k) + \frac{3a - 2}{12a^2} F'(w_k)$$
$$\left. + \frac{2 + \gamma}{4a^2} \left(F'(x_k) - F'(y_k)\right) (I - t_k)\right), \tag{5.120}$$

satisfies (5.119) for any a and γ.

Proof By virtue of (5.88) and (5.119) we have

$$bF'(x_k) + cF'(y_k) + dF'(w_k)$$
$$= F'(x_k)\left((b + c + d)I - 2c(c - d)\Theta_k - 3a^2(c + d)d_k\right) + O\left(h^3\right). \tag{5.121}$$

From (5.119) and (5.121) it clear that the fourth-order convergence condition (5.102) fulfilled if

$$b + c + d = 1, \quad -2a(c - d) = 1 + \gamma, \quad -3a^2(c + d) = 1 + \frac{3a}{2}\gamma.$$

The solution of last system is

$$b = 1 + \frac{2 + 3a\gamma}{6a^2}, \quad c = -\frac{3a + 2 + 6a\gamma}{12a^2}, \quad d = \frac{3a - 2}{12a^2}.$$

Using (5.102) and substituting the linear combination (5.121) with obtained coefficients into (5.119) we obtain (5.120). $\qquad\square$

Let $\gamma = -2$ and $a = 2/3$. Then (5.120) gives

$$T_k = \frac{1}{2}\left(3F'(y_k) - F'(x_k)\right)^{-1}\left(3F'(y_k) + F'(x_k)\right),$$

that coincides with ones obtained by Cordero et al. [19]. When $\gamma = -2$ the condition (5.119) leads to

$$T_k = \left(I - \gamma\frac{I - t_k}{a}\right)^{-1}(I - \Theta_k + (1 - 3a)d_k). \tag{5.122}$$

Similarly, the following theorem can be proved

Theorem 5.12 *Let z_k be defined as:*

$$z_k - x_k = \Omega_k, \tag{5.123}$$

where

$$\Omega_k = \left(\left(1 - \frac{1}{a}\right)F'(x_k) + \frac{1}{a}F'(y_k)\right)^{-1}$$
$$\times \left(\left(\left(1 + \frac{2 - 3a}{a^2}\right)F(x_k) + \frac{1 - 4a}{2a^3}F(y_k) + \frac{1 - 2a}{2a^3}F(w_k)\right)\right). \tag{5.124}$$

Then $F(z_k) = O\left(h^4\right)$.

Proof As in Theorem 5.10, we consider linear combination of $bF(x_k) + cF(y_k) + dF(w_k)$. Taking into account (5.110) and (5.111), we have

$$bF(x_k) + cF(y_k) + dF(w_k) = F'(x_k)\big((b + c(1 - a) + d(a + 1)) \cdot I$$
$$+ (c + d)a^2\Theta_k + (c - d)a^3d_k\big)\left(F'(x_k)\right)^{-1}F(x_k) + O\left(h^3\right). \tag{5.125}$$

We find the coefficient in (5.125) such that

$$(b + c(1 - a) + d(1 + a))I + (c + d)a^2\Theta_k + (c - d)a^3d_k = I - \Theta_k + (1 - 3a)d_k. \tag{5.126}$$

This gives

$$b = 1 + \frac{2 - 3a}{a^2}, \quad c = \frac{1 - 4a}{2a^3}, \quad d = \frac{2a - 1}{2a^3}.$$

Then from (5.123)–(5.126) we get

$$z_k = x_k + \Omega_k = x_k + T_k \left(F'(x_k)\right)^{-1} F(x_k), \qquad (5.127)$$

where

$$T_k = \left(\left(1 - \frac{1}{a}\right) F'(x_k) + \frac{1}{a} F'(y_k)\right)^{-1} F'(x_k)\left(I - \Theta_k + (1 - 3a)d_k\right). \quad (5.128)$$

According to (5.122), T_k given by (5.128), satisfies the condition (5.102). Hence $F(z_k) = O\left(h^4\right)$. $\qquad\qquad\square$

Note that Theorem 5.12 was proven by Cordero et al. in [8] only when $a = 1/2$.

Remark 5.4 In this case the matrix T_k in (5.127) is not known. But, the factor Ω_k is known. The matrix T_k is the solution of system

$$T_k\left(\frac{y_k - x_k}{a}\right) = \Omega_k,$$

and T_k can be found using by division in R^n algorithm discovered by Chun et al. [25]. The above mentioned algorithm can be applied also for finding matrix $A_k = \Theta_k + d_k$. Indeed, from the expansions (5.110) and (5.111) we obtain

$$A_k(y_k - x_k) = \xi_k, \qquad (5.129)$$

where ξ_k determined by solving system

$$F'(x_k)\xi_k = \frac{(1 + a)F(y_k) + (a - 1)F(w_k)}{2a^2}.$$

Since the vectors $y_k - x_k$ and ξ_k are known, the unknown matrix A_k in (5.129) can be found using by division in R^n algorithm. Then the formula (5.102) leads to

$$T_k = I + \frac{(I - t_k)^2}{2a^2} A_k.$$

But this algorithm is more complicated than (5.120) and (5.108), (5.113). Thus, we propose four different choices for T_k, satisfying the sufficient convergence condition (5.102). The main advantage of those choices is that they work well for any $a \neq 1$, while the methods constructed in [8, 13, 16, 19, 20, 26] work only for $a = 2/3$ or $a = 1/2$. Our families of iterations (5.104) with T_k given by one of the formulas (5.108), (5.113), (5.114), (5.120), (5.124) and with $\tilde{\alpha}_k$ given by (5.99) included existing methods as particular cases. We list some these in Table 5.7.

Remark 5.5 The high-order convergence of iterations often proved when $F(x)$ is the sufficiently differentiable function. Of course, the convergence order reduces when the corresponding high-order derivatives do not exist. The proposed methods may be extended to Banach space and local convergence analysis may be proceed based

Table 5.7 Choices of parameters a, T_k and $\tilde{\alpha}_k$

Methods	a	T_k	$\tilde{\alpha}_k$
Sharma et al. [20] Montazeri et al. [28]	2/3	(5.108)	(5.99), $\delta = 0$
Cordero et al. [8]	1/2	(5.124)	(5.99), $\delta = -2$
Cordero et al. [19]	2/3	(5.120), $\gamma = -2$	(5.99), $\delta = -2$
Xiao et al. [13]	2/3	(5.114)	(5.99), $\delta = -2$
Chun et al. [29] ($b_5 = 0$)	2/3	(5.108)	(5.99), $\delta = 0$
Abbasbandy et al. [30]	2/3	(5.108)	(5.99), $\delta = 0$
Madhu et al. [16]	2/3	(5.108)	(5.99), $\delta = 0$

on the mild conditions such as Lipschitz constants and hypothesis on the divided difference of order one as in [27].

During the comparison analysis we reveal redundant terms in some methods, see Table 5.8. From (5.85) we see that it suffice to find $\tilde{\alpha}_k$ with accuracy $O\left(h^2\right)$. Since $I - t_k = O(h)$ the term with $(I - t_k)^2$ in $\tilde{\alpha}_k$ has no influence in the order of convergence. Hence these terms must be canceled. Dropping the redundant terms in T_k and $\tilde{\alpha}_k$ we reduce the computational cost and time, thereby the corresponding algorithms are simplified.

If we use the idea of [27], then one can easy to show that Theorem 5.8 is directly extended to the i-step iterations:

$$
\begin{aligned}
\psi_1^k &= x_k - a\left(F'(x_k)\right)^{-1} F(x_k), \\
\psi_2^k &= x_k - T_k\left(F'(x_k)\right)^{-1} F(x_k), \\
\psi_3^k &= \psi_2^k - \tilde{\alpha}_k(\delta)\left(F'(x_k)\right)^{-1} F(\psi_2^k), \\
\psi_4^k &= \psi_3^k - \tilde{\alpha}_k(\delta)\left(F'(x_k)\right)^{-1} F(\psi_3^k), \\
&\ \ \cdots \\
\psi_i^k &= \psi_{i-1}^k - \tilde{\alpha}_k(\delta)\left(F'(x_k)\right)^{-1} F(\psi_{i-1}^k), \quad i = 3, 4, \ldots, l,
\end{aligned}
\tag{5.130}
$$

and T_k and $\tilde{\alpha}_k$ satisfy the conditions (5.102) and (5.86), respectively.

Theorem 5.13 *The i-step iterations (5.130) has order of convergence $2i$.*

Proof By Theorem 5.8 we have $F(\psi_2^k) = O\left(h^4\right)$, $F(\psi_3^k) = O\left(h^6\right)$. Then, by Theorem 5.8 we have $F(\psi_4^k) = O\left(h^8\right)$, i.e., the repeating application of Theorem 5.8 gives $F(\psi_i^k) = O\left(h^{2i}\right)$. $\qquad\square$

Remark 5.6 The coefficients $\tilde{\alpha}_k$ may be depend on own δ for each i. But the choice $\delta_i = \delta$ for all i is best from the computational cost and complexity of view.

Table 5.8 Choices of parameters T_k and $\tilde{\alpha}_k$

Methods	T_k	$\tilde{\alpha}_k$
Chun et al. [29]	–	$b_5(t_k - I)^2$
Abbasbandy et al. [30]	$\frac{15}{8}(t_k - I)^3$	$\frac{1}{2}(t_k - I)^2$
Madhu et al. [16]	–	$\frac{1}{2}(t_k - I)^2$

5.2.4 Computational Efficiency

To compare the different iterative methods we will use the computational efficiency index [2, 31] $CE = \rho^{1/C}$, where ρ is order of convergence and $C = d + op$ is the total computational cost of each iteration. Here, d is the number of function evaluations F and F', and op is the number of operations[1] (products and quotients) required per iteration.

Computational cost as follows:

- n and n^2 scalar functions are calculated to evaluate F and F', respectively.
- Further to calculate an inverse matrix, a $n \times n$ linear system be solved, we have to do $(n^3 - n)/3$ and n^2 operations for get LU factorization and solve the obtained triangular systems, respectively. For solving m linear systems with this matrix, we will do $(n^3 + 3mn^2 - n)/3$ operations.
- On other hand, we will need n^2 product for each matrix-vector multiplication and n product for scalar vector multiplication.

We discuss the computational cost and efficiency of the methods. The values of C were calculated and shown in Table 5.9.

In Table 5.9 we denote by NLS1 the number of linear systems with same coefficients matrix $F'(x_k)$ and by NLS2 the number of linear systems with other matrices. From Table 5.9 see that the most effective and cheapest methods are (5.85) with choices T_k given by (5.108), (5.113) and (5.120) and $\tilde{\alpha}_k$ given by (5.99) with $\delta = 0$ and for which the cost total is $n^3/3$. The methods with choices T_k given by (5.114) and (5.118) and $\tilde{\alpha}_k$ given by (5.99) with $\delta \neq 0$, $\delta \neq -2$, $\delta \neq -2a$ cost n^3.

The calculation algorithm for iteration (5.85) depends on the form of T_k. For example, the parameter T_k is given by (5.108), (5.113) then the algorithm is described as follows:

1. $F'(x_k)w_k = -F(x_k), \quad y_k = x_k + aw_k,$
2. $F'(x_k)\xi_k = (F'(x_k) - F'(y_k))w_k,$
3. $F'(x_k)\eta_k = -F(x_k) + \frac{1}{2a}(F'(x_k) - F'(y_k))w_k + \left(1 - \frac{3a}{2}\right)d_k w_k$
   $+\frac{1}{2a^2}(F'(x_k) - F'(y_k))\xi_k, \quad z_k = x_k + \eta_k,$
4. $F'(x_k)\Omega_k = -F(z_k),$
5. $\left(\left(1 + \frac{\delta}{2a}\right)F'(x_k) - \frac{\delta}{2a}F'(y_k)\right)d_k = \left(\left(1 + \frac{2+\delta}{2a}\right)F'(x_k) - \frac{2+\delta}{2a}F'(y_k)\right)\Omega_k,$
   $x_{k+1} = z_k + d_k.$

---

[1] without taking into account the operations of subtraction and addition.

Table 5.9 Values of C

T_k	$\tilde{\alpha}_k$	NLS1	NLS2	d	op	C
(5.108), (5.113)	(5.99), $\delta = 0$	5	–	$3n^2 + 2n$	$\frac{1}{3}n^3 + 9n^2 + \frac{14}{3}n$	$\frac{1}{3}n^3 + 12n^2 + \frac{20}{3}n$
	(5.99), $\delta \neq 0$	4	1		$\frac{2}{3}n^3 + 10n^2 + \frac{13}{3}n$	$\frac{2}{3}n^3 + 13n^2 + \frac{19}{3}n$
(5.114)	(5.99), $\delta = 0$, $\delta = -2a$	4	1	$3n^2 + 2n$	$\frac{2}{3}n^3 + 9n^2 + \frac{16}{3}n$	$\frac{2}{3}n^3 + 12n^2 + \frac{22}{3}n$
	(5.99), $\delta \neq 0$, $\delta \neq -2a$	3	2		$n^3 + 10n^2 + 5n$	$n^3 + 13n^2 + 7n$
(5.120)	(5.99), $\delta = \gamma = 0$	5	–	$3n^2 + 2n$	$\frac{1}{3}n^3 + 9n^2 + \frac{14}{3}n$	$\frac{1}{3}n^3 + 12n^2 + \frac{20}{3}n$
	(5.99), $\delta = \gamma \neq 0$	3	2		$\frac{2}{3}n^3 + 11n^2 + \frac{13}{3}n$	$\frac{2}{3}n^3 + 14n^2 + \frac{19}{3}n$
(5.124)	(5.99), $\delta = 0$, $\delta = -2$	2	2	$2n^2 + 3n$	$\frac{2}{3}n^3 + 6n^2 + \frac{13}{3}n$	$\frac{2}{3}n^3 + 8n^2 + \frac{22}{3}n$
	(5.99), $\delta \neq 0$, $\delta \neq -2$	2	2		$n^3 + 7n^2 + 4n$	$n^3 + 9n^2 + 7n$

In this case, we must solve four (five, when $\delta = 0$) linear system with same matrix $F'(x_k)$ and one linear system (when $\delta \neq 0$) with other matrix. For calculation of the matrix and of the right hand sides of these systems we need five matrix-vector multiplications and five scalar vector multiplications.

An addition, we need two evaluation F and three evaluation F'. Hence $op = (2n^3 + 30n^2 + 13n)/3$ and $d = 3n^2 + 2n$. In a similar way, calculated the total computational cost for other cases.

We developed a new methods like Jarratt with high-order of convergence. They include some popular methods as special cases. We suggest four different selections for parameter matrix T_k. The main advantage of the proposed methods is that they work well for any value of parameter "a" that extend essentially the domain of applicability of the original ones.

5.3 Novel Multi-parametric Newton-Jarratt-Type Family of Higher Order Methods for Solving Nonlinear Systems

In this section, we develop new multi-parametric families of higher order ($6 \leq p \leq 8$) iterative methods based on the generating function one. The families contain some iterative methods known already. In fact, the proposed classes of higher order methods are very wide in the sense that they exhaust almost all the existing ones with the same order of convergence. Several applied problems are solved to check the performance of our methods and other existing ones and to verify the theoretical results, and we found that our methods show better results than the other compared ones. Moreover, the basins of attraction of the methods for systems are introduced to confirm the same results for the performance of the proposed methods [32].

We consider the following iterative scheme:

$$
\begin{aligned}
y_k &= x_k - a \left(F'(x_k) \right)^{-1} F(x_k), \quad a \in R \setminus \{0\}, \\
z_k &= x_k - T_k \left(F'(x_k) \right)^{-1} F(x_k), \\
x_{k+1} &= z_k - \tilde{\alpha}_k \left(F'(x_k) \right)^{-1} F(z_k).
\end{aligned}
\tag{5.131}
$$

In [23, 33, 34] were found some sufficient conditions for T_k and $\tilde{\alpha}_k$ such that the iteration (5.131) have p-th ($p = 6, 7, 8$) order of convergence. In order to describe these conditions we needed the following notations:

$$
\begin{aligned}
t_k &= \left(F'(x_k) \right)^{-1} F'(y_k), \quad s_k = \left(F'(y_k) \right)^{-1} F'(x_k), \\
\gamma_k &= \left(F'(x_k) \right)^{-1} F'(w_k), \quad w_k = x_k + a \left(F'(x_k) \right)^{-1} F(x_k).
\end{aligned}
\tag{5.132}
$$

Then it is easy to show that [33] (see also Sect. 5.2)

$$
\begin{aligned}
t_k &= I - 2a\Theta_k - 3a^2 d_k + O\left(h^3\right), \\
\gamma_k &= I + 2a\Theta_k - 3a^2 d_k + O\left(h^3\right), \\
s_k &= I + 2a\Theta_k + 3a^2 d_k + 4a^2\Theta_k^2 + O\left(h^3\right),
\end{aligned}
\tag{5.133}
$$

where I is identity matrix and Θ_k and d_k are defined by (5.87) and (5.89), respectively. We also use the assumption (5.105). By summarizing the results of [23, 33, 34], we deduce that

$$
F(x_{k+1}) = O\left(h^p\right),
\tag{5.134}
$$

if the parametric matrices T_k, $\tilde{T}_k$ and $\tilde{\alpha}_k$ satisfy one of the following relations I-VI in Table 5.10.

Let

$$
g(t) = 1 + at + bt^2 + \cdots, \quad a, b \in R, \quad |t| < 1.
\tag{5.135}
$$

Table 5.10 The parametric matrices. Here $M_k = I - t_k$

No	T_k	$\tilde{\alpha}_k$	a	p
I	$I + \frac{M_k}{2a} + \frac{M_k^2}{2a^2} + (1 - \frac{3}{2}a)d_k + O\left(h^3\right)$	$I + \frac{M_k}{a} + O\left(h^2\right)$		6
II	$I + \frac{M_k}{2a} + O\left(h^2\right)$	$I + \frac{M_k}{a} + \frac{3M_k^2}{2a^2} + 3(1 - a)d_k + O\left(h^3\right)$		
III	$I + \frac{M_k}{2} + O\left(h^2\right)$	$I + M_k$		
IV	$I + \frac{M_k}{2} + \frac{5}{4}M_k^2 + O\left(h^3\right)$	I	$a = 1$	
V	$\tilde{T}_k = I + (a + 1)\Theta_k + (a^2 + 2a + 2)\Theta_k^2 + (a^2 + a + 1)d_k + O\left(h^3\right)$	$I + 2\Theta_k + 3d_k + 6\Theta_k^2 + O\left(h^3\right)$		7
VI	$\tilde{T}_k = I + 2\Theta_k + 5\Theta_k^2 + 3d_k + O\left(h^3\right)$	$I + 2\Theta_k + 3d_k + 6\Theta_k^2 + O\left(h^3\right)$	$a = 1$	8

The parametrization of function $g(t)$ can be performed, for example, using GFM [7]. The GFM allows us to represent $g(t)$ as:

$$g(t) = (1 - \alpha t)^{-1} \left(1 + (a - \alpha)t + (b - \alpha a)t^2\right) + O\left(t^3\right), \quad \alpha \in R. \qquad (5.136)$$

Two-parametric representation of (5.135) is

$$g(t) = \left(1 + \alpha t + \beta t^2\right) \times \left(1 + (a - \alpha)t + (b - \beta - \alpha(a - \alpha))t^2\right) + O\left(t^3\right), \quad \alpha, \beta \in R. \qquad (5.137)$$

If $\beta = \alpha^2$ then (5.137) leads to (5.136) within of accuracy $O\left(t^3\right)$. Using (5.134) and GFM (or (5.135), (5.136)) we can formulate the following:

Theorem 5.14 *Let $F(x) : D \subset R^n \to R^n$ be a sufficiently Fréchet differentiable function in a convex set $D \subset R^n$ containing a zero x^* of $F(x)$. Suppose that $F'(x)$ is continuous and nonsingular in x^*. Then, for an initial value x_0 sufficiently close to the solution, the iterations (5.131) have sixth-order convergence when T_k and $\tilde{\alpha}_k$ are satisfied by one of the following pairs:*

$$T_k = (I - \alpha M_k)^{-1} \left(I + \left(\frac{1}{2a} - \alpha\right) M_k + \frac{1 - a\alpha}{2a^2} M_k^2\right)$$

$$+ \left(1 - \frac{3}{2}a\right) d_k + O\left(h^3\right), \quad \alpha \in R, \qquad (5.138a)$$

$$\tilde{\alpha}_k = (I - \beta M_k)^{-1} \left(I + \left(\frac{1}{a} - \beta\right) M_k\right) + O\left(h^2\right), \quad \beta \in R, \qquad (5.138b)$$

and

$$T_k = (I - \gamma M_k)^{-1} \left(I + \left(\frac{1}{2a} - \gamma \right) M_k \right) + O\left(h^2\right), \quad \gamma \in R, \quad (5.139a)$$

$$\tilde{\alpha}_k = (I - \delta M_k)^{-1} \left(I + \left(\frac{1}{a} - \delta \right) M_k + \frac{3 - 2a\delta}{2a^2} M_k^2 \right)$$
$$+ 3(1 - a)d_k + O\left(h^3\right), \quad \delta \in R. \tag{5.139b}$$

Proof From Table 5.10 we see that $F(x_{k+1}) = O\left(h^6\right)$ under conditions I and II. If we take GFM or (5.135), (5.136) into account, then the condition I can be rewritten in parameterized from as (5.138a, 5.138b). In a similar way, the condition II can be rewritten as (5.139a, 5.139b). $\square$

Thus, we have two-parametric families of sixth-order methods. It should be pointed out that in [33] was proposed sufficient conditions for convergence of sixth-order iterations. Our new sufficient conditions for convergence of methods (5.138a, 5.138b), (5.139a, 5.139b) are more general than those of [33]. That is, when $\alpha = 0$ and $\beta = -\delta/(2a)$ conditions (5.138) lead to the conditions (29) and (18) given in [33]. Therefore, our choices of (5.138a, 5.138b), (5.139a, 5.139b) are more general than those of [33]. In [33] the attention was focused on the possible choices of iteration matrices, while in this study, our attention is mainly on introducing parameters into the iteration matrices and obtain more general sufficient conditions for convergence.

Realization of this general theory can be done by different ways. As a result, we will obtain concrete sixth-order methods. For simplicity, we consider the cases $a = 2/3$ and $a = 1$ below in details.

5.3.1 Sixth-Order Jarratt-Type Methods

Let $a = 2/3$ in (5.131). Then (5.131) leads to

$$y_k = x_k - \frac{2}{3} \left(F'(x_k) \right)^{-1} F(x_k),$$
$$z_k = x_k - T_k \left(F'(x_k) \right)^{-1} F(x_k), \tag{5.140}$$
$$x_{k+1} = z_k - \tilde{\alpha}_k \left(F'(x_k) \right)^{-1} F(z_k).$$

The sixth-order convergence condition (5.138) reads as

$$T_k = (I - \alpha M_k)^{-1} \left(I + \left(\frac{3}{4} - \alpha \right) M_k + \left(\frac{9}{8} - \frac{3}{4}\alpha \right) M_k^2 \right) + O\left(h^3\right), \quad \alpha \in R, \tag{5.141}$$

$$\tilde{\alpha}_k = (I - \beta M_k)^{-1} \left(I + \left(\frac{3}{2} - \beta \right) M_k \right) + O\left(h^2\right), \quad \beta \in R. \tag{5.142}$$

Table 5.11 Particular cases of iterations (5.140)

No	Methods	T_k	(5.141),α	$\tilde{\alpha}_k$	(5.142),β
1	Sharma et al. [12] Montazeri et al. [28]	$\frac{23}{8}I - 3t_k + \frac{9}{8}t_k^2$	0	$I + \frac{3}{2}M_k$	0
2	Hueso et al. [36]	$\frac{5}{8}I + \frac{9}{24}s_k^2 + a_2\big(-I + s_k + \frac{1}{3}t_k - \frac{1}{3}s_k^2\big)\big)$	0	$I + \frac{3}{2}M_k + \frac{9-8b}{8}M_k^2$	0
3	Abbasbandy et al. [30]	$I + \frac{3}{4}M_k + \frac{9}{8}M_k^2 - \frac{15}{8}M_k^3$	0	$I + \frac{3}{2}M_k + \frac{1}{2}M_k^2$	0
4	Behl et al. [37]	$\frac{23}{8}I - 3t_k + \frac{9}{8}t_k^2$	0	$I + \frac{3}{2}M_k + \Theta_6 M_k^2$	0
5	Chun et al. [29]	$\frac{23}{8}I - 3t_k + \frac{9}{8}t_k^2$	0	$I + \frac{3}{2}M_k + b_5 M_k^2$	0
		$\frac{157}{64}I - \frac{117}{64}s_k - \frac{39}{64}t_k + \frac{63}{64}s_k^2$	0	$I + \frac{3}{2}M_k + 3b_5(-I + M_k) + (\frac{28}{8} + 3b_5)M_k^2$	0
6	Bahl et al. [38]	$I + \frac{3}{4}M_k\big(I + 6(4I - 3\alpha M_k)^{-1}M_k\big)$	0	$\frac{2-3\lambda}{5}\big((F'(x_k))^{-1} + \lambda F'(y_k))^{-1} \times \big(F'(x_k) + \frac{2-3\lambda}{5}(F'(y_k))\big)$	$\frac{5\lambda}{2(1+\lambda)}$
7	Narang et al. [26]	$\big(I + \frac{1}{2\beta}\big(I - \frac{\lambda}{\beta}M_k\big)^{-1}M_k\big)H_k$ $H_k = I + \frac{3\beta-2}{4\beta}M_k + \frac{9\beta^2 - 3\beta - 4\lambda + 2}{8\beta^2}M_k^2$	$\frac{\lambda}{\beta}$	$I + \frac{3}{2}M_k$	0
8	Wang et al. [39]	$\frac{1}{2}(3F'(y_k) - F'(x_k))^{-1} \times (F'(x_k) + 3F'(y_k))$	$\frac{3}{2}$	$I + \frac{3}{2}M_k$	0
9	Cordro et al. [3]	$\frac{1}{2}(3F'(y_k) - F'(x_k))^{-1} \times (3F'(y_k) + F'(x_k))$	$\frac{3}{2}$	$I + \frac{3}{2}M_k$	0
10	Soleymani et al. [40]	$\frac{1}{2}(3F'(y_k) - F'(x_k))^{-1} \times (3F'(y_k) + F'(x_k))$	$\frac{3}{2}$	$\big(I + \frac{3}{2}M_k + \frac{9}{8}M_k^2\big)^2$	0
11	Kansal et al. [35] $a = 1$	$\big(I - \frac{3}{2}M_k\big)^{-1} \times \big(I - \frac{3}{4}M_k\big)$	$\frac{3}{2}$	$I + \frac{3}{2}M_k + rM_k^2$	0
12	Behl et al. [41]	$\frac{5}{8}I + \frac{3}{2}s_k^2$	0	$-2(1 + 3b_1)F'(x_k) - (3 + 5b_1)(F'(y_k))^{-1} \times \big(F'(x_k) + b_1 F'(y_k)\big)$	$\beta = \frac{3+5b_1}{2(1+b_1)}$
13	Zhanlav et al. [33]	$I + \frac{3}{4}M_k + \frac{9}{8}M_k^2$	0	$\big(I + \frac{3\delta}{4}M_k\big)^{-1} \times \big(I + \frac{3(2+\delta)}{4}M_k\big)$	$-\frac{3\delta}{4}$

That is, we have two-parametric (α, β) sixth-order family of iterations (5.140) with T_k and $\tilde{\alpha}_k$ given by (5.141) and (5.142), respectively. Our family (5.140)–(5.142) includes many of methods as particular cases. We list some of these in Table 5.11. Note that all the three-step methods listed below can be written as (5.140) with own T_k and $\tilde{\alpha}_k$.

It is worthy to note, that some of the above mentioned methods contain redundant terms, which are underlined by red line in Table 5.11. These terms have no influence on the convergence of the corresponding method. Dropping these terms reduces the complexity of methods. From Table 5.11 it is clear that the second scheme of [29] has sixth-order of convergence, only when $b_5 = 0$. As far as concerned to the scheme of Kansal et al. [35], it pointed out that the original scheme has a very complicated expression for $T_k = v_k^{-1} \mu_k$. Both v_k and μ_k are polynomials of four degree with respect to t_k and each coefficient of them is a cubic polynomial of parameter a. But these difficulties have been removed after some long calculations, and we obtain

$$v_k = -8 \left(I - \frac{3(3-2a)}{2} M_k - \frac{9}{2}(a-1)M_k^2 - \frac{27}{4} A_8 M_k^3 \right),$$

$$\mu_k = -8 \left(I + \frac{3(4a-5)}{4} M_k - \frac{9}{4}(a-1)M_k^2 - \frac{27}{4} A_4 M_k^3 \right). \quad (5.143)$$

So, the parameter matrix T_k in scheme of [35] has a form

$$T_k = \left(I + \frac{3(3-2a)}{2} M_k - \frac{9}{4}(4a^2 - 10a + 7)M_k^2 \right) \left(I + \frac{3(4a-5)}{4} M_k \right)$$

$$= I + \frac{3}{4} M_k - \frac{9}{8}(a-1)M_k^2 + O\left(h^3\right), \quad (5.144)$$

for $a \neq 1$. Comparison of (5.137) and (5.144) gives

$$\alpha = \frac{3(3-2a)}{2}, \quad \beta = \frac{9}{4}(4a^2 - 10a + 7). \quad (5.145)$$

That is, (5.144) is a special case of (5.137) with α and β given by (5.145).

5.3.2 Sixth-Order Newton-Type Methods

Let $a = 1$ in (5.131). Then (5.131) leads to

$$y_k = x_k - \left(F'(x_k)\right)^{-1} F(x_k),$$
$$z_k = x_k - T_k \left(F'(x_k)\right)^{-1} F(x_k), \quad (5.146)$$
$$x_{k+1} = z_k - \tilde{\alpha}_k \left(F'(x_k)\right)^{-1} F(z_k).$$

The sixth-order convergence condition (5.139) reads as

$$T_k = (I - \gamma M_k)^{-1} \left(I + (\tfrac{1}{2} - \gamma)M_k \right) + O\left(h^2\right), \quad \gamma \in R, \quad (5.147)$$

$$\tilde{\alpha}_k = (I - \delta M_k)^{-1} \left(I + (1-\delta)M_k + \frac{3-2\delta}{2} M_k^2 \right) + O\left(h^3\right), \quad \delta \in R. \quad (5.148)$$

Thus, we have two-parametric (γ, δ) sixth-order family of iterations (5.146) with T_k and $\tilde{\alpha}_k$ given by (5.147) and (5.148), respectively. The proposed family of iterations

Table 5.12 Particular cases of iterations (5.146)

Methods	T_k	(5.147),γ	$\tilde{\alpha}_k$	(5.148),δ
Behl et al. [42]	$2(F'(x_k) + F'(y_k))^{-1} \times F'(x_k)$	$\frac{1}{2}$	$(3F'(y_k) - F'(x_k))^{-1} \times (F'(x_k) + F'(y_k))$	$\frac{3}{2}$
Lotfi et al. [43]	$2(F'(x_k) + F'(y_k))^{-1} \times F'(x_k)$	$\frac{1}{2}$	$\frac{7}{2}I - 4t_k + \frac{1}{2}t_k^2$	0
Wang et al. [35]	$I + \frac{1}{2}M_k$	0	$I + M_k$	0
New1	$I + M_k + \frac{5}{4}M_k^2$	(5.149) $\sigma = 0$	I	
New2	$\left(I - \frac{5}{4}M_k\right)^{-1} \times \left(I - \frac{1}{4}M_k\right)$	(5.149) $\sigma = \frac{5}{4}$	I	

(5.146)–(5.148) includes also many of existing methods as particular cases. We list some of these methods in Table 5.12.

As above, instead of the IV condition in Table 5.10 one can use its parameterized form

$$T_k = (I - \sigma M_k)^{-1}\left(I + (1 - \sigma)M_k + \left(\frac{5}{4} - \sigma\right)M_k^2\right) + O\left(h^3\right), \quad \sigma \in R. \tag{5.149}$$

The New 1 and New 2 methods in Table 5.12 are obtained from (5.149) when $\sigma = 0$ and $\sigma = 5/4$, respectively. Note that the new schemes are more simple than the other ones from the view of the computational cost.

Repeating applications of (5.146)–(5.148) gives i-step iterations

$$
\begin{aligned}
\psi_1^k &= x_k - \left(F'(x_k)\right)^{-1} F(x_k), \\
\psi_2^k &= x_k - T_k \left(F'(x_k)\right)^{-1} F(x_k), \\
\psi_3^k &= \psi_2^k - \tilde{\alpha}_k \left(F'(x_k)\right)^{-1} F(\psi_2^k), \\
&\cdots \\
\psi_i^k &= \psi_{i-1}^k - \tilde{\alpha}_k \left(F'(x_k)\right)^{-1} F(\psi_{i-1}^k), \quad i = 3, 4, \ldots,
\end{aligned}
\tag{5.150}
$$

where T_k and $\tilde{\alpha}_k$ are given by (5.147) and (5.148), respectively. For (5.150) the following holds.

Theorem 5.15 *Suppose that all the assumptions of Theorem 5.14 are fulfilled. Then the order of local convergence of iterations (5.150) equals to $3(i - 1)$.*

Proof According to (5.146) we have $\psi_2^k = O\left(h^3\right)$, $\psi_3^k = O\left(h^6\right)$. Repeating applications of this result gives $\psi_i^k = O\left(h^{3(i-1)}\right)$. $\qquad\square$

The particular cases of (5.150) with $\gamma = 1/2$, $\delta = 3/2$ and $\gamma = 1/2$, $\delta = 0$ are the schemes of Behl et al. [42] and Lotfi et al. [43].

In the similar way, based on (5.149) one can construct $i - step$ iterations

$$\psi_1^k = x_k - \left(F'(x_k)\right)^{-1} F(x_k),$$
$$\psi_2^k = \psi_1^k - \tilde{T}_k \left(F'(x_k)\right)^{-1} F(\psi_1^k),$$
$$\psi_3^k = \psi_2^k - \tilde{\alpha}_k \left(F'(x_k)\right)^{-1} F(\psi_2^k),$$
$$\cdots$$
$$\psi_i^k = \psi_{i-1}^k - \tilde{\alpha}_k \left(F'(x_k)\right)^{-1} F(\psi_{i-1}^k), \quad i = 3, 4, \ldots,$$

(5.151)

where $\tilde{T}_k$ and $\tilde{\alpha}_k$ are described as (5.149) and (5.148). For (5.151) we establish the following result:

Theorem 5.16 *Let all the assumptions of Theorem 5.14 be satisfied. Then, the order of local convergence of iterations (5.151) equals to* $3(i-1) + 2$.

Proof $\psi_1^k = O\left(h^2\right)$, $\psi_2^k = O\left(h^5\right)$ by virtue of (5.149). As a consequence $\psi_3^k = O\left(h^8\right) = O\left(h^{3(3-1)+2}\right)$. By mathematical induction we get $\psi_i^k = O\left(h^{3(i-1)+2}\right)$. $\qquad\square$

In [33] was proposed also $i - step$ iterations similar to (5.150) and (5.151). But in our case, the iteration matrices T_k and $\tilde{\alpha}_k$ are more general than those of [33].

5.3.3   Development of New Two-Parametric Seventh and Eighth-Order Methods

Now we consider another version of (5.131)

$$y_k = x_k - a \left(F'(x_k)\right)^{-1} F(x_k),$$
$$z_k = y_k - \tilde{T}_k \left(F'(x_k)\right)^{-1} F(y_k),$$
$$x_{k+1} = z_k - \tilde{\alpha}_k \left(F'(x_k)\right)^{-1} F(z_k).$$

(5.152)

From Table 5.10 we see that the iterations (5.152) have seventh-order of convergence under condition V. The condition V can be rewritten in terms of M_k as

$$\tilde{T}_k = I + \frac{a+1}{2a} M_k + \frac{a^2 + 2a + 2}{4a^2} M_k^2 - \frac{(a-1)(a+2)}{2} d_k + O\left(h^3\right), \quad (5.153)$$
$$\tilde{\alpha}_k = I + \frac{1}{a} M_k + \frac{3}{2a^2} M_k^2 - 3(a-1)d_k + O\left(h^3\right). \quad (5.154)$$

According to (5.135), (5.136), the parametrization of $\tilde{T}_k$ and $\tilde{\alpha}_k$ by GFM gives

$$\tilde{T}_k = (I - \beta_1 M_k)^{-1} \left(I + \left(\frac{a+1}{2a} - \beta_1\right) M_k + \frac{a^2 + 2a + 2 - 2a(a+1)\beta_1}{4a^2} M_k^2\right)$$
$$- \frac{(a-1)(a+2)}{2} d_k, \quad \beta_1 \in R,$$

(5.155)

and

$$\tilde{\alpha}_k = (I - \beta_2 M_k)^{-1} \left(I + \left(\frac{1}{a} - \beta_2 \right) M_k + \frac{3 - 2a\beta_2}{2a^2} M_k^2 \right)$$
$$- 3(a - 1)d_k, \quad \beta_2 \in R. \tag{5.156}$$

Then we have the following result.

Theorem 5.17 *The order of local convergence of iterations (5.152) equals to seven, when $\tilde{T}_k$ and $\tilde{\alpha}_k$ are given by (5.155) and (5.156), respectively. In particular, the methods (5.152) for $a = 1$ has eighth-order of convergence.*

The simple cases of (5.155), (5.156) are obtained when $\beta_1 = \beta_2 = 0$. In these cases formulae (5.155), (5.156) lead to (5.153), (5.154), respectively. Another easy case is

$$\beta_1 = \frac{a^2 + 2a + 2}{2a(a + 1)}, \quad \beta_2 = \frac{3}{2a}. \tag{5.157}$$

In this case the formulae (5.155) and (5.156) lead to

$$\tilde{T}_k = \left(I - \frac{a^2 + 2a + 2}{2a(a + 1)} M_k \right)^{-1} \left(I - \frac{1}{2a(a + 1)} M_k \right) - \frac{(a - 1)(a + 2)}{2} d_k, \tag{5.158}$$

and

$$\tilde{\alpha}_k = \left(I - \frac{3}{2a} M_k \right)^{-1} \left(I - \frac{1}{2a} M_k \right) - 3(a - 1)d_k, \tag{5.159}$$

respectively.

Remark 5.7 Of course, the convergence order of iteration (5.152) reduce by one unit, if we ignore the terms of order $O\left(h^2\right)$ in one of the formulas (5.155) and (5.156). Our family of iterations (5.152), (5.155), (5.156) contain many of iterations as particular cases. We exhibit some of them in Table 5.13. Note that we use the relation $s_k = 2I - t_k + (I - t_k)^2 + O\left(h^3\right)$ there.

The i-th step version of iterations (5.152) is

$$\begin{aligned}
\psi_1^k &= x_k - a \left(F'(x_k) \right)^{-1} F(x_k), \\
\psi_2^k &= \psi_1^k - \tilde{T}_k \left(F'(x_k) \right)^{-1} F(\psi_1^k), \\
\psi_3^k &= \psi_2^k - \tilde{\alpha}_k \left(F'(x_k) \right)^{-1} F(\psi_2^k), \\
&\cdots \\
\psi_i^k &= \psi_{i-1}^k - \tilde{\alpha}_k \left(F'(x_k) \right)^{-1} F(\psi_{i-1}^k), \quad i = 3, 4, \ldots,
\end{aligned} \tag{5.160}$$

where $\tilde{T}_k$ and $\tilde{\alpha}_k$ are given by (5.155) and (5.156), respectively. The order of local convergence of iterations (5.160) equals to $3(i - 1) + 1$, which becomes $3(i - 1) + 2$, when $a = 1$.

174 5 Higher Order Newton-Type Iterations

Table 5.13 Particular cases of iterations (5.152)

Methods	a	$\tilde{T}_k$	(5.155) β_1	$\tilde{\alpha}_k$	(5.156) β_2	p
Sharma et al.'s NLM$_8$ [44]	1	$\frac{13}{4}I - \frac{7}{2}t_k + \frac{5}{4}t_k^2$	0	$\frac{7}{2}I - 4t_k + \frac{3}{2}t_k^2$	0	8
Cordero et al.'s CCGT1 [45]	1	$\left(\frac{5}{4}I - \frac{1}{2}s_k + \frac{1}{4}s_k^2\right)s_k$	0	$\left(\frac{3}{2}I - s_k + \frac{1}{2}s_k^2\right)s_k$	0	8
Cordero et al.'s CCGT2 [45]	1	$\frac{1}{4} + \frac{1}{2}s_k + \frac{1}{4}s_k^2$	0	$\frac{1}{2}I + \frac{1}{2}s_k^2$	0	8
Zhanlav et al. [46]	1	$\alpha I - (\frac{25}{4} - 3\alpha)t_k + (3\alpha - \frac{17}{2})t_k^2 + (\frac{13}{4} - \alpha)t_k^3$	0	$\beta I + (\frac{13}{2} - 3\beta)t_k + (3\beta - 9)t_k^2 + (\frac{7}{2} - \beta)t_k^3$	0	8
Zhanlav et al. [46]	1	$\frac{5}{4}s_k - \frac{1}{2}s_k^2 + \frac{1}{4}s_k^3 + \sigma(I - s_k)^3$	0	$\frac{3}{2}s_k - s_k^2 + \frac{1}{2}s_k^3 + \gamma(I - s_k)^3$	0	8
Behl et al.'s PM [47]	1	$I + M_k$	0	$I + M_k + \frac{3}{2}M_k^2$	0	7

5.3.4 Total Cost Comparison Between Methods

Here, we consider methods (5.131) with (5.138), (5.131) with (5.139), and (5.152) with (5.155), (5.156). The method (5.131) with (5.138) can be rewritten in the following equivalent form in terms of linear systems:

$$F'(x_k)\phi_1 = F(x_k), \quad y_k = x_k - a\phi_1, \quad w_k = x + a\phi_1,$$
$$(I - \alpha M_k)\phi_2 = \phi_1,$$
$$F'(x_k)\phi_3 = F'(y_k)\phi_1,$$
$$(I - \alpha M_k)\phi_4 = M_k\phi_1 \equiv \phi_1 - \phi_3,$$
$$(I - \alpha M_k)\phi_5 = M_k^2\phi_1 \equiv M_k(\phi_1 - \phi_3),$$
$$F'(x_k)\phi_6 = F'(w_k)\phi_1,$$
$$z_k = x_k - \left(\phi_2 + \left(\frac{1}{2a} - \alpha\right)\phi_4 + \frac{1 - a\alpha}{2a^2}\phi_5 + \left(1 - \frac{3}{2}a\right)\left(\frac{\phi_1}{3a^2} - \frac{\phi_3}{6a^2} - \frac{\phi_6}{6a^2}\right)\right),$$
$$F'(x_k)\phi_7 = F(z_k),$$
$$(I - \beta M_k)\phi_8 = \phi_7,$$
$$F'(x_k)\phi_9 = F'(y_k)\phi_7,$$
$$(I - \beta M_k)\phi_{10} = M_k\phi_7 \equiv \phi_7 - \phi_9,$$
$$x_{k+1} = z_k - \left(\phi_8 + \left(\frac{1}{a} - \beta\right)\phi_{10}\right),$$

and similarly for the other methods.

Table 5.14 The cost of each iteration

Method	Evaluation of F and Jacobian	Scalar vector multiply multiply	Matrix vector multiply	Linear solve	Inverse Matrix + Matrix matrix multiply	Total
(5.131) with (5.138a), (5.138b)	$2n\mu_0 + 3n^2\mu_1$	$14n$	$4n^2$	$3\alpha + 9\beta + \left(\frac{3\beta}{2} + 9n\right)\ell$	$11n^3/3$	$14n^3/3 + \left(3\mu_1 + \frac{23+3\ell}{2}\right)n^2 + (2\mu_0 + \frac{11+15\ell}{2})n$
(5.131) with (5.139a), (5.139b)	$2n\mu_0 + 3n^2\mu_1$	$14n$	$4n^2$	$3\alpha + 9\beta + \left(\frac{3\beta}{2} + 9n\right)\ell$	$11n^3/3$	$14n^3/3 + \left(3\mu_1 + \frac{23+3\ell}{2}\right)n^2 + (2\mu_0 + \frac{1+15\ell}{6})n$
(5.152) with (5.155), (5.156)	$3n\mu_0 + 3n^2\mu_1$	$18n$	$6n^2$	$3\alpha + 13\beta + \left(\frac{3\beta}{2} + 9n\right)\ell$	$11n^3/3$	$14n^3/3 + \left(3\mu_1 + \frac{35+3\ell}{2}\right)n^2 + (3\mu_0 + \frac{11+15\ell}{6})n$

We check Table 5.14 to see the total cost of each iteration for each method, where n is the system dimension, $\alpha = n(n-1)(2n-1)/6$, $\beta = n(n-1)$, μ_0 and μ_1 are relative costs of evaluation of F and Jacobian, respectively, in terms of multiplications and ℓ is the relative cost of division in terms of multiplications. We note that all the methods cost $14n^3/3$. Methods (5.131) with (5.138), (5.131) with (5.139), and (5.152) with (5.155), (5.156) become most inexpensive when $\alpha = 0$ and $\beta = 0$, $\gamma = 0$ and $\delta = 0$, and $\beta_1 = 0$ and $\beta_2 = 0$, respectively.

5.4 Family of Newton-Type Methods with Seventh and Eighth-Order of Convergence for Solving Systems of Nonlinear Equations

In this section we develop a new family of three-step seventh and eighth-order Newton-type iterative methods for solving systems of nonlinear equations. We also propose some different choices of parameter matrices that ensure the convergence order. The proposed family includes some known methods of special cases [46].

5.4.1 Seventh-Order Iterations

We first consider the following three-point iterative method:

$$y_k = x_k - a \left(F'(x_k) \right)^{-1} F(x_k), \quad a \in R \setminus \{0\},$$

$$z_k = \varphi_p(x_k, y_k) = y_k - \bar{\tau}_k \left(F'(x_k) \right)^{-1} F(y_k), \qquad (5.161)$$

$$x_{k+1} = z_k - \alpha_k \left(F'(x_k) \right)^{-1} F(z_k).$$

Here $z_k = \varphi_p(x_k, y_k)$ is an iteration with order $p \geq 3$. The second substep in (5.161) can also be written as:

$$z_k = x_k - \tau_k \left(F'(x_k) \right)^{-1} F(x_k). \qquad (5.162)$$

In [23], it was shown that $F(z_k) = O\left(h^4\right)$ if τ_k satisfies

$$\tau_k = I + \Theta_k + 2\Theta_k^2 + d_k + O\left(h^3\right), \qquad (5.163)$$

where Θ_k and d_k are given by (5.87) and (5.89). To obtain high-order iterations of type (5.161), we need the assumption (5.105).

To proceed the convergence analysis of iteration (5.161), first we will establish the relation of $\bar{\tau}_k$ and τ_k which allows to ensure the fourth-order convergence of $F(z_k)$.

Theorem 5.18 *The fourth-order convergence condition (5.163) for τ_k is equivalent to*

$$\bar{\tau}_k = I + (a+1)\Theta_k + (a^2 + 2a + 2)\Theta_k^2 + (a^2 + a + 1)d_k + O\left(h^3\right). \qquad (5.164)$$

Proof From (5.161), (5.162) we can get

$$x_k - \tau_k \left(F'(x_k) \right)^{-1} F(x_k) = x_k - a \left(F'(x_k) \right)^{-1} F(x_k) - \bar{\tau}_k \left(F'(x_k) \right)^{-1} F(y_k).$$

Then

$$\bar{\tau}_k \left(F'(x_k) \right)^{-1} F(y_k) = (\tau_k - aI) \left(F'(x_k) \right)^{-1} F(x_k). \qquad (5.165)$$

On the other hand, using Taylor expansion of $F(y_k)$ around x_k, this gives

$$\left(F'(x_k) \right)^{-1} F(y_k) = \left((1-a)I + a^2\Theta_k + a^3 d_k \right) \left(F'(x_k) \right)^{-1} F(x_k) + O\left(h^4\right). \qquad (5.166)$$

Using (5.166) in (5.165) we get

$$\bar{\tau}_k(1-a) \left(I + \frac{a^2\Theta_k + a^3 d_k}{1-a} \right) = \tau_k - aI + O\left(h^3\right). \qquad (5.167)$$

Since $\Theta_k = O(h)$, $d_k = O\left(h^2\right)$ by virtue of (5.87), (5.89) and using well-known expansion (5.7) from (5.167) we obtain

$$\bar{\tau}_k = \frac{aI - \tau_k}{a-1}\left(I + \frac{a^2\Theta_k + a^3 d_k}{1-a}\right)^{-1} + O\left(h^3\right)$$

$$= \left(I + \frac{\Theta_k + 2\Theta_k^2 + d_k}{1-a}\right)\left(I - \frac{a^2\Theta_k + a^3 d_k}{1-a} + \left(\frac{a^2\Theta_k + a^3 d_k}{1-a}\right)^2 + \cdots\right) + O\left(h^3\right)$$

$$= I + (a+1)\Theta_k + (a^2 + 2a + 2)\Theta_k^2 + (a^2 + a + 1)d_k + O\left(h^3\right),$$

in which we used (5.163). Thus, we arrive at (5.164). The converse is obvious from (5.164) and (5.165). $\qquad\square$

We now state and prove the below theorem

Theorem 5.19 *Let $F : D \subseteq R^n \to R^n$ be a sufficiently Fréchet differentiable function in an open neighborhood D containing a zero α of $F(x)$. Suppose that $F'(x)$ is continuous and nonsingular in x^*. Then, for an initial approximation x_0 sufficiently close to x^*, the sequence $\{x_k\}$, $x_0 \in D$ obtained by (5.161) has order of convergence $p + 3$ if and only if the parameter matrix α_k in (5.161) satisfies*

$$\alpha_k = I + 2\Theta_k + 3d_k + 6\Theta_k^2 + O\left(h^3\right). \tag{5.168}$$

Proof Taylor's formula of $F(x_{k+1})$ around z_k gives

$$F(x_{k+1}) = \left(I - F'(z_k)\alpha_k\left(F'(x_k)\right)^{-1}\right) F(z_k) + O\left(\|F(z_k)\|^2\right). \tag{5.169}$$

Since $F(z_k) = O\left(h^p\right)$, then from (5.169) it is clear that

$$F(x_{k+1}) = O\left(h^{p+3}\right),$$

under the condition

$$\alpha_k = \left[F'(z_k)\right]^{-1} F'(x_k) + O\left(h^3\right)). \tag{5.170}$$

So, we have to find $\left[F'(z_k)\right]^{-1}$ with accuracy $O\left(h^3\right))$. For this end, we apply Taylor's formula of $F'(z_k)$ at point x_k

$$F'(z_k) = F'(x_k) + F''(x_k)(z_k - x_k) + \frac{F'''(x_k)}{2}(z_k - x_k)^2 + O\left((z_k - x_k)^3\right). \tag{5.171}$$

From the first two steps in (5.161) we can obtain

$$z_k - x_k = -a\left(F'(x_k)\right)^{-1} F(x_k) - \bar{\tau}_k\left(F'(x_k)\right)^{-1} F(y_k). \tag{5.172}$$

Using (5.166) in (5.172) we get

$$z_k - x_k = -\left(aI + (1-a)\bar{\tau}_k + a^2\bar{\tau}_k\Theta_k\right)\left(F'(x_k)\right)^{-1} F(x_k) + O\left(h^3\right). \tag{5.173}$$

Using (5.164) in (5.173) we obtain

$$z_k - x_k = -(I + \Theta_k)\left(F'(x_k)\right)^{-1} F(x_k) + O\left(h^3\right). \qquad (5.174)$$

Substituting (5.174) into (5.171) we get

$$
\begin{aligned}
F'(z_k) &= F'(x_k)\Bigg(I - \left(F'(x_k)\right)^{-1} F''(x_k)(I + \Theta_k)\left(F'(x_k)\right)^{-1} F(x_k) \\
&\quad + \frac{1}{2}\left(F'(x_k)\right)^{-1} F'''(x_k)\left(\left(F'(x_k)\right)^{-1} F(x_k)\right)^2\Bigg) + O\left(h^3\right) \\
&= F'(x_k)(I - 2\Theta_k - 2\Theta_k^2 - 3d_k) + O\left(h^3\right),
\end{aligned}
$$

in which (5.87), (5.89) and (5.105) are used. Using (5.7), we get

$$
\begin{aligned}
\left(F'(z_k)\right)^{-1} &= \left(I - 2\Theta_k - 2\Theta_k^2 - 3d_k\right)^{-1}\left(F'(x_k)\right)^{-1} + O\left(h^3\right) \\
&= \left(I + 2\Theta_k + 2\Theta_k^2 + 3d_k + 4\Theta_k^2\right)\left(F'(x_k)\right)^{-1} + O\left(h^3\right). \qquad (5.175)
\end{aligned}
$$

Substituting (5.175) into (5.170) we arrive at (5.168). $\qquad\square$

Thus, the choice (5.168) ensures $F(x_{k+1}) = O\left(h^{p+3}\right)$.

Remark 5.8 The parameter a in (5.161) belongs to the set $R \setminus \{0\}$. But there is a suitable choice for a from the computational point of view.

From (5.161) we see that the first step in (5.161) can be considered independent Newton method with damping parameter a. According to [48], it is desirable to choose

$$0 < a < 2. \qquad (5.176)$$

Indeed, from (5.166) we have

$$\|F(y_k)\| < |1 - a|\|F(x_k)\| + O\left(h^2\right).$$

That is, such choice (5.176) allows us to decrease the residual $\|F(y_k)\|$ and thereby to speed up the convergence of iteration (5.161).

The direct consequence of Theorems 5.18, 5.19 is:

Theorem 5.20 *Let assumptions of Theorem 5.19 be satisfied. Then the convergence order of iteration (5.161) equal to seven, when $\bar{\tau}_k$ and α_k satisfy (5.164) and (5.168), respectively.*

Proof By Theorem 5.18, $F(z_k) = O\left(h^4\right)$ provided (5.164). Then by virtue of Theorem 5.19, $F(x_{k+1}) = O\left(h^7\right)$ under choice (5.168). $\qquad\square$

Our next goal is to express Θ_k and d_k in (5.164), (5.168) by means of known quantities without loss of accuracy. In order to do this, we use (5.132) and (5.133). From (5.90) and (5.133) we find that

$$\Theta_k = \frac{\gamma_k - t_k}{4a} + O\left(h^3\right),\tag{5.177}$$

and

$$d_k = \frac{2I - \gamma_k - t_k}{6a^2} + O\left(h^3\right).\tag{5.178}$$

Substituting (5.177) and (5.178) into (5.164) and (5.168) we get

$$\bar{\tau}_k = I + (1 + a)\frac{\gamma_k - t_k}{4a} + (a^2 + 2a + 2)\left(\frac{\gamma_k - t_k}{4a}\right)^2$$
$$+(1 + a + a^2)\frac{2I - \gamma_k - t_k}{6a^2} + O\left(h^3\right)\tag{5.179}$$

and

$$\alpha_k = I + \frac{\gamma_k - t_k}{2a} + \frac{2I - \gamma_k - t_k}{2a^2} + 6\left(\frac{\gamma_k - t_k}{4a}\right)^2 + O\left(h^3\right).\tag{5.180}$$

Remark 5.9 One can replace Θ_k^2 in (5.164) and (5.168) by $((I - t_k)/(2a))^2$ without loss of accuracy.

Theorem 5.20 remains true for $\bar{\tau}_k$ and α_k given by (5.179) and (5.180), respectively, which is the first main finding of this subsection.

5.4.2 Eighth-Order Iterations

Let $a = 1$ in (5.161). Then the iteration (5.161) becomes as:

$$y_k = x_k - \left(F'(x_k)\right)^{-1} F(x_k),$$
$$z_k = \varphi_p(x_k, y_k) = y_k - \bar{\tau}_k \left(F'(x_k)\right)^{-1} F(y_k),\tag{5.181}$$
$$x_{k+1} = z_k - \alpha_k \left(F'(x_k)\right)^{-1} F(z_k).$$

When $a = 1$, the formula (5.164) leads to

$$\bar{\tau}_k = I + 2\Theta_k + 5\Theta_k^2 + 3d_k + O\left(h^3\right).\tag{5.182}$$

In this case, one can expect stronger results than Theorem 5.20.

Theorem 5.21 *Let the assumptions of Theorem 5.19 be satisfied. Then the convergence order of iteration (5.181) equal to eight, when $\bar{\tau}_k$ and α_k satisfy (5.182) and (5.168), respectively.*

Proof Taylor expansion of $F(z_k)$ around y_k gives

$$F(z_k) = F(y_k) - F'(y_k)\bar{\tau}_k \left(F'(x_k)\right)^{-1} F(y_k)$$
$$+ \frac{F''(y_k)}{2}(\bar{\tau}_k \left(F'(x_k)\right)^{-1} F(y_k))^2 + O\left((F(y_k))^3\right). \tag{5.183}$$

Keeping in mind (5.87), (5.105), (5.182), and $F''(y_k) = F''(x_k) + O(h)$, $F(y_k) = O\left(h^2\right)$ and neglecting small term $O\left(h^5\right)$ in (5.183) we can reach

$$F(z_k) = F'(x_k)(I - t_k\bar{\tau}_k + \Theta_k^2)\left(F'(x_k)\right)^{-1} F(y_k) + O\left(h^5\right). \tag{5.184}$$

Since $F(y_k) = O\left(h^2\right)$, then from (5.184) it is clear that

$$F(z_k) = O\left(h^5\right), \tag{5.185}$$

provided

$$I - t_k\bar{\tau}_k + \Theta_k^2 = O\left(h^3\right),$$

i.e.,

$$\bar{\tau}_k = t_k^{-1}\left(I + \Theta_k^2\right) + O\left(h^3\right). \tag{5.186}$$

Using (5.93) with $a = 1$ in (5.186) we get (5.182). Thus, (5.185) holds true under (5.182). Due to Theorem 5.19, the order of iteration (5.181) equals to $p + 3 = 8$. $\square$

Remark 5.10 Repeating application of Theorems 5.19 and 5.21 gives an iterative methods of arbitrary order of convergence as in [49, 50].

When $a = 1$, the formula (5.90), and (5.93) lead to

$$t_k = I - 2\Theta_k - 3d_k + O\left(h^3\right), \quad s_k = I + 2\Theta_k + 3d_k + 4\Theta_k^2 + O\left(h^3\right), \tag{5.187}$$

and

$$\gamma_k = I + 2\Theta_k - 3d_k + O\left(h^3\right).$$

From (5.187) we find

$$\Theta_k = \frac{I - t_k}{2} + O\left(h^2\right). \tag{5.188}$$

Using (5.187), (5.188), one can rewrite (5.182) and (5.168) in term of t_k as:

$$\bar{\tau}_k = \alpha I + \left(\frac{25}{4} - 3\alpha\right) t_k + \left(3\alpha - \frac{17}{2}\right) t_k^2 + \left(\frac{13}{4} - \alpha\right) t_k^3$$
$$= \frac{25}{4} t_k - \frac{17}{2} t_k^2 + \frac{13}{4} t_k^3 + \alpha(I - t_k)^3, \quad \alpha \in R, \tag{5.189}$$

and

$$\alpha_k = \bar{\tau}_k + \Theta_k^2 = \beta I + \left(\frac{13}{2} - 3\beta\right) t_k + (3\beta - 9)t_k^2 + \left(\frac{7}{2} - \beta\right) t_k^3$$

$$= \frac{13}{2} t_k - 9t_k^2 + \frac{7}{2} t_k^3 + \beta(I - t_k)^3, \quad \beta \in R. \tag{5.190}$$

When $\alpha = 13/4$ and $\beta = 7/2$ the formulas (5.189) and (5.190) lead to

$$\bar{\tau}_k = \frac{13}{4} I - \frac{7}{2} t_k + \frac{5}{4} t_k^2,$$

and

$$\alpha_k = \frac{7}{2} I - 4t_k + \frac{3}{2} t_k^2,$$

respectively. In this case, the iteration (5.181) converted to NLM8 given by Sharma et al. in [44]. When $\alpha = \beta = 0$ then the formulas (5.189) and (5.190) lead to

$$\bar{\tau}_k = \frac{25}{4} t_k - \frac{17}{2} t_k^2 + \frac{13}{4} t_k^3,$$

and

$$\alpha_k = \frac{13}{2} t_k - 9t_k^2 + \frac{7}{2} t_k^3,$$

respectively.

Analogously, the conditions (5.168) and (5.182) can be rewritten in terms of s_k as:

$$\bar{\tau}_k = \sigma I + \left(\frac{5}{4} - 3\sigma\right) s_k + \left(3\sigma - \frac{1}{2}\right) s_k^2 + \left(\frac{1}{4} - \sigma\right) s_k^3$$

$$= \frac{5}{4} s_k - \frac{1}{2} s_k^2 + \frac{1}{4} s_k^3 + \sigma(I - s_k)^3, \quad \sigma \in R, \tag{5.191}$$

and

$$\alpha_k = \gamma I + \left(\frac{3}{2} - 3\gamma\right) s_k + (3\gamma - 1) s_k^2 + \left(\frac{1}{2} - \gamma\right) s_k^3$$

$$= \frac{3}{2} s_k - s_k^2 + \frac{1}{2} s_k^3 + \gamma(I - s_k)^3, \quad \gamma \in R, \tag{5.192}$$

respectively. When $\sigma = \gamma = 0$ in (5.191) and (5.192) then we get

$$\bar{\tau}_k = \frac{5}{4} s_k - \frac{1}{2} s_k^2 + \frac{1}{4} s_k^3,$$

$$\alpha_k = \frac{3}{2} s_k - s_k^2 + \frac{1}{2} s_k^3.$$

In this case the iteration (5.181) converted to CCLT1 given by Cordero et al. [45]. When $\sigma = 1/4$, $\gamma = 1/2$ in (5.191) and (5.192), then we get

$$\bar{\tau}_k = \frac{1}{4}I + \frac{1}{2}s_k + \frac{1}{4}s_k^2,$$

$$\alpha_k = \frac{1}{2}I + \frac{1}{2}s_k^2.$$

In this case the iteration (5.181) converted to CCLT2 given by Cordero et al. [45].

Using, (5.177), (5.178), the condition (5.182) can be rewritten in terms of γ_k and t_k as:

$$\bar{\tau}_k = \eta I + \left(\frac{9}{4} - 3\eta\right)\gamma_k + \left(3\eta - \frac{5}{2}\right)\gamma_k^2 + \left(\frac{5}{4} - \eta\right)\gamma_k^3 + 6d_k + O\left(h^3\right), \quad \eta \in R. \tag{5.193}$$

When $\eta = 5/4$, we have

$$\bar{\tau}_k = \frac{13}{4}I - \frac{7}{2}\gamma_k + \frac{5}{4}\gamma_k^2 + (\gamma_k - t_k) + O\left(h^3\right). \tag{5.194}$$

Analogously, we have

$$\alpha_k = \xi I + \left(\frac{5}{2} - 3\xi\right)\gamma_k + (3\xi - 3)\gamma_k^2 + \left(\frac{3}{2} - \xi\right)\gamma_k^3 + 6d_k + O\left(h^3\right), \quad \xi \in R. \tag{5.195}$$

When $\xi = 3/2$, we obtain

$$\alpha_k = \frac{7}{2}I - 4\gamma_k + \frac{3}{2}\gamma_k^2 + (\gamma_k - t_k) + O\left(h^3\right). \tag{5.196}$$

Thus, we also propose eighth-order convergence iteration (5.181) with $\bar{\tau}_k$ and α_k given by (5.194) and (5.196), respectively. We denote it by ZMO1.

Analogously, it is easy to show that

$$\bar{\tau}_k = -\frac{1}{2}I + \frac{5}{4}s_k + \frac{1}{4}t_k, \tag{5.197}$$

$$\alpha_k = -I + \frac{3}{2}s_k + \frac{1}{2}t_k, \tag{5.198}$$

which satisfy the conditions (5.168) and (5.182), respectively. The iteration (5.181) with $\bar{\tau}_k$ and α_k given by (5.197) and (5.198), respectively has also an eighth-order of convergence, which denoted by ZMO2.

Another way to find $\bar{\tau}_k$ and α_k is to use generating function method [7]. Thus, we can easily show that $\bar{\tau}_k$ and α_k given by the following formulas:

$$\bar{\tau}_k = (I + \delta\Theta_k)^{-1}\left(I + (\delta + 2)\Theta_k + 3d_k + (5 + 2\delta)\Theta_k^2\right) \tag{5.199}$$

Table 5.15 Choices of parameters

$\bar{\tau}_k$	(5.189)	(5.191)	(5.193)	(5.197)	(5.199)
α_k	(5.190)	(5.192)	(5.195)	(5.198)	(5.200)

Table 5.16 Computational cost

Methods	d	NLS1	NLS2	$M \times V$	C
NLM8	$2n^2 + 3n$	7	0	4	$\frac{1}{3}n^3 + 13n^2 + \frac{8}{3}n$
CCLT1	$2n^2 + 3n$	1	6	4	$\frac{2}{3}n^3 + 13n^2 + \frac{7}{3}n$
CCLT2	$2n^2 + 3n$	3	4	2	$\frac{2}{3}n^3 + 11n^2 + \frac{7}{3}n$
ZMO1	$3n^2 + 3n$	7	0	6	$\frac{1}{3}n^3 + 15n^2 + \frac{8}{3}n$
ZMO2	$2n^2 + 3n$	5	2	4	$\frac{2}{3}n^3 + 13n^2 + \frac{7}{3}n$
ZMO3*	$3n^2 + 3n$	7	2	6	$\frac{2}{3}n^3 + 18n^2 + \frac{7}{3}n$

$*$ when $\delta = 0$ ZMO3 leads to NLM8

and

$$\alpha_k = (I + \delta\Theta_k)^{-1}\left(I + (\delta + 2)\Theta_k + 3d_k + (6 + 2\delta)\Theta_k^2\right) \tag{5.200}$$

satisfy conditions (5.168) and (5.182), respectively. The iteration (5.181) with $\bar{\tau}_k$ and α_k given by (5.199) and (5.200), respectively, has all eighth-order of convergence, which denoted by ZMO3.

Furthermore, we propose a family of eighth-order iterative methods (5.181) with parameter matrices $\bar{\tau}_k$ and α_k given by formulas listed in Table 5.15. The proposed family includes some known methods as special cases. The choices for $\bar{\tau}_k$ and α_k in Table 5.15 ensure the eighth-order of convergence of methods (5.181), which is the second main finding of this subsection.

From Table 5.16, we see that the most cheapest are NLM8 and ZMO1, the total computational cost is $n^3/3$ (not including lower powers of the dimension n of the system). All other methods cost $2n^2/3$. So, NLM8 and ZMO1 have a higher CE than other methods.

We develop a new family of three-step seventh and eighth-order Newton-type iterative methods. The proposed family includes some existing methods as particular cases. We suggest different choices of parameter matrices of iterations that ensure the expected order of convergence.

5.5   A Novel Families of Higher Order Multi-step Iterative Methods for Solving Nonlinear Systems

We consider the following two-step iterative methods

$$y_k = x_k - a \left(F'(x_k)\right)^{-1} F(x_k), \quad a \in R \setminus \{0\}, \quad x_0 \in D,$$

$$z_k = y_k - \bar{\tau}_k \left(F'(x_k)\right)^{-1} F(y_k), \quad (x_{k+1} = z_k), \tag{5.201}$$

where $\bar{\tau}_k$ is parameter matrix to be determined properly. Our aim is to find the values of parameter matrix of (5.201) in such a way that the proposed scheme (5.201) may produce order of convergence as high as possible [51]. To this end, we assume that $F(x)$ sufficiently differentiable in D and $F'(x)$ is nonsingular in the vicinity of x^*, where x^* is simple root of $F(x) = 0$. We use Taylor expansion of $F(z_k)$ at point y_k

$$F(z_k) = \sum_{j=0}^{5} \frac{F^{(j)}(y_k)}{j!} (z_k - y_k)^j + O\left(F(y_k)^6\right). \tag{5.202}$$

Introducing the notations

$$A_j = \frac{1}{j!} F'(x_j)^{-1} F^{(j)}(y_k), \quad \eta_k = \bar{\tau}_k \left(F'(x_k)\right)^{-1} F(y_k), \tag{5.203}$$

we can write (5.202) as

$$F(z_k) = F'(x_k) \left\{ I - \left[I - (I - t_k) - A_2 \eta_k + A_3 \eta_k^2 - A_4 \eta_k^3 + A_5 \eta_k^4 \right] \bar{\tau}_k \right\}$$
$$\times \left(F'(x_k)\right)^{-1} F(y_k) + O\left(F(y_k)^6\right), \tag{5.204}$$

where I is the $n \times n$ identity matrix and $t_k = \left(F'(x_k)\right)^{-1} F'(y_k)$. If we take into account that $F(y_k) = O(h)$, $I - t_k = O(h)$, $\eta_k = O(h)$, and (5.7) then from (5.204) it follows that

$$F(z_k) = O\left(h^\sigma\right), \quad \sigma = 5, \text{ when } a \neq 1, \text{ and } \sigma = 6, \text{ when } a = 1, \tag{5.205}$$

under condition

$$I - \left[I - (I - t_k) - A_2 \eta_k + A_3 \eta_k^2 - A_4 \eta_k^3 + A_5 \eta_k^4 \right] \bar{\tau}_k = O\left(h^4\right). \tag{5.206}$$

The matrix in square bracket in (5.206) invertible and therefore, from (5.206) we find

$$\bar{\tau}_k = \left(I - (I - t_k) - A_2 \eta_k + A_3 \eta_k^2 - A_4 \eta_k^3 + A_5 \eta_k^4 \right)^{-1} + O\left(h^4\right). \tag{5.207}$$

After some calculations we get

$$D_k^2 = (A_2\eta_k - A_3\eta_k^2 + A_4\eta_k^3)^2$$
$$= \left(A_2 - 2A_2A_3\eta_k + (2A_2A_4 + A_3^2)\eta_k^2\right)\eta_k^2 + O\left(h^5\right),$$
$$D_k^3 = (A_2^3 - 3A_2^3\eta_k)\eta_k^3 + O\left(h^5\right),$$
$$D_k^4 = A_2^4\eta_k^4 + O\left(h^5\right),$$

(5.208)

and

$$s_k = t_k^{-1} = I + (I - t_k) + (I - t_k)^2 + (I - t_k)^3 + \cdots, \quad s_k = \left(F'(y_k)\right)^{-1} F'(x_k).$$

(5.209)

Using (5.7), (5.208), and (5.209) in (5.207) we get

$$\bar{\tau}_k = s_k + A_2\eta_k + (A_2^2 - A_3)\eta_k^2 + (A_4 - 2A_2A_3 + A_2^3)\eta_k^3$$
$$+ (I - t_k)\left(2A_2 + (3A_2^2 - 2A_3)\eta_k\right)\eta_k$$
$$+ 3(I - t_k)^2\left(A_2 + (2A_2^2 - A_3)\eta_k\right)\eta_k + O\left(h^4\right)$$

or

$$\bar{\tau}_k = s_k + (I + 2(I - t_k) + 3(I - t_k)^2)A_2\eta_k$$
$$+ \left((I + 3(I - t_k))A_2^2 - (I + 2(I - t_k))A_3\right)\eta_k^2$$
$$+ (A_4 - 2A_2A_3 + A_2^3)\eta_k^3 + O\left(h^4\right).$$

(5.210)

Further, Taylor expansion of $F(y_k)$ at point x_k gives

$$\left(F'(x_k)\right)^{-1} F(y_k) = \left((1 - a)I + a^2\Theta_k + a^3 d_k + \frac{a^4}{4}c_k\right)\xi_k + O\left(h^5\right),$$

(5.211)

where

$$\Theta_k = \frac{1}{2}\left(F'(x_k)\right)^{-1} F''(x_k)\xi_k, \quad d_k = -\frac{1}{6}\left(F'(x_k)\right)^{-1} F'''(x_k)\xi_k^2,$$

$$c_k = \frac{1}{6}\left(F'(x_k)\right)^{-1} F^{(4)}(x_k)\xi_k^3, \quad \xi_k = \left(F'(x_k)\right)^{-1} F(x_k).$$

(5.212)

Using (5.211) in (5.203) we get

$$\eta_k = \bar{\tau}_k\left((1 - a)I + a^2\Theta_k + a^3 d_k + \frac{a^4}{4}c_k\right)\xi_k + O\left(h^5\right).$$

(5.213)

The fourth-order convergence condition for $F(z_k) = O\left(h^4\right)$ is

$$\bar{\tau}_k = I + (a + 1)\Theta_k + (a^2 + 2a + 2)\Theta_k^2 + (a^2 + a + 1)d_k + O\left(h^3\right).$$

(5.214)

Substituting (5.214) into (5.213), we have

$$\eta_k = \left((1 - a)I + \Theta_k + 2\Theta_k^2 + d_k\right)\xi_k + O\left(h^4\right). \tag{5.215}$$

Hence, using (i), (ii) and notations (a), (b) we have

$$F''(y_k)\eta_k^2 = (1 - a)F''(y_k)\left((1 - a)\xi_k^2 + 2\Theta_k\xi_k^2\right) + O\left(h^4\right), \tag{5.216}$$

and

$$F'''(y_k)\eta_k^3 = (1 - a)^3 F'''(y_k)\xi_k^3 + O\left(h^4\right). \tag{5.217}$$

The Taylor expansion of $F'(y_k)$ at point x_k gives

$$t_k = \left(F'(x_k)\right)^{-1} F'(y_k) = I - 2a\Theta_k - 3a^2 d_k - a^3 c_k + O\left(h^4\right). \tag{5.218}$$

Hence,

$$s_k = t_k^{-1} = I + 2a\Theta_k + 3a^2 d_k + 4a^2\Theta_k^2 + a^3 c_k + 12a^3\Theta_k d_k + 8a^3\Theta_k^3 + O\left(h^4\right). \tag{5.219}$$

According to (5.203) and (5.215) we have

$$A_2\eta_k = (1 - a)\Theta_k + \Theta_k^2 + 3a(1 - a)d_k + 2\Theta_k^3 + (1 + 3a)\Theta_k d_k$$
$$+ \frac{3}{2}a^2(1 - a)c_k + O\left(h^4\right), \tag{5.220}$$
$$(A_2\eta_k)^2 = (1 - a)^2\Theta_k^2 + 2(1 - a)\Theta_k^3 + 6a(1 - a)^2\Theta_k d_k + O\left(h^4\right), \tag{5.221}$$
$$(A_2\eta_k)^3 = (1 - a)^3\Theta_k^3 + O\left(h^4\right), \tag{5.222}$$

and

$$A_3\eta_k^2 = -(1 - a)^2 d_k - 2(1 - a)\Theta_k d_k - a(1 - a)^2 c_k + O\left(h^4\right), \tag{5.223}$$
$$(A_4 - 2A_2 A_3 + A_2^3)\eta_k^3 = (1 - a)^3 \left(\Theta_k^3 + 2\Theta_k d_k + \frac{1}{4}c_k\right) + O\left(h^4\right). \tag{5.224}$$

Substituting (5.218)–(5.224) into (5.210) we obtain

$$\bar{\tau}_k = I + (a + 1)\Theta_k + (a^2 + 2a + 2)\Theta_k^2 + (a^2 + a + 1)d_k$$
$$+ (2a^3 + 4a^2 + 5a + 5)\Theta_k d_k + \frac{1}{4}(a^3 + a^2 + a + 1)c_k$$
$$+ (a^3 + 3a^2 + 5a + 5)\Theta_k^3 + O\left(h^4\right). \tag{5.225}$$

Thus, (5.206) holds under choice (5.225). That is, (5.205) holds when $\bar{\tau}_k$ is given by (5.225). The obtained result can be formulated as:

Theorem 5.22 *Suppose that $F(x)$ is sufficiently differentiable in domain D containing the simple zero x^* of $F(x)$ and $F'(x)$ is continuous and nonsingular in x^*. Let $x_0 \in D$ be an initial approximation sufficiently close to x^*. Then the local order*

Table 5.17 Choices of $\bar{\tau}_k$

$a \backslash p$	2	3	4	5	6
$\neq 1$	(5.226)	(5.227)	(5.228)	(5.225)	–
1	–	(5.226)	(5.227)	(5.228)	(5.225)

of convergence of two-step iterative method equal to five, when parameter matrix $\bar{\tau}_k$ satisfies (5.225). In particular, the method (5.201) for $a = 1$ has sixth-order of convergence.

It is worthy to note that the formula (5.225) consist of several terms. Each of them have a meaning and including next term of order $O\,(h^\sigma)$ increases the convergence order of the method (5.201). In connection with this we split it into the following compositions:

$$\bar{\tau}_k = I, \tag{5.226}$$

$$\bar{\tau}_k = I + (a + 1)\Theta_k, \tag{5.227}$$

$$\bar{\tau}_k = I + (a + 1)\Theta_k + (a^2 + 2a + 2)\Theta_k^2 + (a^2 + a + 1)d_k. \tag{5.228}$$

For completeness, we present in Table 5.17 the convergence order (p) of the iterative method (5.201). When $a = 1$ the formula (5.225) leads to

$$\bar{\tau}_k = I + 2\Theta_k + 5\Theta_k^2 + 3d_k + 14\Theta_k^3 + 16\Theta_k d_k + c_k + O\left(h^4\right). \tag{5.229}$$

Of course, there are many concrete schemes satisfying the condition (5.229). For example, Behl et al. in [52] proposed sixth-order two-step iteration (we denote it 6BCT) with

$$\bar{\tau}_k = \left(\alpha I + \beta t_k + \gamma t_k^2 + t_k(t_k - I)(I - \eta_k)\right) s_k, \tag{5.230}$$

where $\alpha = 7/4$, $\gamma = \alpha - 1$, $\beta = 2(1 - \alpha)$, and $\eta_k = \left(F'(x_k)\right)^{-1} [F; y_k, x_k]$. It is easy to show that (5.230) satisfy the condition (5.229). More simple version of (5.230) is

$$\bar{\tau}_k = \alpha s_k + \beta I + \gamma t_k - (I - t_k)(I - \eta_k). \tag{5.231}$$

Note that (5.230), (5.231) include not only first derivative of $F(x)$, but also first-order divided difference $[F; y_k, x_k]$. The another choices of $\bar{\tau}_k$ satisfying (5.229) are

$$\bar{\tau}_k = \alpha s_k + \beta I + \gamma t_k - \frac{1}{12}(I - t_k)(5(I - t_k) - (I - \gamma_k)),$$

$$\gamma_k = \left(F'(x_k)\right)^{-1} F'(w_k), \quad w_k = x_k + a\left(F'(x_k)\right)^{-1} F(x_k), \tag{5.232}$$

and

$$\bar{\tau}_k = \alpha s_k + \beta I + \gamma t_k - \left(a_1 a_2 \Theta_k^2 + (b_1 a_2 + b_2 a_1)\Theta_k^3 + (c_1 a_2 + c_2 a_1)\Theta_k d_k\right)$$

$$(5.233)$$

with coefficients $\gamma = \alpha - 1$, $\beta = 2(1 - \alpha)$, $4\alpha - a_1 a_2 = 5$, $8\alpha - b_1 a_2 - b_2 a_1 = 14$, $12\alpha - c_1 a_2 - c_2 a_1 = 16$.

Now we will give the implementation algorithm for $\bar{\tau}_k$ given by (5.225). To this end, we find Θ_k from relation (5.218). Then substituting the obtained value of Θ_k and its square Θ_k^2 into (5.225) one can easily shown that

$$\bar{\tau}_k = I + \frac{a+1}{2a}(I - t_k) + \frac{a^2 + 2a + 2}{4a^2}(I - t_k)^2 - \frac{(a-1)(a+2)}{2}d_k$$
$$+ \frac{(a+1)(1 - a^2)}{4}c_k - (a^3 + 2a^2 + a - 5)\Theta_k d_k$$
$$+ \frac{a^3 + 3a^2 + 5a + 5}{8a^3}(I - t_k)^3 + O\left(h^4\right).$$

$$(5.234)$$

Let $w_k = x_k + a\left(F'(x_k)\right)^{-1} F(x_k)$. Then it is easy to show that

$$\gamma_k = \left(F'(x_k)\right)^{-1} F'(w_k) = I + 2a\Theta_k - 3a^2 d_k + a^3 c_k + O\left(h^4\right).$$

$$(5.235)$$

From (5.218) and (5.235) it follows that

$$d_k = \frac{2I - \gamma_k - t_k}{6a^2} + O\left(h^3\right).$$

$$(5.236)$$

As a consequence of (5.218), (5.235) and (5.236) we get

$$\Theta_k d_k = \frac{(I - t_k)^2 - (I - \gamma_k)^2}{24a^3} + O\left(h^4\right).$$

$$(5.237)$$

When $a = \pm 1$ the c_k canceled out from (5.234). So, it is not needed c_k. Otherwise, in order to find c_k it is required additional information on $F'(\Omega_k)$, where

$$\Omega_k = x_k + ma\left(F'(x_k)\right)^{-1} F(x_k), \quad m \neq \pm 1.$$

$$(5.238)$$

Let

$$\delta_k = \left(F'(x_k)\right)^{-1} F'(\Omega_k).$$

$$(5.239)$$

Then, it is easy to show that

$$\delta_k = I + 2ma\Theta_k - 3m^2 a^2 d_k - m^3 a^3 c_k + O\left(h^4\right).$$

$$(5.240)$$

From (5.218) and (5.240), we find

$$c_k = \frac{\delta_k - I - m(I - t_k) + 3ma^2(m+1)d_k}{a^3 m(m^2 - 1)} + O\left(h^4\right). \qquad (5.241)$$

Thus, we find implementation algorithm for $\bar{\tau}_k$ by means of (5.237), (5.241) and (5.234). By using generating function method in [7], one can obtain parameterized variant of (5.234) as:

$$\bar{\tau}_k = (I - \alpha M_k)^{-1}\left(I + \left(\frac{a+1}{2a} - \alpha\right)M_k + \left(\frac{a^2 + 2a + 2}{4a^2} - \frac{a+1}{2a}\alpha\right)M_k^2\right)$$

$$-\frac{(a-1)(a+2)}{2}d_k + \frac{(a+1)(1-a^2)}{4}c_k - (a^3 + 2a^2 + a - 5)\Theta_k d_k$$

$$+\frac{a^3 + 3a^2 + 5a + 5}{8a^3}M_k^3 + O\left(h^4\right), \quad \alpha \in R, \quad M_k = I - t_k. \qquad (5.242)$$

The most simplest cases are $\alpha = 0$ and $\alpha = (a^2 + 2a + 2)/(2a(a+1))$ in (5.242). Thus, we have family of fifth-order two-step iterative methods (5.201), (5.242). When $a = 1$ the formula (5.242) reads as

$$\bar{\tau}_k = (I - \alpha M_k)^{-1}\left(I + (1 - \alpha)M_k + \left(\frac{5}{4} - \alpha\right)M_k^2\right)$$

$$+\Theta_k d_k + \frac{7}{4}M_k^3 + O\left(h^4\right), \quad \alpha \in R, \qquad (5.243)$$

which ensures the sixth-order of convergence.

5.5.1   *Development of New Ninth-Order Three-Step Iterative Methods*

Based on the two-step fifth-order methods (5.201), we consider the following three-step scheme:

$$y_k = x_k - a\left(F'(x_k)\right)^{-1}F(x_k),$$

$$z_k = \phi_p(x_k, y_k) \quad \text{or} \quad z_k = y_k - \bar{\tau}_k\left(F'(x_k)\right)^{-1}F(y_k), \qquad (5.244)$$

$$x_{k+1} = z_k - \alpha_k\left(F'(x_k)\right)^{-1}F(z_k).$$

Here $z_k = \phi_p(x_k, y_k)$ any iteration with p-order of convergence. That is, $\bar{\tau}_k$ in the second-step of (5.244) is determined by one of the formulas (5.225)–(5.228). Our aim in this subsection is to find α_k such that the scheme (5.244) have the order of convergence $O\left(h^{p+4}\right)$. Namely, we shall prove the following [51].

Theorem 5.23 *Suppose that all the assumptions of Theorem 5.22 are fulfilled. Then, the local order of convergence of iterations (5.244) equal to $p + 4$ if α_k satisfies*

$$\alpha_k = I + 2\Theta_k + 6\Theta_k^2 + 3d_k + 20\Theta_k^3 + 20\Theta_k d_k + c_k + O\left(h^4\right),\tag{5.245}$$

where Θ_k, d_k and c_k are defined by (5.212).

Proof The Taylor expansion of $F(x_{k+1})$ around z_k gives

$$F(x_{k+1}) = \left(I - F'(z_k)\alpha_k \left(F'(x_k)\right)^{-1}\right) F(z_k) + O\left(\|F(z_k)\|^2\right).\tag{5.246}$$

Since $F(z_k) = O\left(h^p\right)$, then from (5.246) it is clear that

$$F(x_{k+1}) = O\left(h^{p+4}\right),\tag{5.247}$$

if α_k satisfies

$$\alpha_k = \left(F'(z_k)\right)^{-1} F'(x_k) + O\left(h^4\right).\tag{5.248}$$

So, the aim is to find $\left(F'(z_k)\right)^{-1}$ with accuracy $O\left(h^4\right)$. To this end, first we employ Taylor expansion of $F'(z_k)$ at point x_k

$$F'(z_k) = F'(x_k) + F''(x_k)(z_k - x_k) + \frac{F'''(x_k)}{2}(z_k - x_k)^2$$
$$+ \frac{F^{(4)}(x_k)}{6}(z_k - x_k)^3 + O\left((z_k - x_k)^4\right).\tag{5.249}$$

From (5.244), we get

$$z_k - x_k = -\left(a \left(F'(x_k)\right)^{-1} F(x_k) + \bar{\tau}_k \left(F'(x_k)\right)^{-1} F(y_k)\right).\tag{5.250}$$

Using (5.211) in last expression we get

$$z_k - x_k = -\left(aI + \bar{\tau}_k \left((1-a)I + a^2\Theta_k + a^3 d_k\right)\right)\xi_k + O\left(h^4\right)$$
$$= -(I + \Theta_k + 2\Theta_k^2 + d_k)\xi_k + O\left(h^4\right)\tag{5.251}$$

in which we used (5.225) and $d_k = O\left(h^2\right)$. As a consequence, we have

$$F'''(x_k)(z_k - x_k)^2 = F'''(x_k)(I + 2\Theta_k)\xi_k^2 + O\left(h^4\right),\tag{5.252}$$

and

$$F^{(4)}(x_k)(z_k - x_k)^3 = -F^{(4)}(x_k)\xi_k^3 + O\left(h^4\right).\tag{5.253}$$

Substituting (5.251)–(5.253) into (5.249), we obtain

$$F'(z_k) = F'(x_k)(I - 2\Theta_k - 2\Theta_k^2 - 3d_k - q_k) + O\left(h^4\right),\tag{5.254}$$

where

$$q_k = 4\Theta_k^3 + 8\Theta_k d_k + c_k = O\left(h^3\right). \tag{5.255}$$

Taking (5.7) into account, from (5.254) we get

$$\left(F'(z_k)\right)^{-1} = \left(I + 2\Theta_k + 6\Theta_k^2 + 3d_k + q_k + 16\Theta_k^3 + 12\Theta_k d_k\right)$$
$$\times \left(F'(x_k)\right)^{-1} + O\left(h^4\right). \tag{5.256}$$

So, we find $\left(F'(z_k)\right)^{-1}$ with accuracy $O\left(h^4\right)$. Substituting (5.255), (5.256) into (5.248) we arrive at (5.245). $\qquad\square$

The direct consequence of Theorems 5.22, 5.23 is the following.

Corollary 5.1 *The local order of convergence of iteration (5.244) equals to 9, if $\bar{\tau}_k$ and α_k satisfy (5.225) and (5.245), respectively. In particular, the method (5.244) for $a = 1$ has a tenth-order of convergence.*

Remark 5.11 As $\bar{\tau}_k$ in the preceding section, one can split α_k into the following compositions:

$$\alpha_k = I, \tag{5.257}$$
$$\alpha_k = I + 2\Theta_k, \tag{5.258}$$
$$\alpha_k = I + 2\Theta_k + 6\Theta_k^2 + 3d_k. \tag{5.259}$$

The convergence order of iterations (5.244) defined by

$$F(x_{k+1}) = O\left(h^{p+q}\right),$$

where q is listed in the Table 5.18 [23].

For convenience, summarizing the results given in Theorems 5.22, 5.23 and in Tables 5.17, 5.18, we display the local order of convergence ρ of iterations (5.244) in Table 5.19.

As in the preceding section, we propose the parameterized form of (5.245) as

$$\alpha_k = (I - \beta M_k)^{-1}\left(I + \left(\frac{1}{a} - \beta\right)M_k + \frac{3 - 2a\beta}{2a^2}M_k^2\right) + 3(1 - a)d_k$$
$$+(1 - a^2)c_k + 20\Theta_k^3 + (20 - 18a)\Theta_k d_k + O\left(h^4\right), \quad \beta \in R. \tag{5.260}$$

Thus, we have two-parametric (α, β) ninth-order family of iterative methods (5.244), with $\bar{\tau}_k$ and α_k given by (5.242) and (5.260), respectively, In particular, this family

Table 5.18 Choice of α_k

α_k	(5.257)	(5.258)	(5.259)	(5.245)
q	1	2	3	4

Table 5.19 Choices of pairs $(\bar{\tau}_k, \alpha_k)$

$a \setminus \rho$	4	5	6	7	8	9	10
$a \neq 1$	(5.226), (5.258)	(5.226), (5.259)	(5.226), (5.245)	(5.227), (5.245)	(5.228), (5.245)	(5.225), (5.245)	
	(5.227),(5.257)	(5.227), (5.258)	(5.227), (5.259)	(5.228), (5.259)	(5.225), (5.259)		
		(5.228), (5.257)	(5.228), (5.258)	(5.225), (5.258)			
			(5.225),(5.257)				
$a = 1$	(5.226), (5.257)	(5.226), (5.258)	(5.226), (5.259)	(5.226), (5.245)	(5.227),(5.245)	(5.228), (5.245)	(5.225), (5.245)
		(5.227), (5.257)	(5.227), (5.258)	(5.227), (5.259)	(5.228), (5.259)	(5.225), (5.259)	
			(5.228), (5.257)	(5.228), (5.258)	(5.225), (5.258)		
				(5.225), (5.257)			

for $a = 1$ has a tenth-order of convergence. Based on Theorems 5.22, 5.23 one can propose i-step iterative methods:

$$
\begin{aligned}
\psi_1^k &= x_k - a \left(F'(x_k)\right)^{-1} F(x_k), \\
\psi_2^k &= \psi_1^k - \bar{\tau}_k \left(F'(x_k)\right)^{-1} F(\psi_1^k), \\
\psi_3^k &= \psi_2^k - \alpha_k \left(F'(x_k)\right)^{-1} F(\psi_2^k), \\
&\cdots \\
\psi_i^k &= \psi_{i-1}^k - \alpha_k \left(F'(x_k)\right)^{-1} F(\psi_{i-1}^k), \quad i = 3, 4, \ldots,
\end{aligned}
$$

(5.261)

where $\bar{\tau}_k$ and α_k are given by (5.242) and (5.260), respectively. For (5.261), we have:

Theorem 5.24 *Suppose that all the assumptions of Theorem 5.22 are fulfilled. Then the local order of convergence of the family (5.261) equal to $4i - 3$ ($4i - 2$ for $a = 1$).*

Proof According to Theorems 5.22, 5.23, we have $\psi_3^k = O\left(h^9\right)$, i.e. the assertion is true for $i = 3$. Let $\psi_{i-1}^k = O\left(h^p\right)$, where $p = 4(i - 1) - 3$. Then by Theorem 5.23, we have $\psi_i^k = O\left(h^{p+4}\right) = O\left(h^{4i-3}\right)$. $\qquad\square$

The local order of convergence of (5.261) higher than that of the existing i-step methods proposed in [21, 27, 42, 49].

Remark 5.12 The theory developed in this chapter works well for solving nonlinear operator equations given on Banach space. In particular, the choices of iteration parameters maintained also in this case [53–55].

References

1. E. Cătinaş, A survey on the high convergence orders and computational convergence orders of sequences. Appl. Math. Comput. **343**, 1–20 (2019)
2. J.M. Ortega, W.C. Rheinboldt, *Iterative Solution of Nonlinear Equations in Several Variables* (Academic Press, New York, 1970)
3. A. Cordero, J.L. Hueso, E. Martínez, J.R. Torregrosa, A modified Newton-Jarratt's composition. Numer. Algorithms **55**, 87–99 (2010)
4. T. Zhanlav, V. Ulziibayar, O. Chuluunbaatar, Necessary and sufficient conditions for the convergence of two- and three-point Newton-type iterations. Comput. Math. Math. Phys. **57**, 1090–1100 (2017)
5. F.A. Potra, *Nondiscrete Induction and Iterative Processes* (Pitman, London, 1984)
6. T. Zhanlav, I.V. Puzynin, The convergence of iteration based on a continuous analogue of Newton's method. Comput. Math. Math. Phys. **32**, 729–737 (1992)
7. T. Zhanlav, O. Chuluunbaatar, V. Ulziibayar, Generating function method for constructing new iterations. Appl. Math. Comput. **315**, 414–423 (2017)
8. P. Jarratt, Some fourth-order multipoint methods for solving equations. Math. Comput. **20**, 434–437 (1966)
9. J.R. Sharma, H. Arora, On efficient weighted-Newton methods for sollving systems of nonlinear equations. Appl. Math. Comput. **222**, 497–506 (2013)
10. J.R. Sharma, R.K. Guha, Simple yet efficient Newton-like method for systems of nonlinear equations. Calcolo **53**, 451–473 (2016)
11. A. Cordero, E. Martínez, J.R. Torregrosa, Iterative methods of order four and five for systems of nonlinear equations. Appl. Math. Comput. **231**, 541–551 (2009)
12. J.R. Sharma, P. Gaupta, An efficient fifth-order method for solving systems of nonlinear equations. Comput. Math. Appl. **67**, 591–601 (2014)
13. X.Y. Xiao, H.W. Yin, Increasing the order of convergence for iterative methods to solve nonlinear systems. Calcolo **53**, 285–300 (2016)
14. M. Grau-Sánchez, À. Grau, On the computational efficiency index and some iterative methods for solving systems of nonlinear equations. J. Comput. Appl. Math. **236**, 1259–1266 (2011)
15. M. Grau-Sánchez, À. Grau, M. Noguera, Ostrowski type methods for solving systems of nonlinear equations. Appl. Math. Comput. **218**, 2377–2385 (2011)
16. K. Madhu, J. Jayaraman, Some higher order Newton-like methods for solving system of nonlinear equations and its applications. Int. J. Appl. Comput. Math. **3**(3), 2213–2230 (2017)
17. J.R. Sharma, R.K. Guha, R. Sharma, An efficient fourth-order weighted-Newton method for systems of nonlinear equations. Numer. Algorithms **62**, 307–323 (2013)
18. A. Cordero, J.L. Hueso, E. Martínez, J.R. Torregrosa, Increasing the convergence order of an iterative method for nonlinear systems. Appl. Math. Lett. **25**, 2369–2374 (2012)
19. A. Cordero, J.R. Torregrosa, M.P. Vassileva, Pseudocomposition: a technique to design predictor-corrector methods for systems of nonlinear equations. Appl. Math. Comput. **218**, 11496–11504 (2012)
20. J.R. Sharma, H. Arora, Efficient Jarratt-like methods for solving systems of nonlinear equations. Calcolo **51**, 193–210 (2014)
21. X. Xiao, H. Yin, A new class of methods with higher order of convergence for solving systems of nonlinear equations. Appl. Math. Comput. **264**, 300–309 (2015)
22. X. Xiao, H. Yin, A simple and efficient method with high order convergence for solving systems of nonlinear equations. Comput. Appl. Math. **69**, 1220–1231 (2015)
23. T. Zhanlav, Ch. Chun, K. Otgondorj, V. Ulziibayar, High-order iterations for systems of nonlinear equations. Int. J. Comput. Math. **97**, 1704–1724 (2020)
24. J.R. Sharma, R. Sharma, N. Kalra, A novel family of composite Newton-Traub methods for solving systems of nonlinear equations. Appl. Math. Comput. **269**, 520–535 (2015)
25. B. Ghanbari, C. Chun, A constructive method for solving the equation $Xa = b$ in R^n: a generalization of division in R^n. Appl. Math. Comput. **364**, 124673-1–12 (2020)

26. M. Narang, S. Bhatia, V. Kanwar, New two-parameter Chebyshev-Halley-like family of fourth and sixth-order methods for systems of nonlinear equations. Appl. Math. Comput. **275**, 394–403 (2016)
27. S. Bhalla, S. Kumar, I.K. Argyros, R. Behl, A family of higher order derivative free methods for nonlinear systems with local convergence analysis. Comput. Appl. Math. **37**, 5807–5828 (2018)
28. H. Montazeri, F. Soleymani, S. Shateyi, S.S. Motsa, On a new method for computing the numerical solution of systems of nonlinear equations. J. Appl. Math. **2012**, Article ID: 751975-1–15 (2012)
29. Ch. Chun, B. Neta, Developing high order methods for the solution of systems of nonlinear equations. Appl. Math. Comput. **342**, 178–190 (2019)
30. S. Abbasbandy, P. Bakhtiari, A. Cordero, J.R. Torregrosa, T. Lotfi, New efficient methods for solving nonlinear systems of equations with arbitrary even order. Appl. Math. Comput. **287–288**, 94–103 (2016)
31. M.S. Petković, B. Neta, L.D. Petković, J. Džunić, *Multipoint Methods for Solving Nonlinear Equations* (Elsevier, 2013)
32. T. Zhanlav, Ch. Chun, K. Otgondorj, Construction and dynamics of efficient high-order methods for nonlinear systems. Int. J. Comput. Methods **19**, 2250020-1–31 (2022)
33. T. Zhanlav, K. Otgondorj, Higher order Jarratt-like iterations for solving systems of nonlinear equations. Appl. Math. Comput. **395**, 125849-1–12 (2021)
34. T. Zhanlav, O. Chuluunbaatar, V. Ulziibayar, Two-sided approximation for some Newton's type methods. Appl. Math. Comput. **236**, 239–246 (2014)
35. M. Kansal, A. Cordero, S. Bhalla, J.R. Torregrosa, New fourth- and sixth-order classes of iterative methods for solving systems of nonlinear equations and their stability analysis. Numer. Algorithms (2020)
36. J. Hueso, E. Martínez, C. Teruel, Convergence, efficiency and dynamics of new fourth and sixth order families of iterative methods for nonlinear systems. Comput. Appl. Math. **275**, 412–420 (2015)
37. R. Behl, P. Maroju, S.S. Motsa, Efficient family of sixth-order methods for nonlinear models with its dynamics. Int. J. Comput. Methods **16**, 1840008-1–26 (2019)
38. A. Bahl, A. Cordero, R. Sharma, J.R. Torregrosa, A novel bi-parametric sixth order iterative scheme for solving nonlinear systems and its dynamics. Appl. Math. Comput. **357**, 147–166 (2019)
39. X. Wang, J. Kou, Ch. Gu, Semilocal convergence of a sixth-order Jarratt method in Banach spaces. Numer. Algorithms **57**, 441–456 (2011)
40. F. Soleymani, T. Lofti, P. Bakhtiari, A multistep class of iterative solvers for nonlinear systems. Optim. Lett. **8**, 1001–1015 (2014)
41. R. Behl, I. Sarria, R. Gonzalez, A.A. Magrerian, Highly efficient family of iterative methods for solving nonlinear models. J. Comput. Appl. Math. **346**, 119–132 (2019)
42. R. Behl, S. Bhalla, Á.A. Magreñán, S. Kumar, An efficient high order iterative scheme for large nonlinear systems with dynamics. J. Comput. Appl. Math. **404**, 113249-1–16 (2022)
43. T. Lotfi, P. Bakhtiari, A. Cordero, K. Mahdiani, J.R. Torregrosa, Some new efficient multipoint iterative methods for solving nonlinear systems of equations. Int. J. Comput. Math. **92**(9), 1921–1934 (2015)
44. J.R. Sharma, H. Arora, Improved Newton-like methods for solving systems of nonlinear equations. SeMA J. **74**, 147–163 (2017)
45. A. Cordero, E. Martínez, J.R. Torregrosa, Efficient high-order iterative methods for solving nonlinear systems and their application on heat conduction problems. Complexity **11**, Artical ID: 6457532-1–11 (2017)
46. T. Zhanlav, R. Mijiddorj, K. Otgondorj, A family of Newton-type methods with seventh and eighth-order of convergence for solving systems of nonlinear equations. Hacet. J. Math. Stat. **52**(4), 1006–1021 (2023)
47. R. Behl, H. Arora, CMMSE: a novel scheme having seventh-order convergence for nonlinear systems. J. Comput. Appl. Math. **404**, 113301-1–16 (2020)

48. J.F. Traub, *Iterative Methods for the Solution of Equations* (Prentice-Hall, Englewood Cliffs, NJ, 1964)
49. K. Madhu, D.K.R. Babajee, J. Jayaraman, An improvement to double-step Newton method and its multi-step version for solving system of nonlinear equations and its applications. Numer. Algorithms **74**, 593–607 (2017)
50. P. Sivakumar, J. Jayaraman, Efficient two-step fifth-order and its higher-order algorithms for solving nonlinear systems with applications. Axioms MDPI **8**(2), 37-1–17 (2019)
51. B. Saheya, T. Zhanlav, R. Mijiddorj, A novel families of higher-order multistep iterative methods for solving nonlinear systems. Accepted Math. Methods Appl. Sci. (2024)
52. R. Behl, A. Cordero, J.R. Torregrosa, High order family of multivariate iterative methods: convergence and stability. J. Comput. Appl. Math. **405**, 113053-1–16 (2022)
53. I.K. Argyros, S. George, Expanding the applicability of generalized high convergence order methods for solving equations. Khayyam J. Math. **4**, 167–177 (2018)
54. R. Behl, I.K. Argyros, F.O. Mallawi, C.I. Argyros, Convergence of higher order Jarratt-type schemes for nonlinear equations from applied sciences. Symmetry **13**, 1162-1–19 (2021)
55. T. Zhanlav, D. Khongorzul, Semilocal convergence with R-order three theorems for the Chebyshev method and its applications, in *Optimization, Simulation, and Control, Springer Optimization and Its Applications*, ed. by A. Chinchuluun et al., vol. 76, pp. 331–345 (2013)

Chapter 6
High-Order Derivative-Free
Newton-Type Iterations

Abstract A unified approach to the construction of higher-order two- and three-step derivative-free iterations for solving a system of nonlinear equations is introduced. As in Chap. 5, the necessary and sufficient convergence conditions are written in terms of parameters. The two parametric families of derivative-free iterations include some well-known methods as particular cases. We also discuss the computational cost of the considered methods and select the most efficient and cheapest methods.

6.1   Unified Approach to Construction of Higher-Order Derivative Free Iterative Methods

Unified approach to construction of higher-order derivative free iterative methods for solving systems of nonlinear equations.

The problem of finding solution of the system of nonlinear equations $F(x) = 0$, where $F : D \subset R^n \to R^n$, D is open convex domain in R^n is an important and interesting task in numerical analysis and in engineering sciences [1–3]. The most widely used method for solving this problem is quadratically convergent Newton's method:

$$x_{k+1} = x_k - \left(F'(x_k)\right)^{-1} F(x_k), \quad k = 0, 1, \ldots, \tag{6.1}$$

where x_0 is the initial approximation and $\left(F'(x)\right)^{-1}$ is the inverse of Fréchet derivative $F'(x)$ of the function $F(x)$. In many practical situations, the derivative $F'(x)$ does not exist or tedious to calculate. Instead of (6.1) often used derivative-free methods. For example, the quadratically convergent Traub-Steffensen method [1, 3] is given by

$$x_{k+1} = x_k - M^{-1} F(x_k), \quad k = 0, 1, \ldots, \tag{6.2}$$

The original version of the chapter has been revised. A correction to this chapter can be found at
https://doi.org/10.1007/978-3-031-63361-4_9

T. Zhanlav and O.chuluunbaatar, *New Developments of Newton-Type Iterations for Solving Nonlinear Problems, Mathematical Engineering,*
https://doi.org/10.1007/978-3-031-63361-4_6

where $M = [F; w_k, x_k]$ is the first-order divided difference of F and $w_k = x_k + \gamma F(x_k)$, γ is an arbitrary non-zero constant. M^{-1} is the inverse of matrix M. The divided difference of F is a mapping $M = [F; \cdot, \cdot] : D \times D \subset R^n \times R^n \to L(R^n)$ defined by

$$[F; x, y](x - y) = F(x) - F(y), \quad \forall x, y \in R^n, \quad x \neq y$$

or

$$[F; x + h_1, x] = \int_0^1 F'(x + th_1)dt, \quad \forall x, h_1 \in R^n. \tag{6.3}$$

Expand $F'(x + th_1)$ in Taylor series at the point x and integrate, we obtain

$$[F; x + h_1, x] = F'(x) + \frac{1}{2}F''(x)h_1 + \frac{1}{6}F'''(x)h_1^2 + O\left(h_1^3\right), \tag{6.4}$$

where $h_1^i = \overbrace{(h_1, h_1, \ldots, h_1)}^{i}$. In last decade, many efficient derivative-free high-order methods have been proposed for solving nonlinear systems, see [4–20] and references therein. But each of them is the case study and particular character. Perhaps, such case study may be continued infinitely many times. The general theory or unified approach to the construction of derivative-free iterations is absent until now and so this is needed actually. The purpose of this chapter is to develop a unified approach to the construction high-order derivative-free methods [21]. In Sect. 6.1.1, we propose the necessary and sufficient condition for two-point derivative-free methods to be fourth-order of convergence. In Sect. 6.1.2, we propose the necessary and sufficient condition for three-point derivative-free methods. Based on this conditions we present a new family of derivative-free methods with higher-order of convergence $O\left(h^p\right)$ for $p = 4, 5, 6, 7$. This family of methods contains some known methods as a particular cases. In Sect. 6.1.3, we considered the computational cost of the proposed methods.

6.1.1   Two-Step Iterative Methods and Convergence Analysis

We consider the following two-step derivative-free method

$$\begin{aligned}
y_k &= x_k - M^{-1}F(x_k), \\
x_{k+1} &= y_k - \bar{\tau}_k M^{-1} F(y_k),
\end{aligned} \tag{6.5}$$

where $\bar{\tau}_k$ is parameter matrix to be determined properly. We now prove the theorem.

Theorem 6.1 *Let $F : D \subset R^n \to R^n$ be a sufficiently Fréchet differentiable in an open convex set D containing simple solution α. Suppose that $F'(x)$ is continuous and nonsingular in $x = \alpha$ of $F(x) = 0$. Let, $x^{(0)}$ be an initial approximation, which* *is sufficiently close to α. Then the convergence order of iterations (6.5) equal to four if and only if the parameter matrix satisfies*

$$\bar{\tau}_k = \left(F'(y_k)\right)^{-1} M + O\left(h^2\right).$$ (6.6)

Proof Expanding $F(x_{k+1})$ in Taylor series at the point y_k, we obtain

$$F(x_{k+1}) = \left(I - F'(y_k)\bar{\tau}_k M^{-1}\right) F(y_k) + O\left(F^2(y_k)\right),$$ (6.7)

where I is $n \times n$ identity matrix. Since $F(y_k) = O\left(F^2(x_k)\right) = O\left(h^2\right)$ then from (6.7) it follows that $F(x_{k+1}) = O\left(h^4\right)$ if and only if the parameter matrix $\bar{\tau}_k$ satisfies the condition (6.6). $\square$

Further we will use the Banach lemma of inverse (5.7). Using formulae (5.7), (6.4) and Taylor expansion of $F'(y_k)$ around the point x_k one can obtain

$$[F; w_k, x_k] = F'(x_k)(I + A_k) + O\left(h^2\right),$$ (6.8)

$$M^{-1} = [F; w_k, x_k]^{-1} = (I - A_k)\left(F'(x_k)\right)^{-1} + O\left(h^2\right),$$ (6.9)

and

$$F'(y_k) = F'(x_k)(I - 2\Theta_k) + O\left(h^2\right),$$ (6.10)

$$\left(F'(y_k)\right)^{-1} = (I + 2\Theta_k)\left(F'(x_k)\right)^{-1} + O\left(h^2\right),$$ (6.11)

where

$$\Theta_k = \frac{1}{2}\left(F'(x_k)\right)^{-1} F''(x_k)\left(F'(x_k)\right)^{-1} F(x_k),$$ (6.12)

$$A_k = \frac{1}{2}\left(F'(x_k)\right)^{-1} F''(x_k)\gamma F(x_k).$$ (6.13)

In a similar way, we obtain

$$[F; w_k, y_k] = F'(x_k)(I - \Theta_k + A_k) + O\left(h^2\right),$$ (6.14)

$$[F; y_k, x_k] = F'(x_k)(I - \Theta_k) + O\left(h^2\right).$$ (6.15)

Using the expressions (6.8), (6.9) one can easily shown that, the convergence condition (6.6) is equivalent to

$$\bar{\tau}_k = I + 2\Theta_k + A_k + O\left(h^2\right).$$ (6.16)

Of course, there are many choices of $\bar{\tau}_k$ satisfying the condition (6.16). Assume that $\bar{\tau}_{1k}$ and $\bar{\tau}_{2k}$ are two different choices. Then from (6.16) it clear that the iterations (6.5) with choice

$$\bar{\tau}_k = \alpha \bar{\tau}_{1k} + (1 - \alpha)\bar{\tau}_{2k}, \quad \alpha \in R,$$ (6.17)

have also fourth-order of convergence. Moreover, it is easy to show that if $\bar{\tau}_k$ satisfies (6.16) then

$$\bar{\tau}_k + (\bar{\tau}_k)^{-1} = 2I + O\left(h^2\right), \tag{6.18}$$

and

$$\bar{\tau}^* = \frac{1}{2}\left(2I + \bar{\tau}_k - (\bar{\tau}_k)^{-1}\right), \tag{6.19}$$

also satisfies the condition (6.16). It means that the iteration (6.5) with choice (6.19) has also fourth-order of convergence. Thus, the formulae (6.17) and (6.19) give general rule for choice of parameter matrix $\bar{\tau}_k$ in (6.5). From (6.6) it clear that in order to construct fourth-order methods it suffices to approximate $\left(F'(y_k)\right)^{-1}$ with accuracy $O\left(h^2\right)$. This can be done, for example, as

$$F'(y_k) = -[F; w_k, x_k] + [F; w_k, y_k] + [F; y_k, x_k] + O\left(h^2\right), \tag{6.20}$$

which is obtained using (6.7) and (6.14), (6.15). Using (6.20) in (6.6) we have

$$\bar{\tau}_k = ([F; y_k, x_k] + [F; w_k, y_k] - [F; w_k, x_k])^{-1}[F; w_k, x_k]. \tag{6.21}$$

The iteration (6.5) with choice (6.21) often used in practice [14, 19]. Further we rewrite (6.20) as

$$F'(y_k) = [F; w_k, x_k] + D_k + O\left(h^2\right),$$
$$D_k = [F; w_k, y_k] - [F; w_k, x_k] + [F; y_k, x_k] - [F; w_k, x_k].$$

According to (6.8), (6.14) and (6.15), we have $D_k = O(h)$. Hence

$$\left(F'(y_k)\right)^{-1} = \left(\left(I + D_k[F; w_k, x_k]^{-1}\right)[F; w_k, x_k]\right)^{-1}$$
$$= [F; w_k, x_k]^{-1}\left(I - D_k[F; w_k, x_k]^{-1}\right) + O\left(h^2\right).$$

Then $\bar{\tau}_k$ given by (6.6) leads to

$$\bar{\tau}_k = [F; w_k, x_k]^{-1}\left([F; w_k, x_k] - D_k\right) + O\left(h^2\right)$$
$$= 3I - [F; w_k, x_k]^{-1}\left([F; w_k, y_k] + [F; y_k, x_k]\right). \tag{6.22}$$

The iteration method (6.5) with choice (6.22) coincides with method given in [14]. Analogously, rewriting (6.20) in the form

$$F'(y_k) = \left(I + ([F; y_k, x_k] - [F; w_k, x_k])[F; w_k, y_k]^{-1}\right)[F; w_k, y_k] + O\left(h^2\right),$$

we find that

$$\left(F'(y_k)\right)^{-1} = [F; w_k, y_k]^{-1}$$
$$\times \left(I - ([F; y_k, x_k] - [F; w_k, x_k])[F; w_k, y_k]^{-1}\right) + O\left(h^2\right).$$

Since $[F; y_k, x_k] - [F; w_k, x_k] = O(h)$ and $[F; w_k, y_k]^{-1} = [F; w_k, x_k]^{-1} + O(h)$, then we can replace $[F; w_k, y_k]^{-1}$ by $[F; w_k, x_k]^{-1}$ without loss of accuracy. As a result, we get

$$\bar{\tau}_k = \left(F'(y_k)\right)^{-1}[F; w_k, x_k] = [F; w_k, y_k]^{-1}(2[F; w_k, x_k] - [F; y_k, x_k]). \tag{6.23}$$

In a similar way, (6.20) can be rewritten as

$$F'(y_k) = \left(I + ([F; w_k, y_k] - [F; w_k, x_k])[F; y_k, x_k]^{-1}\right)[F; y_k, x_k] + O\left(h^2\right).$$

Hence

$$\left(F'(y_k)\right)^{-1} = [F; y_k, x_k]^{-1}\left([F; y_k, x_k] - [F; w_k, y_k] + [F; w_k, x_k]\right)$$
$$\times[F; y_k, x_k]^{-1} + O\left(h^2\right). \tag{6.24}$$

The iteration (6.5) with choice $\bar{\tau}_k = \left(F'(y_k)\right)^{-1}$, where $\left(F'(y_k)\right)^{-1}$ is given by (6.24) leads to one given by Liu et al. [9].

Now we consider another approximation for $F'(y_k)$. Let $u_k = y_k + cF(y_k)$ and $c \neq 0$. Then by virtue of (6.4) and $F(y_k) = O\left(h^2\right)$, we have

$$[F; u_k, y_k] = F'(y_k) + O\left(h^2\right),$$

Hence

$$[F; u_k, y_k]^{-1} = \left(F'(y_k)\right)^{-1} + O\left(h^2\right).$$

Therefore, according to (6.6), one can choose $\bar{\tau}_k$ such that

$$\bar{\tau}_k = [F; u_k, y_k]^{-1}[F; w_k, x_k]. \tag{6.25}$$

Using this and (6.18) one can propose the simple choice

$$\bar{\tau}_k = 2I - [F; w_k, x_k]^{-1}[F; y^k + cF(y_k), y_k]. \tag{6.26}$$

In general, one can propose two parameters choice of $\bar{\tau}_k$ of the form

$$\bar{\tau}_k = ((c - b)[F; y_k, x_k] + a[F; w_k, y_k] + b[F; w_k, x_k])^{-1}$$
$$\times ((2a + 2c + b)[F; w_k, x_k] - c[F; w_k, y_k] - (a + b)[F; y_k, x_k]), \tag{6.27}$$
$$a + c \neq 0,$$

satisfying the condition (6.16). Thus, we propose two parameters families iteration (6.5) with choices (6.27). When $c = 0$, $a = 1$, $b = -1$ the (6.27) leads to (6.21), when $b = c = 1$, $a = 0$ the (6.27) leads to (6.22). When $b = c = 0$, $a = 1$ the (6.27) leads to (6.23) and so on. Note that all the fourth-order derivative-free methods can be rewritten as (6.5) with different parameters matrices $\bar{\tau}_k$. But all the parameter matrices $\bar{\tau}_k$ coincide each other within the accuracy $O\left(h^2\right)$. As in [6], we use the notation

$$G_k = [F; w_k, x_k]^{-1}[F; u_k, y_k]. \tag{6.28}$$

It is easy to show that

$$[F; w_k, x_k] = F'(x_k)\left(I + \frac{1}{2}\left(F'(x_k)\right)^{-1} F''(x_k)\gamma F(x_k)\right) + O\left(h^2\right),$$

$$[F; u_k, y_k] = F'(x_k)(I - 2\Theta_k) + O\left(h^2\right).$$

Hence, we have

$$[F; w_k, x_k]^{-1} = \left(I - \frac{1}{2}\left(F'(x_k)\right)^{-1} F''(x_k)\gamma F(x_k)\right)\left(F'(x_k)\right)^{-1} + O\left(h^2\right),$$

$$[F; u_k, y_k]^{-1} = (I + 2\Theta_k)\left(F'(x_k)\right)^{-1} + O\left(h^2\right).$$

Then we get

$$G_k + (G_k)^{-1} = 2I + O\left(h^2\right). \tag{6.29}$$

Using (6.29) in (6.26) we obtain

$$\bar{\tau}_k = 2I - [F; w_k, x_k]^{-1}[F; u_k, y_k] + O\left(h^2\right). \tag{6.30}$$

From (6.29) it follows that

$$(I - G_k)^2 = O\left(h^2\right) \quad \text{or} \quad I - G_k = O(h). \tag{6.31}$$

Adding (6.31) in the right hand side of (6.30) we get

$$\bar{\tau}_k = aI + G_k\left((3 - 2a)I + (a - 2)G_k\right) + O\left(h^2\right). \tag{6.32}$$

When $a = 2$, (6.32) leads to (6.30). The two-step iteration (6.5) with choice (6.32) was given in [6, 16].

Analogously, one can choose $\bar{\tau}_k$ as

$$\bar{\tau}_k = bI + (G_k)^{-1}\left((1 - 2b)I + b(G_k)^{-1}\right). \tag{6.33}$$

When $b = 0$, (6.33) leads to (6.25). Inclusion of parameters a and b in (6.32), (6.33), in our opinion, does not give any advantage for (6.5). But they may be used to speed up the iteration of (6.5) [15, 16]. Instead of (6.5) often used two-point iterations

$$y_k = x_k - [F; w_k, s_k]^{-1} F(x_k),$$
$$x_{k+1} = y_k - \bar{\tau}_k [F; w_k, s_k]^{-1} F(y_k), \tag{6.34}$$

where $w_k = x_k + \gamma F(x_k)$, $s_k = x_k - \gamma F(x_k)$, $\gamma \neq 0$. For the iteration (6.34), we have the following result.

Theorem 6.2 *Assume that all the assumptions of Theorem 6.1 are fulfilled. Then the convergence order of iteration* (6.34) *equals to four if and only if the parameter matrix $\bar{\tau}_k$ satisfies*

$$\bar{\tau}_k = I + 2\Theta_k + O\left(h^2\right). \tag{6.35}$$

Proof The iteration (6.34) is obtained from (6.5) replacing $[F; w_k, x_k]^{-1}$ by $[F; w_k, s_k]$. So the fourth-order convergence condition (6.6) for (6.34) reads as

$$\bar{\tau}_k = F(y_k)^{-1} [F; w_k, s_k] + O\left(h^2\right). \tag{6.36}$$

It is easy to show that

$$[F; w_k, s_k] = \frac{1}{2}([F; w_k, x_k] + [F; s_k, x_k]) = F'(x_k) + O\left(h^2\right), \tag{6.37}$$

$$[F; w_k, s_k]^{-1} = \left(F'(x_k)\right)^{-1} + O\left(h^2\right), \tag{6.38}$$

$$F'(y_k) = F'(x_k) - F''(x_k)[F; w_k, s_k]F(x_k) + O\left(h^2\right)$$
$$= F'(x_k)(I - 2\Theta_k) + O\left(h^2\right), \tag{6.39}$$

$$[F; y_k, x_k] = F'(x_k)(I - \Theta_k) + O\left(h^2\right), \tag{6.40}$$

$$\left(F'(y_k)\right)^{-1} = (I + 2\Theta_k)\left(F'(x_k)\right)^{-1} + O\left(h^2\right)$$
$$= (I + 2\Theta_k)[F; w_k, s_k]^{-1} + O\left(h^2\right). \tag{6.41}$$

Substituting (6.37) and (6.41) into (6.36) we get (6.35) that complete the proof of theorem. $\square$

Since $\Theta_k = O(h)$ then from (6.35) we obtain

$$(\bar{\tau}_k)^{-1} = I - 2\Theta_k + O\left(h^2\right). \tag{6.42}$$

A consequence of (6.35) and (6.42) is

$$\bar{\tau}_k + (\bar{\tau}_k)^{-1} = 2I + O\left(h^2\right). \tag{6.43}$$

As before, the fourth-order convergence condition (6.35) can be expressed through first-order divided differences. Namely, using (6.37) and (6.40) we get

$$I - [F; w_k, s_k]^{-1}[F; y_k, x_k] = \Theta_k + O\left(h^2\right). \tag{6.44}$$

Substituting (6.44) into (6.35) we obtain

$$\bar{\tau}_k = 3I - 2[F; w_k, s_k]^{-1}[F; y_k, x_k]. \tag{6.45}$$

The iteration (6.34) with choice (6.45) are considered by Wang et al. in [17] (see also [15]). In general, it is easy to show that $\bar{\tau}_k$ given by formula

$$\bar{\tau}_k = (2 + a)I - [F; w_k, s_k]^{-1}((2a + b)[F; y_k, x_k] + b[F; y_k, w_k]$$
$$-b[F; w_k, x_k] + (1 - a - b)[F; y_k + cF(y_k), y_k]) + O\left(h^2\right), \quad a, b \in R, \tag{6.46}$$

satisfies the condition (6.35). Thus we propose the two parameters families of iterations (6.34) with choice (6.46). When $a = 1$, $b = 0$ the formula (6.46) leads to (6.45). We list some particular cases of (6.46)

$$\bar{\tau}_k = 2I - [F; w_k, s_k]^{-1}\left([F; y_k, x_k] + [F; w_k, y_k] - [F; w_k, x_k]\right),$$
$$\text{when } a = 0, \quad b = 1, \tag{6.47}$$
$$\bar{\tau}_k = 2I - [F; w_k, s_k]^{-1}[F; y_k + cF(y_k), y_k],$$
$$\text{when } a = b = 0, \tag{6.48}$$
$$\bar{\tau}_k = I - 2[F; w_k, s_k]^{-1}\left([F; w_k, y_k] - [F; w_k, x_k]\right),$$
$$\text{when } a = -1, \ b = 2, \tag{6.49}$$
$$\bar{\tau}_k = [F; w_k, s_k]^{-1}\left([F; y_k, x_k] + 3[F; w_k, x_k] - 3[F; y_k, w_k]\right),$$
$$\text{when } a = -2, \quad b = 3. \tag{6.50}$$

6.1.2 High-Order Three-Step Derivative-Free Methods

Now we consider three-step iterative methods:

$$y_k = x_k - [F; w_k, x_k]^{-1}F(x_k),$$
$$z_k = y_k - \bar{\tau}_k[F; w_k, x_k]^{-1}F(y_k) = \psi_4(x_k, y_k, w_k), \tag{6.51}$$
$$x_{k+1} = z_k - \alpha_k[F; w_k, x_k]^{-1}F(z_k),$$

where $\bar{\tau}_k$ and α_k are some square matrices and $z_k = \psi_4(x_k, y_k, w_k)$ stands for any fourth-order method. Of course, the convergence order of method (6.51) essentially depends on the choice of matrix α_k. In the preceding subsection we prove that $z_k =$

$\psi_4(x_k, y_k, w_k) = O\left(h^4\right)$ under choice (6.16) or (6.27). The convergence order of iterations (6.51) established by following theorem.

Theorem 6.3 *Assume that all the assumptions of Theorem 6.1 are fulfilled. Then the convergence order of iterations (6.51) equal to six if and only if*

$$\alpha_k = \bar{\tau}_k, \tag{6.52}$$

where $\bar{\tau}_k$ satisfy (6.16).

Proof The Taylor expansion of $F(x_{k+1})$ around the point z_k gives

$$F(x_{k+1}) = \left(I - F'(z_k)\alpha_k[F; w_k, x_k]^{-1}\right)F(z_k)) + O\left(h^8\right). \tag{6.53}$$

From (6.53) it clear that

$$F(x_{k+1}) = O\left(h^6\right), \tag{6.54}$$

under condition

$$\alpha_k = \left(F'(z_k)\right)^{-1}[F; w_k, x_k] + O\left(h^2\right). \tag{6.55}$$

On the other hand, expanding $F'(z_k)$ at point y_k, we get

$$F'(z_k) = F'(y_k) - F''(y_k)\bar{\tau}_k[F; w_k, x_k]^{-1}F(y_k) = F'(y_k) + O\left(h^2\right), \tag{6.56}$$

because of $F(y_k) = O\left(h^2\right)$. Then from (6.56) it follows that

$$F^{-1}(z_k) = F^{-1}(y_k) + O\left(h^2\right). \tag{6.57}$$

If we take (6.57) into account, then from (6.6) and (6.55) it follows that the iterations (6.51) has a sixth-order of convergence if and only if $\alpha_k = \bar{\tau}_k$. $\square$

Theorems 6.2 and 6.3 are derivative-free extensions of Theorems 2.1 and Theorem 3.1 in [22]. In the preceding subsection we propose some choices of $\bar{\tau}_k$ as (6.21), (6.22) and in general (6.27). In connection with this, we reveal the following important fact:

Corollary 6.1 *Sixth-order convergence of iterations (6.51) maintained under choices (6.27) and*

$$\begin{aligned} \alpha_k = {}&(c_1 - b_1)[F; y_k, x_k] + a_1[F; w_k, y_k] + b_1[F; w_k, x_k]^{-1} \\ &\times ((2a_1 + 2c_1 + b_1)[F; w_k, x_k] - c_1[F; w_k, y_k] \\ &- (a_1 + b_1)[F; y_k, x_k]), \quad a_1 + c_1 \neq 0. \end{aligned} \tag{6.58}$$

Proof Since the choice (6.27) satisfies the condition (6.6) or (6.16), it is possible to choose α_k in (6.51) as any of (6.27) with different parameters not equal to that of (6.27). Then by virtue Theorem 6.3 the convergence order of iteration (6.51) is equal to six under choices (6.27) and (6.58).   □

The iteration (6.51) with $\bar{\tau}_k$ and α_k given by formulae (6.27) and (6.58) define eight-parameter $(\gamma, \alpha, a, b, c, a_1, b_1, c_1)$ family of sixth-order methods without memory. Of course $\alpha_k = \bar{\tau}_k$ when $a = a_1$, $b = b_1$, $c = c_1$ in (6.27) and (6.58).

Note that, the another choice of $\bar{\tau}_k$ is the formula (6.32). Then by Corollary 6.1 sixth-order convergence of iteration (6.51) maintained under choice (6.32) and

$$\alpha_k = \beta I + G_k \left((3 - 2\beta)I + (\beta - 2)G_k\right). \tag{6.59}$$

Thus, we arrive at the result Theorem 2.1 obtained by Sharma et al. in [16].

Now we consider more symmetric iterations

$$\begin{aligned}
y_k &= x_k - [F; w_k, s_k]^{-1} F(x_k), \\
z_k &= y_k - \bar{\tau}_k [F; w_k, s_k]^{-1} F(y_k), \\
x_{k+1} &= z_k - \alpha_k [F; w_k, s_k]^{-1} F(z_k).
\end{aligned} \tag{6.60}$$

Above, we have shown that $F(z_k) = O\left(h^4\right)$ under choice (6.35). Now we show that Theorem 6.3 and Corollary 6.1 holds also for iteration (6.60). Indeed,

$$F(x_{k+1}) = O\left(h^6\right) \iff \alpha_k = \left(F'(z_k)\right)^{-1} [F; w_k, s_k] + O\left(h^2\right). \tag{6.61}$$

From (6.57), (6.61) and (6.36) it follows that $\alpha_k = \bar{\tau}_k$. Above, we propose some choices of $\bar{\tau}_k$ such as (6.45), (6.47)–(6.50) (in general the choice (6.46)). Note that in [7, 15, 17] was proven similar theorem under special choice (6.45) by using symbolic computation. Grau-Sánchez et al. [8] also developed some efficient derivative-free methods. One of these sixth-order methods is of the form (6.60) with choice

$$\bar{\tau}_k = (2[F; y_k, x_k] - [F; w_k, s_k])^{-1} [F; w_k, s_k], \tag{6.62}$$

satisfying the condition (6.35). Using the relation (6.43) we can find $\bar{\tau}_k$ as

$$\begin{aligned}
\bar{\tau}_k &= 2I - [F; w_k, s_k]^{-1} (2[F; y_k, x_k] - [F; w_k, s_k]) \\
&= [F; w_k, s_k]^{-1} (3[F; w_k, s_k] - 2[F; y_k, x_k]) \\
&= 3I - 2[F; w_k, s_k]^{-1} [F; y_k, x_k].
\end{aligned}$$

It means that the choice $\bar{\tau}_k$ given by (6.62) is equivalent to choice (6.45) within accuracy $O\left(h^2\right)$. They also proved similar theorem under choice (6.62). It is known, that evaluation of inverse of matrix requires a lengthy and cumbersome computation work. Therefore, while constructing a numerical methods it will be more appropriate if the number of matrix inversion is as small as possible. From (6.60) and (6.62) we see that the Grau-Sánchez sixth-order method uses two matrix inversions, whereas if instead of (6.62) we use (6.46) then the obtained scheme requires only one inversion per iteration.

Hence, more appropriate sixth-order method is

$$
\begin{aligned}
y_k &= x_k - [F; w_k, s_k]^{-1} F(x_k), \\
z_k &= y_k - \bar{\tau}_k [F; w_k, s_k]^{-1} F(y_k), \\
x_{k+1} &= z_k - \bar{\tau}_k [F; w_k, s_k]^{-1} F(z_k),
\end{aligned}
\tag{6.63}
$$

where $\bar{\tau}_k$ is defined by (6.46). Note that, recently in [7] was proposed i-step iterative scheme of type (6.63) with $2i$-order of convergence. Perhaps, it is possible to construct modification of this method that requires only one inverse matrix per iteration, as (6.63). Here we give a modification of this method only for $i = 4$ and it has a form

$$
\begin{aligned}
y_k &= x_k - [F; w_k, s_k]^{-1} F(x_k), \\
z_k &= y_k - \bar{\tau}_k [F; w_k, s_k]^{-1} F(y_k), \\
t_k &= z_k - \bar{\tau}_k [F; w_k, s_k]^{-1} F(z_k), \\
x_{k+1} &= t_k - \bar{\tau}_k [F; w_k, s_k]^{-1} F(t_k),
\end{aligned}
\tag{6.64}
$$

where $\bar{\tau}_k$ is given by (6.46). The convergence order of iteration (6.64) is equal to 8. Indeed, according to Theorem 6.3 we have $F(t_k) = O\left(h^6\right)$. Expanding $F(x_{k+1})$ around t_k, we get

$$
F(x_{k+1}) = \left(I - F'(t_k)\bar{\tau}_k[F; w_k, s_k]\right) F(t_k) + O\left(h^{12}\right).
$$

Since $F(z_k) = O\left(h^4\right)$ then from the third step of (6.64) we get

$$
F'(t_k) = F'(z_k) + O\left(h^4\right) \Rightarrow F'(t_k)^{-1} = \left(F'(z_k)\right)^{-1} + O\left(h^4\right).
$$

Hence, by virtue of (6.57), we have

$$
F'(t_k)^{-1} = \left(F'(y_k)\right)^{-1} + O\left(h^2\right).
$$

Since $\bar{\tau}_k$ given by (6.46) satisfies the condition (6.35), then by Theorem 6.2 and 6.3 we have $F(x_{k+1}) = O\left(h^8\right)$. The most easy choices of $\bar{\tau}_k$ are (6.45) and (6.48). Quite recently Narang et al. in [18] also proposed $m-$point method ($m \geq 3$) of type (6.63) with convergence order $m + 1$.

Remark 6.1 It is easy to show that if in (6.51) and (6.60) we choose $\bar{\tau}_k = I$ and $\bar{\alpha}_k$ is given by formula (6.27) and (6.46), respectively, then convergence order of them reduces from 6 to 5.

Now we proceed to construct seventh-order iterations of the form (6.51). In the beginning of Sect. 6.1.2 we shown that

$$F(z_k) = O\left(h^4\right) \iff \bar{\tau}_k = I + 2\Theta_k + A_k + O\left(h^2\right). \tag{6.65}$$

Then from (6.53) it clear that

$$F(x_{k+1}) = O\left(h^7\right) \tag{6.66}$$

under condition

$$\alpha_k = \left(F'(z_k)\right)^{-1} [F; w_k, x_k] + O\left(h^3\right). \tag{6.67}$$

Our aim is to approximate $F'(z_k)^{-1}$ through first divided differences of F with accuracy $O\left(h^3\right)$. Using (6.4) we obtain

$$[F; w_k, x_k] = F'(x_k)\left(I + A_k + \frac{1}{6}\left(F'(x_k)\right)^{-1} F'''(x_k)(\gamma F(x_k))^2\right) + O\left(h^3\right). \tag{6.68}$$

By using (5.7) we have

$$[F; w_k, x_k]^{-1} = \left(I - A_k + A_k^2 - \frac{1}{6}\left(F'(x_k)\right)^{-1} F'''(x_k)(\gamma F(x_k))^2\right)$$
$$\times \left(F'(x_k)\right)^{-1} + O\left(h^3\right). \tag{6.69}$$

Expanding $F'(z_k)$ around the point y_k, we get

$$F'(z_k) = F'(y_k)\left(I - \left(F'(x_k)\right)^{-1} F''(x_k) \left(F'(x_k)\right)^{-1} F(y_k)\right) + O\left(h^3\right), \tag{6.70}$$

in which we have used (6.16), (6.69) and $\left(F'(y_k)\right)^{-1} = \left(F'(x_k)\right)^{-1} + O(h)$, $F''(y_k) = F''(x_k)^{-1} + O(h)$.

Analogously, one can easily shown that

$$F(y_k) = F'(x_k) A_k \left(F'(x_k)\right)^{-1} F(x_k)$$
$$+ \frac{1}{2} F''(x_k)((F'(x_k))^{-1} F(x_k))^2 + O\left(h^3\right), \tag{6.71}$$
$$F'(y_k) = F'(x_k)\left(I - 2\Theta_k + \left(F'(x_k)\right)^{-1} F''(x_k) A_k \left(F'(x_k)\right)^{-1} F(x_k)\right.$$
$$\left. + \frac{1}{2}\left(F'(x_k)\right)^{-1} F'''(x_k)((F'(x_k))^{-1} F(x_k))^2\right) + O\left(h^3\right), \tag{6.72}$$
$$\left(F'(x_k)\right)^{-1} F(y_k) = (A_k + \Theta_k) \left(F'(x_k)\right)^{-1} F(x_k) + O\left(h^3\right), \tag{6.73}$$
$$\left(F'(y_k)\right)^{-1} = \left(I + 2\Theta_k + 4\Theta_k^2 - \left(F'(x_k)\right)^{-1} F''(x_k) A_k \left(F'(x_k)\right)^{-1} F(x_k)\right.$$
$$\left. - \frac{1}{2}\left(F'(x_k)\right)^{-1} F'''(x_k)((F'(x_k))^{-1} F(x_k))^2\right) \left(F'(x_k)\right)^{-1} + O\left(h^3\right). \tag{6.74}$$

Using (6.4), (5.7), (6.16) and (6.69) one can easily shown that

$$[F; y_k, z_k] = F'(z_k) \left(I + \frac{1}{2}\left(F'(x_k)\right)^{-1} F''(x_k)(F'(x_k))\right)^{-1} F(y_k)) + O\left(h^3\right), \quad (6.75)$$

$$[F; x_k, z_k] - [F; y_k, z_k] = F'(x_k)\left(\Theta_k - \frac{1}{2}(F'(x_k))^{-1} F''(x_k)A_k \left(F'(x_k)\right)^{-1} F(x_k)\right.$$
$$\left. -\frac{1}{3}\left(F'(x_k)\right)^{-1} F'''(x_k)((F'(x_k))^{-1} F(x_k))^2\right) + O\left(h^3\right), \quad (6.76)$$

$$[F; y_k, x_k] = F'(x_k)\left(I - \Theta_k + \frac{1}{2}(F'(x_k))^{-1} F''(x_k)A_k \left(F'(x_k)\right)^{-1} F(x_k)\right.$$
$$\left. +\frac{1}{6}\left(F'(x_k)\right)^{-1} F'''(x_k)((F'(x_k))^{-1} F(x_k))^2\right) + O\left(h^3\right), \quad (6.77)$$

$$[F; z_k, x_k] - [F; y_k, x_k] = -\frac{1}{2} F''(x_k)\left(F'(x_k)\right)^{-1} F(y_k) + O\left(h^3\right). \quad (6.78)$$

Using the relations (6.72), (6.73) in (6.70) and (6.78) we obtain

$$F'(z_k) = F'(x_k)\big(I - 2\Theta_k - \left(F'(x_k)\right)^{-1} F''(x_k)\Theta_k \left(F'(x_k)\right)^{-1} F(x_k)$$
$$+\frac{1}{2}\left(F'(x_k)\right)^{-1} F'''(x_k)((F'(x_k))^{-1} F(x_k))^2\big) + O\left(h^3\right), \quad (6.79)$$

and

$$[F; z_k, x_k] - [F; y_k, x_k] = -\frac{1}{2} F''(x_k)(A_k + \Theta_k) \left(F'(x_k)\right)^{-1} F(x_k) + O\left(h^3\right), \quad (6.80)$$

$$[F; y_k, z_k] = F'(z_k) + \frac{1}{2} F''(x_k)(A_k + \Theta_k) \left(F'(x_k)\right)^{-1} F(x_k) + O\left(h^3\right). \quad (6.81)$$

From (6.80) and (6.81) it follows

$$F'(z_k) = [F; y_k, z_k] + [F; z_k, x_k] - [F; y_k, x_k] + O\left(h^3\right). \quad (6.82)$$

Substituting (6.82) into (6.67) we obtain

$$\alpha_k = \left([F; y_k, z_k] + [F; z_k, x_k] - [F; y_k, x_k]\right)^{-1} [F; w_k, x_k]. \quad (6.83)$$

Thus, the iteration (6.51) have seventh-order of convergence provided that $\bar{\tau}_k$ and α_k are given by (6.27) and (6.83), respectively. This result fully coincides with result of Wang et al. [19] that obtained using symbolic computation. Here we give analytic approach for convergence analysis. On the other hand, using the relation

$$\left(F'(x_k)\right)^{-1} = [F; y_k, x_k]^{-1} + O(h), \quad [F; z_k, y_k]^{-1} = [F; z_k, x_k]^{-1} + O(h),$$

from (6.82) one can easily obtain

$$F'(z_k) = [F; y_k, z_k][F; y_k, x_k]^{-1}[F; z_k, x_k] + O\left(h^3\right), \quad (6.84)$$
$$F'(z_k) = [F; z_k, x_k][F; y_k, x_k]^{-1}[F; y_k, z_k] + O\left(h^3\right). \quad (6.85)$$

Rewriting (6.82) as

$$F'(z_k) = \left(I + ([F; z_k, x_k] - [F; y_k, x_k])[F; y_k, z_k]^{-1}\right)[F; y_k, z_k] + O\left(h^3\right),$$

and using $[F; y_k, z_k]^{-1} = [F; w_k, x_k]^{-1} + O(h)$ we get

$$\begin{aligned}
\left(F'(z_k)\right)^{-1} &= [F; y_k, z_k]^{-1} \\
&\times \left(I - ([F; z_k, x_k] - [F; y_k, x_k])[F; w_k, x_k]^{-1}\right) + O\left(h^3\right) \\
&= [F; y_k, z_k]^{-1}\left([F; w_k, x_k] - [F; z_k, x_k] + [F; y_k, x_k]\right) \\
&\times [F; w_k, x_k]^{-1} + O\left(h^3\right).
\end{aligned} \tag{6.86}$$

Substituting (6.86) into (6.67) we obtain

$$\alpha_k = [F; y_k, z_k]^{-1}\left([F; w_k, x_k] - [F; z_k, x_k] + [F; y_k, x_k]\right). \tag{6.87}$$

Thus, the iterations (6.51) with choices (6.27), (6.87) has seventh-order of convergence. The obtained results can be formulated as:

Theorem 6.4 *Assume that all the assumptions of Theorem 6.1 are fulfilled. Then the convergence order of iteration (6.51) equal to seven if the parameter matrix $\bar{\tau}_k$ is given by (6.27) and α_k is given by one of two formulae (6.83), (6.87).*

Seventh-order iteration (6.51) with choices (6.22), (6.87) proposed by Sharma et al. in [12]. Therefore, our iteration (6.51) with choices (6.27), (6.87) are more general and can be regarded as extension of iterations proposed in [12]. The most easy choices are (6.22) and (6.26) for $\bar{\tau}_k$. Moreover, it is possible many other approximations for $\left(F'(z_k)\right)^{-1}$ with accuracy $O\left(h^3\right)$. Analogously, from (6.82) we obtain

$$\begin{aligned}
\left(F'(z_k)\right)^{-1} &= [F; w_k, x_k]^{-1} \\
&\times ([F; w_k, x_k] - [F; z_k, x_k] + [F; y_k, x_k])[F; y_k, z_k]^{-1} + O\left(h^3\right),
\end{aligned} \tag{6.88}$$

$$\begin{aligned}
\left(F'(z_k)\right)^{-1} &= [F; y_k, z_k]^{-1} \\
&\times ([F; y_k, z_k] - [F; z_k, x_k] + [F; y_k, x_k])[F; y_k, z_k]^{-1} + O\left(h^3\right).
\end{aligned} \tag{6.89}$$

Using the relation $[F; y_k, z_k]^{-1} = [F; y_k, x_k]^{-1} + O(h)$ in (6.82) we get

$$\begin{aligned}
\left(F'(z_k)\right)^{-1} &= [F; y_k, z_k]^{-1} \\
&\times (2[F; y_k, x_k] - [F; z_k, x_k])[F; y_k, z_k]^{-1} + O\left(h^3\right),
\end{aligned} \tag{6.90}$$

$$\begin{aligned}
\left(F'(z_k)\right)^{-1} &= [F; y_k, z_k]^{-1} \\
&\times (2[F; y_k, x_k] - [F; z_k, x_k])[F; y_k, z_k]^{-1} + O\left(h^3\right).
\end{aligned} \tag{6.91}$$

Such approximations (6.86), (6.88)–(6.91) are obtained using the property (6.80). Let $\bar{\tau}_k$ in (6.60) satisfies the condition (6.35) Then $F(z_k) = O\left(h^4\right)$. From the Taylor expansion of $F(x_{k+1})$

$$F(x_{k+1}) = \left(I - F'(z_k)\alpha_k[F; w_k, s_k]^{-1}\right)F(z_k) + O\left(F^2(z_k)\right), \quad (6.92)$$

we see that

$$F(x_{k+1}) = O\left(h^7\right), \quad (6.93)$$

if and only if

$$\alpha_k = \left(F'(z_k)\right)^{-1}[F; w_k, s_k]) + O\left(h^3\right). \quad (6.94)$$

Thus, to obtain α_k with accuracy $O\left(h^3\right)$ it suffices to approximate $\left(F'(z_k)\right)^{-1}$ with accuracy $O\left(h^3\right)$. It can be approximated by means of linear combination of first-order divided differences $[F; y_k, z_k], [F; x_k, z_k], [F; y_k, x_k], [F; w_k, x_k], [F; w_k, s_k]$. First of all it is easy to show that

$$[F; w_k, s_k] = F'(x_k)\left(I + \frac{1}{6}\left(F'(x_k)\right)^{-1} F'''(x_k)\gamma^2 F^2(x_k)\right) + O\left(h^3\right), \quad (6.95)$$

$$F'(z_k) = F'(y_k)\left(I - \left(F'(x_k)\right)^{-1} F''(x_k)\left(F'(x_k)\right)^{-1} F(y_k)\right) + O\left(h^3\right), \quad (6.96)$$

$$\left(F'(x_k)\right)^{-1} F(y_k) = \Theta_k \left(F'(x_k)\right)^{-1} F(x_k) + O\left(h^3\right), \quad (6.97)$$

$$F'(y_k) = F'(x_k)\left(I - 2\Theta_k + \frac{1}{2}\left(F'(x_k)\right)^{-1} F'''(x_k)((F'(x_k))^{-1} F(x_k))^2\right)$$
$$+ O\left(h^3\right). \quad (6.98)$$

Using (6.98) and (6.97) in (6.96) we have

$$\left(F'(z_k)\right)^{-1} = \left(I + 2\Theta_k + 4\Theta_k^2 - \frac{1}{2}\left(F'(x_k)\right)^{-1} F'''(x_k)\left((F'(x_k))^{-1} F(x_k)\right)^2\right.$$
$$\left. + \left(F'(x_k)\right)^{-1} F''(x_k)\Theta_k \left(F'(x_k)\right)^{-1} F(x_k)\right)\left(F'(x_k)\right)^{-1} + O\left(h^3\right). \quad (6.99)$$

Substituting (6.95) and (6.99) into (6.94) we obtain

$$\alpha_k = I + 2\Theta_k + 4\Theta_k^2 - C_k + D_k + O\left(h^3\right), \quad (6.100)$$

where

$$C_k = \frac{1}{6}\left(F'(x_k)\right)^{-1} F'''(x_k)\left(3((F'(x_k))^{-1} F(x_k))^2 - \gamma^2 F^2(x_k)\right), \quad (6.101)$$

$$D_k = \left(F'(x_k)\right)^{-1} F''(x_k)\Theta_k \left(F'(x_k)\right)^{-1} F(x_k). \quad (6.102)$$

Thus, we have

Theorem 6.5 *Assume that all the assumptions of Theorem 6.1 are fulfilled. Then the convergence order of iteration (6.60) equal to seven, if parameter matrices $\bar{\tau}_k$ and α_k satisfy the conditions (6.35) and (6.100), respectively.*

We assume that $\gamma = 0$ and

$$D_k = 2\Theta_k^2 + O\left(h^3\right). \tag{6.103}$$

Then the formula (6.100) leads to

$$\alpha_k = I + 2\Theta_k + 3d_k + 6\Theta_k^2 + O\left(h^3\right),$$
$$d_k = -\frac{1}{6}\left(F'(x_k)\right)^{-1} F'''(x_k)\left(\left(F'(x_k)\right)^{-1} F(x_k)\right)^2.$$

Hence, Theorem 6.5 is a derivative-free version of Theorem 3.1 for $p = 7$, see Table 1 in [22]. According to (6.4) we have

$$[F; y_k, x_k] = F'(x_k) - \frac{1}{2}F''(x_k)[F; w_k, s_k]^{-1} F(x_k)$$
$$+ \frac{F'''(x_k)}{6}\left([F; w_k, s_k]^{-1} F(x_k)\right)^2 + O\left(h^3\right). \tag{6.104}$$

From (6.95) we get

$$[F; w_k, s_k]^{-1} = \left(I - \frac{1}{6}\left(F'(x_k)\right)^{-1} F'''(x_k)\gamma^2 F^2(x_k)\right)$$
$$\times \left(F'(x_k)\right)^{-1} + O\left(h^3\right). \tag{6.105}$$

Substituting (6.105) into (6.104) we obtain

$$[F; y_k, x_k] = F'(x_k)$$
$$\times \left(I - \Theta_k + \frac{1}{6}\left(F'(x_k)\right)^{-1} F'''(x_k)((F'(x_k))^{-1} F(x_k))^2\right) + O\left(h^3\right). \tag{6.106}$$

Analogously, we obtain

$$[F; z_k, x_k] = F'(x_k) - \frac{1}{2}F''(x_k)[F; w_k, s_k]^{-1} (F(x_k) + F(y_k))$$
$$+ \frac{F'''(x_k)}{6}\left([F; w_k, s_k]^{-1}(F(x_k) + F(y_k))\right)^2 + O\left(h^3\right),$$

in which we have used (6.35). Using (6.97) and (6.105) in last expression we have

$$[F; z_k, x_k] = F'(x_k) \left(I - \Theta_k - \frac{1}{2} \left(F'(x_k) \right)^{-1} F''(x_k) \Theta_k \left(F'(x_k) \right)^{-1} F(x_k) \right.$$

$$\left. + \frac{1}{6} \left(F'(x_k) \right)^{-1} F'''(x_k) \left(\left(F'(x_k) \right)^{-1} F(x_k) \right)^2 \right) + O\left(h^3 \right). \qquad (6.107)$$

From (6.106) and (6.107) we get

$$[F; z_k, x_k] - [F; y_k, x_k] = -\frac{1}{2} F''(x_k) \Theta_k \left(F'(x_k) \right)^{-1} F(x_k) + O\left(h^3 \right). \qquad (6.108)$$

Analogously, we obtain

$$[F; y_k, z_k] = F'(z_k) + \frac{1}{2} F''(x_k) \Theta_k \left(F'(x_k) \right)^{-1} F(x_k) + O\left(h^3 \right), \qquad (6.109)$$

in which we have used (6.35), (6.97), (6.105). From (6.108) and (6.109) it follows that

$$F'(z_k) = [F; y_k, z_k] + [F; x_k, z_k] - [F; y_k, x_k] + O\left(h^3 \right). \qquad (6.110)$$

The same formula (6.110) was obtained for the iteration (6.51). As a consequence of (6.110) the following hold:

$$\alpha_k = ([F; y_k, z_k] + [F; z_k, x_k] - [F; y_k, x_k])^{-1} [F; w_k, s_k] + O\left(h^3 \right), \qquad (6.111)$$

$$\alpha_k = \left(F'(z_k) \right)^{-1} [F; w_k, s_k]$$
$$= [F; y_k, z_k]^{-1} ([F; w_k, s_k] - [F; z_k, x_k] + [F; y_k, z_k]) + O\left(h^3 \right), \qquad (6.112)$$

and

$$\alpha_k = [F; y_k, z_k]^{-1} ([F; y_k, z_k] - [F; z_k, x_k] + [F; y_k, x_k])$$
$$\times [F; y_k, z_k]^{-1} [F; w_k, s_k] + O\left(h^3 \right). \qquad (6.113)$$

It means that formulae (6.83) and (6.87) hold true for iterations (6.60). The only difference is that $[F; w_k, x_k]$ in (6.83) and (6.87) is replaced by $[F; w_k, s_k]$. Thus, we obtain [23]

Theorem 6.6 *Assume that all the assumptions of Theorem 6.1 are fulfilled. Then the iterations* (6.60) *have seventh-order convergence provided that* $\bar{\tau}_k$ *satisfies the condition* (6.35) *and* α_k *is given by* (6.111) *or* (6.112).

Another way to obtain seventh-order iterations (6.60) is direct application of the sufficient convergence condition (6.100) and $[F; y_k, z_k]$ given by

$$[F; y_k, z_k] = F'(y_k)$$
$$\times \left(I - \frac{1}{2} \left(F'(x_k) \right)^{-1} F''(x_k) \Theta_k (F'(x_k))^{-1} F(x_k) \right) + O\left(h^3 \right), \quad (6.114)$$

for which we have used (6.4), (6.35), (6.105). Substituting (6.98) into (6.114) and using (6.105) we get

$$\tilde{A}_k = [F; w_k, s_k]^{-1}[F; z_k, y_k] = I - 2\Theta_k + C_k - \frac{1}{2}D_k + O\left(h^3\right), \quad (6.115)$$

where C_k and D_k are determined by (6.102), (6.101). From (6.115) it follows that

$$\tilde{A}_k^{-1} = [F; z_k, y_k]^{-1}[F; w_k, s_k] = I + 2\Theta_k - C_k + \frac{1}{2}D_k + 4\Theta_k^2 + O\left(h^3\right). \quad (6.116)$$

From (6.100) and (6.116) it follows that

$$\tilde{A}_k^{-1} = \alpha_k - \frac{1}{2}D_k + O\left(h^3\right).$$

Using formula (6.108) in last expression we have

$$\tilde{A}_k^{-1} = \alpha_k + \left(F'(x_k)\right)^{-1}\left([F; z_k, x_k] - [F; y_k, x_k]\right) + O\left(h^3\right).$$

We can replace $\left(F'(x_k)\right)^{-1}$ by $[F; w_k, s_k]^{-1}$ due to (6.108) and hence from last expression we find

$$\begin{aligned}
\alpha_k &= [F; z_k, y_k]^{-1}[F; w_k, s_k] \\
&\quad -[F; w_k, s_k]^{-1}\left([F; z_k, x_k] - [F; y_k, x_k]\right) + O\left(h^3\right). \quad (6.117)
\end{aligned}$$

Since $\tilde{A}_k^2 = I - 4\Theta_k + 4\Theta_k^2 + 2C_k - D_k + O\left(h^3\right)$ then $\tilde{A}_k^{-1}$ can be seeking for

$$\tilde{A}_k^{-1} = pI + q\tilde{A}_k + s\tilde{A}_k^2 + O\left(h^3\right). \quad (6.118)$$

Substituting $\tilde{A}_k^{-1}$, $\tilde{A}_k$ and $\tilde{A}_k^2$ into last equality we have

$$\begin{aligned}
I + 2\Theta_k - C_k + \frac{1}{2}D_k + 4\Theta_k^2 + O\left(h^3\right) &= (p + q + s)I - 2(q + 2s)\Theta_k \\
+4s\Theta_k^2 + (q + 2s)C_k - \frac{1}{2}(q + 2s)D_k &+ O\left(h^3\right).
\end{aligned}$$

From this we obtain

$$q + 2s = -1, \quad s = 1, \quad p + q + s = 1 \Rightarrow p = 3, \quad q = -3, \quad s = 1.$$

Thus, substituting $\tilde{A}_k^{-1} = 3I - 3\tilde{A}_k + \tilde{A}_k^2$ into (6.117) we obtain

$$\alpha_k = 3I - 3\tilde{A}_k + \tilde{A}_k^2 - [F; w_k, s_k]^{-1}\left([F; z_k, x_k] - [F; y_k, x_k]\right) + O\left(h^3\right). \quad (6.119)$$

The obtained result can be formulated as:

Theorem 6.7 *Assume that all the assumptions of Theorem 6.1 are fulfilled. Then the iteration (6.60) has a seventh-order of convergence when $\bar{\tau}_k$ satisfies the condition (6.46) and α_k is given by (6.119).*

To determine α_k given by (6.119) needed only one inverse matrix, whereas in (6.117) needed two inverse matrices. Similar formula to (6.119) was obtained by Wang et al. [20]. Assume that (6.103) holds. Then (6.100) and (6.115) leads to

$$\alpha_k = I + 2\Theta_k + 6\Theta_k^2 - C_k + O\left(h^3\right), \tag{6.120}$$

and

$$\tilde{A}_k = I - 2\Theta_k - \Theta_k^2 + C_k + O\left(h^3\right), \tag{6.121}$$

respectively, In this case α_k given by (6.120) can be expressed as

$$\alpha_k = p_1 I + q_1 \tilde{A}_k + s_1 \tilde{A}_k^2 + O\left(h^3\right). \tag{6.122}$$

Substituting (6.121) into (6.122) we obtain

$$I + 2\Theta_k + 6\Theta_k^2 - C_k = (p_1 + q_1 + s_1)I - 2\Theta_k(q_1 + 2s_1)$$
$$+(-q_1 + 2s_1)\Theta_k^2 + (q_1 + 2s_1)C_k + O\left(h^3\right),$$

which holds when

$$p_1 + q_1 + s_1 = 1, \; q_1 + 2s_1 = -1, \; -q_1 + 2s_1 = 6 \Rightarrow s_1 = \frac{5}{4}, \; q_1 = -\frac{7}{2}, \; p_1 = \frac{13}{4}.$$

Thus, we get

$$\alpha_k = \frac{13}{4}I - \frac{7}{2}\tilde{A}_k + \frac{5}{4}\tilde{A}_k^2 + O\left(h^3\right). \tag{6.123}$$

The obtained result can be formulated as:

Theorem 6.8 *Assume that all the assumptions of Theorem 6.1 are fulfilled. Then the iteration (6.60) has a seventh-order of convergence when $\bar{\tau}_k$ satisfies the condition (6.46) and α_k is given by (6.123).*

Note that the easy choices of (6.46) are (6.45) and (6.48). Whang et al. in [20] were obtained seventh-order derivative free iteration

$$y_k = x_k - B^{-1}F(x_k),$$
$$z_k = y_k - \left(3I - 2[F; w_k, s_k]^{-1}[F; y_k, x_k]\right)B^{-1}F(y_k), \tag{6.124}$$
$$x_{k+1} = z_k - \alpha_k B^{-1}F(z_k),$$

where $B = [F; w_k, s_k]$ and α_k given by (6.123). Thus, our iteration (6.60) with choices $\bar{\tau}_k$ and α_k given by (6.46) and (6.123), respectively includes the iteration (6.124) as special cases.

Now we consider another first-order divided difference

$$Q = [F; u_k, v_k], \quad u_k = z_k + bF(z_k), \quad v_k = z_k - bF(z_k). \tag{6.125}$$

It is easy to show that

$$Q = F'(z_k) + O\left(h^3\right), \tag{6.126}$$

and

$$B^{-1}Q = B^{-1}F'(z_k) = I - 2\Theta_k + C_k - D_k + O\left(h^3\right). \tag{6.127}$$

Similarly, α_k can be expressed by linear combination of the form

$$\alpha_k = p_2 I + q_2 B^{-1}Q + r_2 \left(B^{-1}Q\right)^2 + s_2 \left(B^{-1}Q\right)^3 + O\left(h^3\right). \tag{6.128}$$

Substituting (6.100) and (6.127) into (6.128) we obtain

$$\begin{aligned}
I + 2\Theta_k + 4\Theta_k^2 - C_k + D_k &= p_2 I + q_2(I - 2\Theta_k + C_k - D_k) \\
&\quad + r_2(I - 4\Theta_k + 4\Theta_k^2 + 2(C_k - D_k)) \\
&\quad + s_2(I - 6\Theta_k + 12\Theta_k^2 + 3(C_k - D_k)) + O\left(h^3\right),
\end{aligned}$$

which holds when

$$\begin{aligned}
&p_2 + q_2 + r_2 + s_2 = 1, \quad q_2 + 2r_2 + 3s_2 = -1, \quad r_2 + 3s_2 = 1 \Rightarrow \\
&q_2 = -3 + 3s_2, \quad r_2 = 1 - 3s_2, \quad p_2 = 3 - s_2.
\end{aligned}$$

Hence

$$\begin{aligned}
\alpha_k &= (3 - s_2)I - 3(1 - s_2)B^{-1}Q + (1 - 3s_2)\left(B^{-1}Q\right)^2 \\
&\quad + s_2 \left(B^{-1}Q\right)^3 + O\left(h^3\right). \tag{6.129}
\end{aligned}$$

Thus, we obtain the family of seventh-order derivative-free iteration (6.60) with α_k given by (6.129). The result can be formulated as:

Theorem 6.9 *Assume that all the assumptions of Theorem 6.1 are fulfilled. Then the iteration (6.60) has a seventh-order of convergence when parameter matrix $\bar{\tau}_k$ satisfies the conditions (6.35) and α_k is given by (6.129).*

Thus, we obtain the family of seventh-order derivative-free iterations (6.60) with α_k given by (6.129). In particular, when $s_2 = 0$ the formula (6.129) leads to

Table 6.1 Choices of parameters

Methods	$\bar{\tau}_k$	α_k	$\bar{\tau}_k$	α_k	$\bar{\tau}_k$	α_k
(6.51)	(6.27)	(6.27)	(6.58)	(6.27)	(6.27)	(6.83), (6.87), (6.129), (6.130)
(6.60)	(6.46)	(6.46)	(6.46)	(6.46)	(6.46)	(6.100), (6.111), (6.112), (6.119), (6.123), (6.129)
p	6				7	

$$\alpha_k = 3I - 3B^{-1}Q + (B^{-1}Q)^2. \tag{6.130}$$

When $s_2 = -5/4$ the formula (6.129) leads to

$$\alpha_k = \frac{17}{4}I - \frac{27}{4}B^{-1}Q + \frac{19}{4}\left(B^{-1}Q\right)^2 - \frac{5}{4}\left(B^{-1}Q\right)^3. \tag{6.131}$$

In this case the third-step of iteration (6.60) coincides with third-step of iteration proposed by Narang et al. in [10]. The scheme proposed by Narang et al. in [10] differ from (6.60) only by second step with $\bar{\tau}_k = I$.

Remark 6.2 For seventh-order iterations (6.60) one can choose $\bar{\tau}_k$ by formula (6.46).

Similar results to Theorem 6.8 and Theorem 6.9 can be obtained for iteration (6.51) using $[F; w_k, x_k]$ instead of $[F; w_k, s_k]$. For example, the Theorem 6.9 can be formulated for iteration (6.51) as:

Theorem 6.10 *Assume that all the assumptions of Theorem 6.1 are fulfilled. Then the iteration (6.51) has a seventh-order of convergence when $\bar{\tau}_k$ satisfies the condition (6.16) and α_k is given by (6.130), where*

$$B^{-1}Q = [F; w_k, x_k]^{-1}[F; v_k, z_k], \quad v_k = z_k + bF(z_k).$$

The proof of Theorem 6.10 is the same as proof of Theorem 6.8 and hence it omitted here. Such theorem was proven by Sharma et al. in [13] by means of symbolic computation. The selected two fourth-order iterations in [13] are obtained from (6.32) with $a = 2$ and $a = 0$, respectively.

Thus, we obtained many iterative methods with convergence order six and seven. For convenience, we present in Table 6.1 the choices of parameter matrices and the convergence order p of proposed three-step methods.

6.1.3 Computational Cost of Proposed Methods and Comparison with Known Methods

We now discuss the computational cost of the considered methods. In Tables 6.2, 6.3 and 6.4, we denote by NLS1 the number of linear system with some coefficient

Table 6.2 Values of C for sixth-order methods

Method	$\bar{\tau}_k, \alpha_k$	NFE	NLS1	NLS2	$M \times V$	C
SA6 [15]		$2n^2 + n$	5	–	2	$\frac{1}{3}n^3 + 11n^2 + \frac{2}{3}n$
Grau-Sánchez [8]		$2n^2 + 3n$	1	2	–	$\frac{2}{3}n^3 + 7n^2 + \frac{7}{3}n$
(6.60)	(6.45), (6.45)	$3n^2 + 3n$	5	–	2	$\frac{1}{3}n^3 + 11n^2 + \frac{8}{3}n$
	(6.45), (6.48)	$3n^2 + 3n$	5	–	2	$\frac{1}{3}n^3 + 13n^2 + \frac{8}{3}n$
	(6.48), (6.45)	$3n^2 + 3n$	5	–	2	$\frac{1}{3}n^3 + 13n^2 + \frac{8}{3}n$
	(6.48), (6.48)	$2n^2 + 4n$	5	–	2	$\frac{1}{3}n^3 + 11n^2 + \frac{11}{3}n$

Table 6.3 Values of C for sixth-order methods

method	$\bar{\tau}_k, \alpha_k$	NFE	NLS1	NLS2	$M \times V$	C
M61 [16]		$2n^2 + 3n$	6	–	4	$\frac{1}{3}n^3 + 14n^2 + \frac{20}{3}n$
M62 [16]		$2n^2 + 3n$	5	–	2	$\frac{1}{3}n^3 + 11n^2 + \frac{8}{3}n$
(6.51)	(6.22), (6.22)	$3n^2 + n$	5	–	2	$\frac{1}{3}n^3 + 13n^2 + \frac{2}{3}n$
	(6.22), (6.25)	$4n^2 + n$	5	–	2	$\frac{1}{3}n^3 + 15n^2 + \frac{2}{3}n$
	(6.25), (6.22)	$4n^2 + n$	5	–	2	$\frac{1}{3}n^3 + 15n^2 + \frac{2}{3}n$
	(6.25), (6.25)	$2n^2 + 3n$	5	–	2	$\frac{1}{3}n^3 + 11n^2 + \frac{8}{3}n$

Table 6.4 Values of C for seventh-order methods

Method	$\bar{\tau}_k, \alpha_k$	NFE	NLS1	NLS2	$M \times V$	C
S7 [20]		$3n^2$	6	–	3	$\frac{1}{3}n^3 + 14n^2 - \frac{1}{3}n$
NM7 [10]		$2n^2 + 3n$	5	–	3	$\frac{1}{3}n^3 + 11n^2 + \frac{2}{3}n$
Whang et al. [19]		$5n^2 - n$	4	1	1	$\frac{2}{3}n^3 + 16n^2 - \frac{5}{3}n$
Sharma et al. [12]		$5n^2 + n$	4	1	2	$\frac{2}{3}n^3 + 17n^2 + \frac{1}{3}n$
(6.51)	(6.22), (6.83)	$5n^2 + n$	4	1	1	$\frac{2}{3}n^3 + 16n^2 + \frac{1}{3}n$
(6.51)	(6.22), (6.130)	$5n^2 + n$	4	1	1	$\frac{2}{3}n^3 + 16n^2 + \frac{1}{3}n$
(6.60)	(6.45), (6.111)	$4n^2 + n$	4	1	1	$\frac{2}{3}n^3 + 14n^2 + \frac{1}{3}n$
	(6.45), (6.112)	$4n^2 + n$	4	1	4	$\frac{2}{3}n^3 + 15n^2 + \frac{1}{3}n$
	(6.45), (6.119)	$4n^2 + n$	4	1	4	$\frac{1}{3}n^3 + 18n^2 + \frac{2}{3}n$
	(6.48), (6.119)	$5n^2$	6	–	3	$\frac{1}{3}n^3 + 16n^2 - \frac{1}{3}n$
	(6.45), (6.130)	$2n^2 + 5n$	6	–	3	$\frac{1}{3}n^3 + 13n^2 + \frac{17}{3}n$
	(6.45), (6.123)	$3n^2 + 2n$	6	–	3	$\frac{1}{3}n^3 + 15n^2 + \frac{5}{3}n$

matrices $[F; w_k, x_k]$ and $[F; w_k, s_k]$ for methods (6.51) and (6.60), respectively, and by NLS2 the number of linear system with other coefficient matrices. NFE– the number of function evaluations, $M \times V$–the matrix by vector multiplications, C–total computation cost per iterations. From these Tables we see that the most effective and cheapest methods of order six are SA6, M61, M62 and (6.51) and (6.60) for which the total cost is $n^3/3$, and S7, NM7 and (6.60) among the considered methods of order seven.

References

1. J.M. Ortega, W.C. Rheinboldt, *Iterative Solution of Nonlinear Equations in Several Variables* (Academic Press, New-York, 1970)
2. F.A. Potra, *Nondiscrete Induction and Iterative Processes* (Pitman, London, 1984)
3. J.F. Traub, *Iterative Methods for the Solution of Equations* (Englewood Cliffs, NJ, Prentice-Hall, 1964)
4. F. Ahmad, F. Soleymani, F.K. Haghani, S. Serra-Capizzano, Higher order derivative-free iterative methods with and without memory for systems of nonlinear equations. Appl. Math. Comput. **314**, 199–211 (2017)
5. A. Amiri, A. Cordero, M.T. Darvishi, J.R. Torregrosa, A fast algorithm to solve systems of nonlinear equations. J. Comput. Appl. Math. **354**, 242–258 (2019)
6. J.R. Sharma, H. Arora, M.S. Petković, An efficient derivative free family of fourth order methods for solving systems of nonlinear equations. Appl. Math. Comput. **235**, 383–393 (2014)
7. S. Bhalla, S. Kumar, I.K. Argyros, R. Behl, A family of higher order derivative free methods for nonlinear systems with local convergence analysis. Comput. Appl. Math. **37**, 5807–5828 (2018)
8. M. Grau-Sánchez, M. Noguera, S. Amat, On the approximation of derivatives using divided difference operators preserving the local convergence order of iterative methods. J. Comput. Appl. Math. **237**, 363–372 (2013)
9. Z. Liu, Q. Zheng, P. Zhao, A variant of Steffensen's method of fourth-order convergence and its applications. Appl. Math. Comput. **216**, 1978–1983 (2010)
10. M. Narang, S. Bhatia, V. Kanwar, New efficient derivative free family of seventh-order methods for solving systems of nonlinear equations. Numer. Algorithms **76**, 283–307 (2017)
11. M.S. Petković, J.R. Sharma, On some efficient derivative-free iterative methods with memory for solving systems of nonlinear equations. Numer. Algorithms **71**, 457–474 (2016)
12. J.R. Sharma, H. Arora, A novel derivative free algorithm with seventh-order convergence for solving systems of nonlinear equations. Numer. Algorithms **67**, 917–933 (2014)
13. J.R. Sharma, H. Arora, A simple yet efficient derivative free family of seventh order methods for systems of nonlinear equations. SeMA J. **73**, 59–75 (2016)
14. J.R. Sharma, H. Arora, An efficient derivative free iterative method for solving systems of nonlinear equations. Appl. Anal. Discret. Math. **7**, 390–403 (2013)
15. J.R. Sharma, H. Arora, Efficient derivative-free numerical methods for solving systems of nonlinear equations. Comput. Appl. Math. **35**, 269–284 (2016)
16. J.R. Sharma, H. Arora, Efficient higher order derivative-free multipoint methods with and without memory for systems of nonlinear equations. Int. J. Comput. Math. **95**, 920–938 (2018)
17. X. Wang, X. Fan, Two efficient derivative-free iterative methods for solving nonlinear systems. Algorithms **9**, 14 (2016)
18. M. Narang, S. Bhatia, A.S. Alshomrani, V. Kanwar, General efficient class of Steffensen type methods with memory for solving systems of nonlinear equations. J. Comput. Appl. Math. **352**, 23–39 (2019)
19. X. Wang, T. Zhang, A family of Steffensen type methods with seventh-order convergence. Numer. Algorithms **62**, 429–464 (2013)
20. X. Wang, T. Zhang, W. Qian, M. Teng, Seventh-order derivative-free iterative method for solving nonlinear systems. Numer. Algorithms **70**, 545–558 (2015)
21. Z. Tugal, O. Khuder, M. Renchin-Ochir, S. Lkhagvadash, Optimal choice of parameters in higher-order derivative-free iterative methods for systems of nonlinear equations, in *Optimization, Simulation and Control, ICOSC 2022*, ed. by R. Enkhbat, A. Chinchuluun, P.M. Pardalos. Springer Proceedings in Mathematics and Statistics, vol. 434 (2023), pp. 165–185
22. T. Zhanlav, Ch. Chun, Kh. Otgondorj, V. Ulziibayar, High-order iterations for systems of nonlinear equations. Int. J. Comput. Math. **97**, 1704–1724 (2020)
23. T. Zhanlav, Kh. Otgondorj, L. Saruul, R. Mijiddorj, A unified approach to the construction of higher-order derivative-free iterative methods for solving systems of nonlinear equations. Proc Mongolian Acad Scie (accepted)

Part III
Applications

Chapter 7
Newton-Type Iterations for Solving Some Problems in Linear Algebra

Abstract A continuous analogue of Newton's methods for solving a system of linear algebraic equations and an algebraic eigenvalue problem are presented. The convergence of iteration processes with optimal iteration parameters is proved. The considered algorithms are also applied to a generalized eigenvalue problem with real- or complex-valued symmetric matrices and to a nonlinear eigenvalue problem.

7.1 CANM for Solving a System of Linear Algebraic Equations

We propose a continuous analogue of Newton's method with inner iteration for solving a system of linear algebraic equations. Implementation of inner iterations is carried out in two ways. The former is to fix the number of inner iterations in advance. The latter is to use the inexact Newton method for solution of the linear system of equations that arises at each stage of outer iterations. We give some new choices of iteration parameter and of forcing term, that ensure the convergence of iterations. The performance and efficiency of the proposed iteration is illustrated by numerous examples that represent a wide range of typical systems.

We consider a system of linear algebraic equations

$$A x = f. \tag{7.1}$$

For a numerical solving of Eq. (7.1) we consider an iterative process:

$$B \frac{x^{(k+1)} - x^{(k)}}{\tau_k} = -r_k, \quad r_k = A x^{(k)} - f, \quad k = 0, 1, \dots,$$

which consists of two steps:

$$B v_k = -r_k,$$
$$x^{(k+1)} = x^{(k)} + \tau_k v_k, \quad k = 0, 1, \dots. \tag{7.2}$$

T. Zhanlav and O. Chuluunbaatar, *New Developments of Newton-Type Iterations for Solving Nonlinear Problems*, Mathematical Engineering, https://doi.org/10.1007/978-3-031-63361-4_7

Of course, the quality of iterative process Eq. (7.2) essentially depends on the choices of matrix B and of iteration parameter $\tau_k \neq 0$.

Let H be a linear space of n-dimensional vectors. We will denote the Euclidean vector norm in H by $||\cdot||$, as well as the corresponding norm of matrices.

Theorem 7.1 *Let $(Av_k, r_k) \neq 0$. Then a necessary and sufficient condition for the $||r_k||$ to be decreasing $(||r_{k+1}|| < ||r_k||)$ is that*

$$\tau_k \in I_k = \begin{cases} \left(0, -\frac{2(Av_k, r_k)}{||Av_k||^2}\right), & \text{when} \quad (Av_k, r_k) < 0, \\ \left(-\frac{2(Av_k, r_k)}{||Av_k||^2}, 0\right), & \text{when} \quad (Av_k, r_k) > 0. \end{cases}$$

Proof From Eq. (7.2) we obtain

$$r_{k+1} = r_k + \tau_k Av_k.$$

Hence we have

$$||r_{k+1}||^2 = ||r_k||^2 + \tau_k[\tau_k||Av_k||^2 + 2(Av_k, r_k)]. \tag{7.3}$$

The assertion follows from (7.3). $\qquad\qquad\square$

The interval I_k we call τ-region of convergence of iteration method (7.2). Thus we have to choose τ_k from this region. Moreover, it is desirable that the τ_k to be optimal in some sense. Further we will use well-known assertions [1] to study the convergence of (7.2).

Theorem 7.2 *Let S be an $n \times n$ matrix. Then the successive approximations*

$$x^{(k+1)} = Sx^{(k)} + z, \quad k = 0, 1, 2, \ldots \tag{7.4}$$

converge for each $z \in R^n$ and each $x^{(0)} \in R^n$ if and only if

$$\rho(S) < 1,$$

where $\rho(S)$ is a spectral radius of the matrix S.

It is easy to show that the iteration process (7.2) can be rewritten as (7.4) with iteration matrix

$$S = I - \tau_k B^{-1} A \quad \text{and} \quad z = \tau_k B^{-1} f.$$

Here I is an $n \times n$ identity matrix.

Theorem 7.3 *The iteration process (7.2) with parameter τ_k given by*

$$\tau_k^* = -\frac{(Av_k, r_k)}{||Av_k||^2} \tag{7.5}$$

converges to x^* for any $x^{(0)} \in H$. The following a posteriori estimate holds true [2]:

$$\|r_k\| = q_{k-1}q_{k-2}\cdots q_0\|r_0\|,$$

where

$$q_i = \sqrt{1 - \frac{(Av_i, r_i)^2}{\|Av_i\|^2\|r_i\|^2}} < 1, \quad i = 0, 1, \ldots, k-1.$$

We call the nonzero value of τ_k^* defined by (7.5) the optimal one in the sense that it yields the minimum value of functional $\|r_{k+1}\|$.

7.2 CANM for Solving Eigenvalues and Eigenvectors of Matrices

Finding the eigenvalues and eigenvectors of a given matrix is one of the complex problems of linear algebra. Currently, there are many direct and iterative methods for solving this problem [3–7]. Among them, an important role is played by methods of reverse iteration with a shift, and combined methods (using reverse iterations and the Rayleigh relation) that have a high convergence rate. However, they require good enough initial approximations to calculate the desired eigenvalue and eigenvector. Therefore, they are usually used in cases where it is necessary to clarify the eigenvalue obtained by some other methods and the corresponding eigenvector.

We consider the possibility of using a continuous analogue of the Newton method (CANM) [8–10] to find an isolated and simple eigenvalue and the corresponding eigenvector of a real square matrix.

Let $\mathbf{R}^n$ be the space of n-dimensional real vectors. We introduce the scalar product of vectors and the norm formulas

$$(u, v) = \sum_{i=1}^{n} u_i v_i, \qquad \|u\| = \sqrt{(u, u)}.$$

Consider the eigenvalue problem in $\mathbf{R}^n$:

$$(A - \lambda I)x = 0, \tag{7.6}$$

where A is a real matrix of dimension $n \times n$. Because the eigenvector corresponding eigenvalue λ is accurate to a constant factor, we introduce the normalization condition

$$(x, x) = 1. \tag{7.7}$$

Our goal is to find an isolated and simple eigenvalue and the corresponding eigenvector (λ, x) of the real matrix A. As we known [8], the problem (7.6), (7.7) can be considered as a nonlinear problem respect to unknown (λ, x) and for this we apply CANM:

$$(A - \lambda I)\frac{dx}{dt} - x\frac{d\lambda}{dt} = -(A - \lambda I)x, \tag{7.8}$$

$$2\left(x, \frac{dx}{dt}\right) = 1 - (x, x). \tag{7.9}$$

Discretization of the Eqs. (7.8) and (7.9) is carried out by the Euler method:

$$\frac{dx}{dt}\bigg|_{t_k} \approx \frac{x_{k+1} - x_k}{\tau_k} = v_k, \quad \frac{d\lambda}{dt}\bigg|_{t_k} \approx \frac{\lambda_{k+1} - \lambda_k}{\tau_k} = \mu_k. \tag{7.10}$$

As a result, we have equations with unknowns v_k and μ_k

$$(A - \lambda_k I)v_k - \mu_k x_k = -r_k, \tag{7.11}$$

$$2(v_k, x_k) = 1 - (x_k, x_k), \tag{7.12}$$

where r_k is the residual of the Eq. (7.6), i.e.

$$r_k = (A - \lambda_k I)x_k. \tag{7.13}$$

If we search for v_k in the form

$$v_k = -x_k + \mu_k \theta_k, \tag{7.14}$$

then substituting (7.14) into (7.11) and (7.12) gives us

$$(A - \lambda_k I)\theta_k = x_k, \tag{7.15}$$

$$2\mu_k(\theta_k, x_k) = 1 + (x_k, x_k). \tag{7.16}$$

Since the right side of the Eq. (7.16) is greater than unity, then $(\theta_k, x_k) \neq 0$ and so from (7.16) we have

$$\mu_k = \frac{1 + (x_k, x_k)}{2(\theta_k, x_k)}. \tag{7.17}$$

The system of linear algebraic Eq. (7.15) is solved with regard to θ_k by one of the standard methods. After computing θ_k, the value μ_k is determined by the formula (7.17). According to (7.10), the following approximations x_{k+1}, λ_{k+1} are calculated using the formulas

$$\begin{aligned} x_{k+1} &= x_k + \tau_k v_k = (1 - \tau_k)x_k + \tau_k \mu_k \theta_k, \\ \lambda_{k+1} &= \lambda_k + \tau_k \mu_k, \quad k = 0, 1, \ldots, \end{aligned} \tag{7.18}$$

where k is the iteration number, $\tau_k > 0$ the iteration parameter, and (λ_0, x_0) the initial approximation from the neighborhood of the desired solution. As it can be seen from the formulas (7.18) for $\tau_k = 1$ the proposed algorithm coincides with the usual Newton method for problems (7.6), (7.7). Moreover, it requires, like the reverse iterations with Rayleigh relation, good enough initial approximations. Generally, as the convergence of an iterative process (7.18), and the rate of convergence depends on choice iteration parameter $\tau_k > 0$.

7.2.1 *Convergence of Method and Choice of Iteration Parameter*

We study the convergence of the proposed iterative process (7.18). Using (7.11), (7.12), and (7.13), it is easy to show that

$$r_{k+1} = (1 - \tau_k)r_k - \tau_k^2 \mu_k v_k. \tag{7.19}$$

Therefore, we have

$$\|r_{k+1}\|^2 = (1 - \tau_k)^2 \|r_k\|^2 - 2(1 - \tau_k)\tau_k^2 \mu_k (r_k, v_k) + \tau_k^4 \mu_k^2 \|v_k\|^2. \tag{7.20}$$

Then, the following theorem holds.

Theorem 7.4 *A necessary and sufficient condition for a decrease of the residual in norm from iteration to iteration is the fulfillment of the inequality*

$$\varphi(\tau_k) = (\tau_k - 2)\|r_k\|^2 - 2(1 - \tau_k)\tau_k \mu_k (r_k, v_k) + \tau_k^3 \mu_k^2 \|v_k\|^2 < 0. \tag{7.21a}$$

Proof The proof of the theorem follows from the relation

$$\|r_{k+1}\|^2 = \|r_k\|^2 + \tau_k \varphi(\tau_k), \tag{7.21b}$$

which follows from (7.20). $\square$

From this, it is clear that the iterative parameter τ_k must be chosen so that the condition (7.21a) is satisfied. However, it is difficult to find such values of the iterative parameter τ_k for which the inequality (7.21a) holds.

On the other hand, the right-hand side of the relation (7.20) is a positive quadratic function of $\alpha_k = \mu_k \tau_k^2$ and it reaches the minimum values

$$\min_{\alpha_k} \|r_{k+1}\|^2 = \|r_{k+1}(\alpha_k^*)\|^2 = (1 - \tau_k)^2 \left(1 - \frac{(r_k, v_k)^2}{\|r_k\|^2 \|v_k\|^2}\right) \|r_k\|^2 \tag{7.22}$$

for $\alpha_k = \alpha_k^*$, where

$$\alpha_k^* = \frac{(1 - \tau_k)(r_k, v_k)}{\|v_k\|^2}.$$ (7.23)

It is easy to see that (7.23) is equivalent to the equation

$$D(\tau_k) \equiv \mu_k \tau_k^2 \|v_k\|^2 - (1 - \tau_k)(r_k, v_k) = 0,$$ (7.24)

whose roots are calculated using the formula

$$\tilde{\tau}_k = \frac{-(r_k, v_k) \pm \sqrt{\Delta_k}}{2\mu_k \|v_k\|^2},$$ (7.25)

where

$$\Delta_k = (r_k, v_k)^2 + 4\mu_k (r_k, v_k)\|v_k\|^2.$$

We state the main result as the following theorem.

Theorem 7.5 *Assume that the iterative parameter τ_k is selected according to the formula:*

$$\tau_k = \begin{cases} \max\{0.1; \tilde{\tau}_k\} \quad \text{or} \quad \min\{\tilde{\tau}_k; 1.9\}, & \text{if} \quad \Delta_k \geq 0, \\ 1, & \text{if} \quad \mu_k^2 \|v_k\|^2 < \|r_k\|^2 \quad \text{and} \quad \Delta_k < 0, \\ \overline{\tau}_k, & \text{if} \quad \mu_k^2 \|v_k\|^2 \geq \|r_k\|^2 \quad \text{and} \quad \Delta_k < 0. \end{cases}$$ (7.26)

Here $\tilde{\tau}_k$ is the root of Eq. (7.24) lying in the interval $(0, 2)$, and $\overline{\tau}_k$ – the root of equation $g(\tau_k) = 0$ lying in the interval $(0, 1)$, where

$$g(\tau_k) = (\tau_k - 2)\|r_k\|^2 - 2(1 - \tau_k)\tau_k \mu_k (r_k, v_k) + \tau_k^2 \mu_k^2 \|v_k\|^2.$$ (7.27)

Then, the next equality is valid

$$\|r_{k+1}\| = q_k \|r_k\|, \quad q_k < 1, \quad k = 0, 1, \dots.$$ (7.28)

Moreover with $\Delta_k > 0$, we have

$$q_k = |1 - \tau_k| \sqrt{1 - \frac{(r_k, v_k)^2}{\|r_k\|^2 \|v_k\|^2}}.$$ (7.29)

Proof Let $\Delta_k > 0$. Then one root of Eq. (7.24) lies in the interval $(0, 2)$, since $D(0)D(2) = -\Delta_k < 0$. More precisely, this root belongs to the interval $(0, 1)$ for $\mu_k(r_k, v_k) > 0$ and belongs to the interval $(1, 2)$ for $\mu_k(r_k, v_k) < 0$. The case $\mu_k(r_k, v_k) = 0$ is excluded because $\tau_k^* = 0$. Then, (7.22) follows from (7.28). Let $\Delta_k < 0$. In this case, real roots of Eq. (7.24) do not exist and cannot find the minimum value of $\|r_{k+1}\|$ in the auxiliary parameter α_k. However, we can found τ_k for which the inequality (7.21a) is satisfied. In fact, we use the fact that

$$\varphi(0) = -2\|r_k\|^2 < 0,$$
$$\varphi(1) = -\|r_k\|^2 + \mu_k^2\|v_k\|^2. \tag{7.30}$$

If $\mu_k^2\|v_k\|^2 < \|r_k\|^2$, then $\varphi(1) < 0$ and, therefore, we can choose $\tau_k = 1$. Then, the inequality $\|r_{k+1}\| < \|r_k\|$ follows from (7.21b). This means that condition (7.28) is satisfied. If $\mu_k^2\|v_k\|^2 \geq \|r_k\|^2$, then $\varphi(1) \geq 0$ and $g(1) \geq 0$. Then, from (7.27) and (7.30) we have the relations $\varphi(0)\varphi(1) < 0$ and $g(0)g(1) < 0$, and the inequality holds

$$\varphi(\tau_k) < g(\tau_k)$$

on the interval $(0, 1)$. This means that if we select $g(\bar{\tau}_k) = 0$, then $\varphi(\bar{\tau}_k) < 0$. In this case, the equality (7.28) with the constant $q_k < 1$ follows from the equality (7.21b). Thus, the choice by formula (7.26) provides a monotone decreasing of the residual in norm.

Let $\Delta_k = 0$, which is equivalent to $\tau_k = 0$ or $\tau_k = 2$. In this case, the convergence rate of the iteration slows down. The way out of this situation is to restrict the lower and upper bounds in choosing the value τ_k, for example, $0.1 \leq \tau_k \leq 1.9$. The theorem is completely proved. $\qquad\square$

Remark 7.1 In the case of convergence of iterative process (7.18) there exists a number N such that for all $k \geq N$ the quantity $\varepsilon_k = \mu_k\|v_k\|^2$ becomes arbitrary small and thus, we have

$$\sqrt{\Delta_k} = |(r_k, v_k)|\sqrt{1 + \frac{4\varepsilon_k}{(r_k, v_k)}} \approx |(r_k, v_k)|\left(1 + \frac{2\varepsilon_k}{(r_k, v_k)}\right).$$

Then, by virtue of formula (7.25), we have the following asymptotical behavior

$$\tilde{\tau}_k = \frac{-(r_k, v_k) \pm \sqrt{\Delta_k}}{2\mu_k\|v_k\|^2} \to 1 \quad \text{when} \quad k \to \infty$$

On the other hand, as noted above, when $\tau_k \to 1$ the iterative process (7.18) turns into the Newton method with a quadratic convergence rate. Therefore, we can expect that the iterative process (7.18) converges quadratically in a sufficiently small neighborhood of the desired solution (λ, x).

Remark 7.2 The estimate (7.28) is almost a posteriori, for $\Delta_k > 0$, because we can accurately calculate the value q_k using the formula (7.29).

For the selection of the iterative parameter τ_k the next remark is important.

Remark 7.3 Let us denote $f(\tau_k) = \|r_{k+1}\|^2$ and examine its derivative

$$f'(\tau_k) = 2(\tau_k - 1)\|r_k\|^2 - 2(2\tau_k - 3\tau_k^2)\mu_k(r_k, v_k) + 4\tau_k^3\mu_k^2\|v_k\|^2. \tag{7.31}$$

From (7.31), it is evident that $f'(0) = -2\|r_k\|^2 < 0$ and $f'(2) = 2f(2) > 0$. Hence, it follows that the function $f(\tau_k)$ has at least one minimum in interval $(0,2)$. If it has three real roots, then the $f(\tau_k)$ has minimums at the smallest and largest roots and should choose the root that give the smallest value. Thus, it is possible to find the best optimal value of the iterative parameter $\tau_k = \tau_k^*$ by calculating the roots of the cubic equation $f'(\tau_k^*) = 0$. It is shown the method of optimal selection of the iterative parameter τ_k^* speeds up the convergence.

We consider another option for choosing the parameter τ_k. To do this we rewrite the relation (7.20) in the form

$$\|r_{k+1}\|^2 = (1 - \tau_k)^2 \|r_k\|^2 + \tau_k^2 \psi(\tau_k), \tag{7.32}$$

where

$$\psi(\tau_k) = \mu_k^2 \|v_k\|^2 \tau_k^2 - 2\mu_k(1 - \tau_k)(r_k, v_k).$$

Suppose

$$\widetilde{\Delta}_k = (r_k, v_k)^2 + 2\mu_k \|v_k\|^2 (r_k, v_k) > 0.$$

Then, the quadratic function $\psi(\tau_k)$ with respect to τ_k has two real roots τ_k^-, τ_k^+ $(\tau_k^- < \tau_k^+)$. It is obvious that

$$\psi(\tau_k) < 0, \quad \text{if} \quad \tau_k^- < \tau_k < \tau_k^+.$$

The choice of the iterative parameter τ_k depends on the expression $\mu_k(r_k, v_k)$:

(i) Let $\mu_k(r_k, v_k) > 0$. Then, $\tau_k^- < 0, 0 < \tau_k^+ < 1$ and therefore

$$\psi(\tau_k) < 0, \quad \text{if} \quad \tau_k \in (0, \tau_k^+). \tag{7.33}$$

(ii) Let $\mu_k(r_k, v_k) < 0$. Then, $1 < \tau_k^- < 2 < \tau_k^+$ and therefore

$$\psi(\tau_k) < 0, \quad \text{if} \quad \tau_k \in (\tau_k^-, 2). \tag{7.34}$$

The case $\mu_k(r_k, v_k) = 0$ is excluded because $\tau_k^- = \tau_k^+ = 0$ and $\psi(\tau_k^\pm) = 0$. From (7.32), (7.33), and (7.34), it follows that

$$\|r_{k+1}\| < q_k \|r_k\|, \quad q_k = |1 - \tau_k| < 1, \quad k = 0, 1, \ldots.$$

provided $\tau_k \in (0, 2) \bigcap (\tau_k^-, \tau_k^+)$. The iterative parameter τ_k is selected from the condition

$$q_k < q_{k-1}. \tag{7.35}$$

Lemma 7.1 *A sufficient condition for inequality* (7.35) *to hold is the choice given by*

$$\tau_k = \begin{cases} \min \left(\frac{\|r_{k-1}\|}{\|r_k\|} \tau_{k-1}; 1 \right), & \text{when} \quad (i), \quad (\tau_0 \approx 0.1), \\ \max \left(\frac{\|r_k\|}{\|r_{k-1}\|} \tau_{k-1}; 1 \right), & \text{when} \quad (ii), \quad (\tau_0 \approx 1.9). \end{cases}$$

Proof Consider the case of (i). Then, $\tau_{k-1} \in (0, \tau_{k-1}^{+}) \subset (0, 1)$ and therefore $\|r_k\| < \|r_{k-1}\|$. Let $\tau_k = \|r_{k-1}\|/\|r_k\|\tau_{k-1}$. Then, we have $1 > \tau_k > \tau_{k-1}$, therefore, condition (7.35) will be satisfied. Let $\tau_k \equiv 1$. Then, $q_k = |1 - \tau_k| = 0$, at the same time $q_{k-1} = |1 - \tau_{k-1}| > 0$, and also holds (7.35). In a similar way, we prove (7.35) in the case of (ii). $\qquad\square$

Let $\widetilde{\Delta}_k < 0$. In this case, the function $\psi(\tau_k)$ has no real roots, and therefore, as before, we need to use the inequality (7.21a) in order to choose an iterative parameter that reduces the norm $\|r_k\|$ from iteration to iteration. Since the inequality $\widetilde{\Delta}_k < 0$ implies that $\mu_k(r_k, v_k) < 0$ and $\Delta_k < 0$, then the following theorem holds.

Theorem 7.6 *Let the iterative parameter τ_k be selected as:*

$$
\tau_k = \begin{cases}
\min\left(1; \dfrac{\|r_{k-1}\|}{\|r_k\|}\tau_{k-1}\right), & \text{if } \widetilde{\Delta}_k > 0 \text{ and } \mu_k(r_k, v_k) > 0, \quad \tau_0 \approx 0.1, \\[2mm]
\max\left(1; \dfrac{\|r_k\|}{\|r_{k-1}\|}\tau_{k-1})\right), & \text{if } \widetilde{\Delta}_k > 0 \text{ and } \mu_k(r_k, v_k) < 0, \quad \tau_0 \approx 1.9, \\[2mm]
1, & \text{if } \mu_k^2\|v_k\|^2 < \|r_k\|^2 \text{ and } \widetilde{\Delta}_k < 0, \\[2mm]
\overline{\tau}_k, & \text{if } \mu_k^2\|v_k\|^2 \geq \|r_k\|^2 \text{ and } \widetilde{\Delta}_k < 0,
\end{cases}
$$

where $\overline{\tau}_k$ – the root of the equation $g(\tau_k) = 0$ belonging to the interval $(0, 1)$. Then, the inequality

$$
\|r_{k+1}\| \leq q_k\|r_k\|, \quad q < 1, \quad k = 0, 1, \ldots.
$$

Moreover, with $\widetilde{\Delta}_k > 0$, we have

$$
q_k = |1 - \tau_k|.
$$

The proof of this theorem is similar to the proof of Theorem 7.5.

7.2.2 *Generalized Eigenvalue Problem with Symmetric Matrices*

Consider the generalized eigenvalue problem

$$
\begin{aligned}
(A - \lambda B)x &= 0, \\
(x, Bx) - 1 &= 0.
\end{aligned}
\tag{7.36}
$$

where A is a symmetric real or complex matrix, B is a symmetric positive definite real matrix.

In this case, we have equations with unknowns v_k and μ_k, similarly as (7.11) and (7.12)

$$(A - \lambda_k B)v_k - \mu_k B x_k = -r_k, \tag{7.37}$$

$$2(v_k, B x_k) = 1 - (x_k, B x_k). \tag{7.38}$$

The residual r_k of the Eq. (7.36) and iteration corrections v_k and μ_k of (7.18) are given by formulas

$$r_k = (A - \lambda_k B)\, x_k, \quad v_k = -x_k + \mu_k \theta_k. \tag{7.39}$$

The iteration corrections θ_k and μ_k to the eigenvector x_k and to the eigenvalue λ_k are calculated from the algebraic problem

$$\begin{cases} (A - \lambda_k B)\, \theta_k = B x_k, \\ 2\mu_k\, (\theta_k, B x_k) = 1 + (x_k, B x_k), \end{cases} \tag{7.40}$$

whereas the iteration corrections μ_k is calculated by means of the formula

$$\mu_k = \frac{1 + (x_k, B x_k)}{2\, (\theta_k, B x_k)}. \tag{7.41}$$

Let A, λ, x be complex. Then μ_k, v_k, r_k are complex and the discrepancy is determined by the relation

$$\|r_{k+1}\|^2 = (1 - \tau_k)^2 \|r_k\|^2 - 2(1 - \tau_k)\tau_k^2 \Re(r_k^*, \mu_k B v_k) + \tau_k^4 \|\mu_k B v_k\|^2 \tag{7.42}$$

Here b^* and $\Re b$ denote the complex conjugate and real part of b, respectively.

We denote again $f(\tau_k) = \|r_{k+1}\|^2$ and calculate its derivative

$$f'(\tau_k) = 2(\tau_k - 1)\|r_k\|^2 - 2(2\tau_k - 3\tau_k^2)\mu_k \Re(r_k^*, \mu_k B v_k) + 4\tau_k^3 \mu_k^2 \|B v_k\|^2 \tag{7.43}$$

We also see that $f'(0) = -2\|r_k\|^2 < 0$ and $f'(2) = 2f(2) > 0$. Hence, the function $f(\tau_k)$ has at least one minimum in the interval $(0,2)$.

Also the iteration corrections θ_k and μ_k can be calculated instead of (7.40) from the following algebraic problem [11]:

$$\begin{cases} (A - \lambda_k B)\, \theta_k = B x_k, \\ (x_k, B x_k)\mu_k = (x_k, (A - \lambda_k B)\, x_k), \end{cases} \tag{7.44}$$

whereas the iteration corrections μ_k is calculated using the formula

$$(x_k, B x_k) = 1, \quad \mu_k = \frac{(x_k, (A - \lambda_k B)\, x_k)}{(x_k, B x_k)} = \frac{(x_k, r_k)}{(x_k, B x_k)} = (x_k, r_k). \tag{7.45}$$

The transition from x_k, λ_k at the k-th step to x_{k+1}, λ_{k+1} at the $k + 1$-th step is executed by means of the formulas

$$\begin{cases} \tilde{x}_{k+1} = x_k + \tau_k v_k = (1 - \tau_k)x_k + \tau_k \mu_k \theta_k, \quad x_{k+1} = \dfrac{\tilde{x}_{k+1}}{(\tilde{x}_{k+1}, B\tilde{x}_{k+1})}, \\ \lambda_{k+1} = \lambda_k + \tau_k \mu_k, \end{cases} \quad (7.46)$$

with the iteration step τ_k calculated using the formula [12]

$$\tau_k = \frac{\|r_k\|^2}{\|r_k\|^2 + \|\tilde{r}_{k+1}\|^2} \le 1, \tag{7.47}$$

where

$$\tilde{r}_{k+1} = (A - \lambda_{k+1}B)\,x_{k+1}, \tag{7.48}$$

and x_{k+1}, λ_{k+1} calculated by formula (7.46) at $\tau_k = 1$.

7.2.3 *Nonlinear Eigenvalue Problem*

Consider a nonlinear eigenvalue problem

$$A(\lambda)x = 0, \tag{7.49}$$
$$(x, x) - 1 = 0, \tag{7.50}$$

where A is a matrix nonlinearly dependent on the spectral parameter λ. CANM is also used for problem (7.49) and (7.50), which reduces to the equation

$$A(\lambda_k)\theta_k = -A'(\lambda_k)x_k.$$

Here, $A'(\lambda_k)$ is the derivative of the matrix $A(\lambda)$ with respect to the spectral parameter λ. Suppose that the correction μ_k is calculated by formula (7.17) and the following approximations for x_{k+1}, λ_{k+1} are found by formula (7.18), and the discrepancy is determined by the expressions $\bar{r}_k = A(\lambda_k)x_k$. Then, it is easy to show that

$$\bar{r}_{k+1} = (1 - \tau_k)\bar{r}_k + \tau_k^2 \mu_k A'(\lambda_k)v_k + O\left(\tau_k^2 \mu_k^2\right). \tag{7.51}$$

A comparison of (7.51) with (7.19) shows that the discrepancies r_{k+1}, $\bar{r}_{k+1}$ have the same main part. Therefore, Theorems 7.5 and 7.6 are valid if v_k is replaced by $-A'(\lambda_k)v_k$.

References

1. R. Kress, *Numerical Analysis* (Springer-Verlag, 1998)
2. T. Zhanlav, On the iteration method with minimal defect for solving a system of linear algebraic equations. Sci. Trans. **8**, 59–64 (2001) (National University of Mongolia)
3. N.S. Bakhvalov, N.P. Zhidkov, G.M. Kobelkov, *Numerical Methods* (Nauka, Moscow, 1987). (in Russian)
4. H.H. Kalitkin, *Numerical Methods* (Nauka, Moscow, 1978). (in Russian)
5. B. Parlett, *Symmetric Eigenvalue Problem* (Prentice-Hall, Englewood Cliffs, 1980)
6. V.V. Voevodin, Y.A. Kuznetsov, Matritsy i vychisleniya (Matrices and computations) (Moscow, Nauka, 1984). (in Russian)
7. J.H. Wilkinson, *Algebraic Eigenvalue Problem* (Nauka, Moscow, 1970). (in Russian)
8. I.V. Puzynin, Continuous analog of Newton's method for the calculation of the quantum mechanical problem. Thesis, 11-12016, Dubna, 1978
9. T. Zhanlav, R. Mijiddorj, O. Chuluunbaatar, The continuous analogue of Newton's method for solving eigenvalues and eigenvectors of the matrices. Tver University Vestnik **14**, 27–37 (2008). (in Russian)
10. T. Zhanlav, I.V. Puzynin, The convergence of iteration based on a continuous analogue of Newton's method. Comput. Math. Math. Phys. **32**, 729–737 (1992)
11. A.A. Gusev, L.L. Hai, O. Chuluunbaatar, V. Ulziibayar, S.I. Vinitsky, V.L. Derbov, A. Góźdź, V.A. Rostovtsev, Symbolic-numeric solution of boundary-value problems for the Schrödinger equation using the finite element method: scattering problem and resonance states. Lecture Notes in Computer Science, vol. 9301 (2015), pp. 182–197
12. V.V. Ermakov, N.N. Kalitkin, The optimal step and regularization for Newton's method. USSR Comput. Math. Math. Phys. **21**, 235–242 (1981)

Chapter 8
Applications of Newton-Type Iterations for Computational Physics

Abstract Methods of computational physics developed for investigating models of complex physical processes in different fields of theoretical physics are considered. A general mathematical formulation of equations for the models under study is given, and numerical methods are described. Within the quantum mechanical few-body system, the following problems are considered in the adiabatic representation: calculations of bound and semi-bound state energies and scattering matrices of three bosons on a straight line with zero radius pair interactions, high-precision calculation of the ground state energy of the helium atom, calculations of multiple differential cross sections for the ionization processes of the ground state of the helium atom by fast electrons and protons, calculations of the ground state energies (including metastable state energies) of the beryllium dimer and high-precision calculation of the ground state energy of the Dirac electron in the field of two-center Coulomb field. The results of their numerical study are presented and compared with the results of other authors and some experimental data.

8.1 General Characterization of Problems

In the general case, the class of equations occurring in mathematical models of the complex physical processes under study can be described using systems of nonlinear integro-differential equations of the form [1]

$$\Gamma \frac{\partial^\alpha \mathbf{u}(\mathbf{x}, t)}{\partial t^\alpha} = \left\{ -\Theta[(\nabla_\mathbf{x} I + A(\boldsymbol{\rho}; \mathbf{x}, t))^2 + V(\boldsymbol{\rho}; \mathbf{x}, \mathbf{u}(\mathbf{x}, t))] \right.$$
$$\left. + \Pi \int_\Omega G(\boldsymbol{\rho}; \mathbf{x}, \mathbf{x}', \mathbf{u}(\mathbf{x}', t)) d\mathbf{x}' \right\} \mathbf{u}(\mathbf{x}, t), \qquad (8.1)$$

where t is time of the evolution process, $\mathbf{x} \in \Omega$, Ω is the domain of the coordinate space, $\boldsymbol{\rho}$ is the vector of the model parameters, A is the external field, V and G are the local and nonlocal interaction potentials, and Γ, Θ, Π are operators defined

---

The original version of the chapter has been revised. A correction to this chapter can be found at
https://doi.org/10.1007/978-3-031-63361-4_9

T. Zhanlav and O.chuluunbaatar, *New Developments of Newton-Type Iterations for Solving Nonlinear Problems, Mathematical Engineering*,
https://doi.org/10.1007/978-3-031-63361-4_8

depending on the model. For each model, system (8.1) is complemented by initial and boundary conditions and, possibly, by normalization conditions of the sought solutions.

The general characteristics of the class of Eqs. (8.1) are its multiparameter character with respect to the model parameters, the multidimensionality of the coordinate space and the presence of singular points in it, and the possibility of a nonunique solution (a spectrum of solutions). This class of nonlinear problems describes the evolution of complex systems with possible bifurcation and critical modes.

Stationary problems ($\Gamma = 0$) play a special role.

The problem of stability of solutions to system (8.1) is solved in a special way in the models considered. Namely, the stability of stationary solutions to system (8.1) for $\Gamma = 0$ is investigated. For the calculation of stationary solutions, the problem of their evolution in a short time interval under small perturbations of a special form is formulated. As a result, a spectral problem is formulated, and this spectral problem, together with stationary boundary value problem (8.1), forms a new system. A conclusion on the character of local stability of the modeled process is drawn on the basis of properties of a part of the system spectrum.

Stationary problems can be reduced to a unified statement in the form of the equation

$$\varphi(\mathbf{a}, \lambda, y) = 0, \tag{8.2}$$

where y is the element of some domain of the Banach space Y; $\mathbf{a} \in R_l$ and $\lambda \in R_n$ are the vectors of Euclidean spaces of the corresponding dimensions. The nonlinear function φ for the given vector $\mathbf{a}$ transforms the elements $z = \{\lambda, y\}$ from the domain $R_n \times Y$ into the space $R_n \times U$, where U is the B—space and $U \supseteq Y$. It is assumed that for each given vector $\mathbf{a}$ Eq. (8.2) has an enumerable (or finite) number of solutions $\{y_k^*\}$, $k = 0, 1, 2, \ldots$, and each solution y_k^* can correspond to the vector of eigenvalues λ_k^*. The solution $z_k^* = \{\lambda_k^*, y_k^*\}$ to (8.2) is a function of the parameter vector $\mathbf{a}$.

The method of dimensional reduction by expansion of the sought solutions in special bases and reduction of the original problem to systems of one-dimensional equations (the Kantorovich method [2]) is widely used for the solution of stationary problems.

The problems under study have the following specific features.

1. There exists particular information on the existence and qualitative behavior of the sought solutions that can be obtained from the nature of the studied processes or from investigation of simplified models, for example, in regions of asymptotic variation of parameters.
2. In problems of low dimension, representing approximations of more complex multidimensional problems, and upon transfer of asymptotic conditions for a solution to finite domains, problems in estimating the accuracy of the applied approximations occur.

It is natural to extend the vector of physical parameters $\mathbf{a}$ in problem (8.2) by parameters of approximation of the problem and the numerical scheme. Numerical investigation of the model is usually reduced to mass calculations in a wide parameter range, which simultaneously provides a possibility of investigation of the properties of the models considered, i.e., the behavior of solutions depending on the "physical" parameters, and the accuracy of the obtained results depending on the parameters of approximation of the original problems. Therefore, upon mass calculations, it is reasonable to apply continuation methods with respect to a parameter, and iterative methods providing the use of all a priori information for refining the calculation results.

Newton's method, which is among the simplest one-level iterative methods, under certain conditions has the fastest quadratic convergence in the vicinity of an isolated solution and provides the minimum linear part of the residual at each step. Newton's method has been further developed on the basis of the generalization of its continuous analogue [3].

8.2 Modified Newton Schemes

8.2.1 Generalized the CANM and Modified Iterative Schemes

In [3], the systematic description of a class of iterative schemes of numerical solution of boundary value problems for differential, integro-differential, and integral equations with additional conditions for the sought solutions, is given. For all these problems, the unified statement in the form of nonlinear equation (8.2)

$$\varphi(z) = 0,$$

is considered.

The basis for the construction of iterative schemes of the class of problems under study is the CANM described by the evolution equation

$$\frac{d}{dt}\varphi(z) = -\varphi(z), \quad 0 \le t \le \infty, \quad z(0) = z_0, \tag{8.3}$$

where t is the additional parameter and z_0 is the initial approximation of the sought solution z^* to (8.1).

The iterative scheme

$$\varphi'(z_k)\Delta z_k = -\varphi(z_k), \quad z_{k+1} = z_k + \tau_k \Delta z_k \tag{8.4}$$

with the additional parameter of optimization of convergence τ_k obtained using the Euler method of solution of Eq. (8.3), is substantiated.

Further generalization of the developed method is based on parameterization of the initial function φ in (8.2) with respect to the additional parameter t with an explicit dependence of φ on t. Following the idea of Davidenko [4], the continuous parameter

$0 \leq t < \infty$ is introduced into the function $\varphi = \varphi(t, z(t))$ in such a way that for $t = 0$ the following simple equation is obtained:

$$\varphi(0, z(0)) \equiv \varphi_0(z_0) = 0 \tag{8.5}$$

and $\lim_{t \to \infty} \varphi(t, z(t)) = \varphi(z)$. For the parameterized function, the generalized equation of the CANM is considered,

$$\frac{d}{dt}\varphi(t, z(t)) = -\varphi(t, z(t)). \tag{8.6}$$

Since the integral of Eq. (8.6) is $\varphi(t, z(t)) = e^{-t}\varphi(0, z_0)$, we have $||\varphi(t, z(t))|| \to 0$ at $t \to \infty$, and the asymptotically stable convergence of $z(t)$ to the sought solution z^* should be expected.

If z_0 is the exact solution to Eq. (8.5), we obtain the Cauchy problem defining the Davidenko method on the half-axis $0 \leq t < \infty$,

$$\frac{dz}{dt} = -(\varphi'_z(t, z(t)))^{-1}\varphi'_t(t, z(t)), \quad z(0) = z_0.$$

If z_0 is the approximate solution to Eq. (8.5), we obtain from Eq. (8.6), by denoting $A(t, z(t)) = \varphi'_z(t, z(t))$, the modified CANM,

$$\frac{dz}{dt} = -A^{-1}(t, z(t))[\varphi(t, z(t)) + \varphi'_t(t, z(t))] \tag{8.7}$$

with the initial condition $z(0) = z_0$.

If Eq. (8.7) is approximated by the Euler scheme, the following sequence of iterations is obtained ($z_k = z(t_k); \;\; B_k = A^{-1}(t_k, z_k)$)

$$V_k = -B_k[\varphi(t_k, z_k) + \varphi'_t(t_k, z_k)], \tag{8.8}$$

$$z_{k+1} = z_k + \tau_k V_k. \tag{8.9}$$

The following additive representation of the function $\varphi(z)$ is often considered for parameterization of $\varphi(t, z(t))$:

$$\varphi(z) = \varphi_0(z) + \varphi_1(z),$$

where $\varphi_0(z)$ is the regular part, and $\varphi_1(z)$ is its perturbation. It is assumed that for the equation $\varphi_0(z) = 0$ it is easy to find the approximate solution z_0, and the operator $\varphi'_0(z)$ is easily invertible.

The parameterization can be performed using the scalar function $g(t)$, the so-called function of inclusion of perturbations, such that, $g(0) = g(\infty) - 1 = g'(\infty) = 0$, for example, $g(t) = 1 - e^{-t}$, and the representation of the function $\varphi(t, z(t))$ in the form of the sum

$$\varphi(t, z(t)) = \varphi_0(z(t)) + g(t)\varphi_1(z(t)). \tag{8.10}$$

The advantage of this approach is in the construction of modified iterative schemes, where instead of inversion of the operator $\varphi'(z)$ at each iteration, it is necessary to invert the derivative of the specially chosen operator φ_0 with a simple structure. Note that iterative schemes on the basis of representation (8.10) are applied for high order multiparameter difference approximations [5]. They preserve the three-diagonal structure of the matrix of the operator φ' in Newton iterations. From the point of view of computer implementation, the preservation of the relatively simple structure of this operator providing a high accuracy of the approximation of the solved equation is of great importance. In the framework of generalization of CANM, iterative schemes with retardation for integro-differential equations can be given as another example.

The next step is the formulation of the functional-operator equation (B is the unknown operator $(\varphi'(z))^{-1}$)

$$\phi(B; z) = \begin{pmatrix} \varphi(z) \\ \varphi'(z)B - I \end{pmatrix} = 0. \tag{8.11}$$

Application of the above approach to this equation provides the construction of iterative schemes without inverting the operator $\varphi'(z_k)$.

The following iterative formulas are used to determine B more precisely:

$$W_k = -B_k[\varphi_z'(t_k, z_k)B_k - I], \tag{8.12}$$

$$B_{k+1} = B_k + \tau_k W_k. \tag{8.13}$$

These formulas are the corollary of the application of CANM to Eq. (8.11). The convergence of this process was proved for $\tau_k = 1$, for example, in [6].

Iterative scheme (8.8), (8.9), (8.12), (8.13) does not include inversion of the operator φ_z', and in this scheme the parameter τ_k minimizes the residual of the original equation. Thus, for the initial approximation z_0 and B_0 all approximations z_k and B_k can be found successively. Practical calculations showed that $B_0 = A^{-1}(z_0)$ is the best initial approximation for B (i.e., in this case the inversion of the operator $\varphi_z'(t, z(t))$ is performed only once for $t = 0$). The advantage of this iterative scheme is the absence of operations of division during all calculations. This excludes cases of division by a small number possible upon the inversion of poorly conditioned matrices. Thus, the stability and accuracy of the calculations are increased. Upon vectorization of operations [7], multiplication of matrices is more preferable than inversion of a matrix, and modified algorithm (8.8), (8.9), (8.12), (8.13) provides a gain in time for a vector computer system. However, this gain is obtained at the expense of the larger memory capacity required for storage of the additional matrices.

Thus, modifications of CANM that increase its efficiency for particular classes of problems and extend the region of its applicability have been developed, and are widely used at present. The problem of the choice of initial approximations is in some sense solved in the developed iterative schemes, and the solution of the linear problem with respect to iterative corrections is simplified. It is also possible to construct an iterative process without inversion of the linear Fréchet operator in this problem.

8.2.2 Estimates of Accuracy of Numerical Results

After reduction using the Kantorovich method, the original multidimensional non-linear stationary boundary value problem

$$\varphi(\mathbf{a}, z) = 0 \tag{8.14}$$

is transformed into the system of N ($N \to \infty$) one-dimensional equations

$$\varphi_N(\mathbf{a}, N, z_N) = 0. \tag{8.15}$$

Taking into account that boundary conditions are set at finite intervals characterized by the boundary points γ, it has the form

$$\varphi_{N,\gamma}(\mathbf{a}, N, \gamma, z_{N,\gamma}) = 0. \tag{8.16}$$

After discretization with the parameter h, we obtain the set of corresponding equations on a grid

$$\varphi_{N,\gamma,h}(\mathbf{a}, N, \gamma, h, z_{N,\gamma,h}) = 0. \tag{8.17}$$

Newton iterative process (8.4) is realized for Eq. (8.17) until the following condition is satisfied

$$\delta_K = ||\varphi_{N,\gamma,h}(\mathbf{a}, N, \gamma, h, z_{N,\gamma,h,K})||_h \le \varepsilon, \tag{8.18}$$

where K is the number of the iteration at which condition (8.18) is satisfied, and $\varepsilon > 0$ is the given small number.

It is necessary to estimate the expression $||z^* - z_{N,\gamma,h,K}||_h$, where z^* is the solution to Eq. (8.14). If $z^*_{N,\gamma,h}$ is the exact solution to Eq. (8.17), the theoretical estimate for condition (8.18) has the form

$$||z^*_{N,\gamma,h} - z_{N,\gamma,h,K}||_h \le B\delta_K \le B\varepsilon.$$

For the exact solution $z^*_{N,\gamma}$ to Eq. (8.16), we have the following theoretical estimate:

$$||z^*_{N,\gamma} - z^*_{N,\gamma,h}||_h \le Ch^p,$$

where p is the order of approximation for discretization (8.17).

Then, the following inequality is satisfied:

$$||z^*_{N,\gamma} - z_{N,\gamma,h,K}||_h \leq B\varepsilon + Ch^p.$$

If $\varepsilon \ll h$ the following estimate is valid:

$$||z^*_{N,\gamma} - z^*_{N,\gamma,h}||_h \sim \tilde{C}h^p. \tag{8.19}$$

This relation should be verified on finer grids $(h \to 0)$, and extrapolation formulas should be used to increase the accuracy of the results.

The contribution of the errors $||z^* - z^*_N||_h$ and $||z^*_N - z^*_{N,\gamma}||_h$, where z^*_N and $z^*_{N,\gamma}$ are the solutions to Eqs. (8.15) and (8.16), respectively, can be indirectly estimated by calculations on sequences of expanding intervals $\{\gamma \to \infty\}$ and for an increasing number $N\{N \to \infty\}$ of equations of system (8.15).

If the values of shifts $||\bar{\gamma}||$ and $\bar{N}$ in the corresponding sequences $\{\gamma\}$ and $\{N\}$ are sufficiently large, and the values $||z^*_{N,\gamma,h,K} - z^*_{N,\gamma+\bar{\gamma},h,K}||_h$ and $||z^*_{N,\gamma,h,K} - z^*_{N+\bar{N},\gamma,h,K}||_h$ are so small that relation (8.19) is satisfied, it can be assumed that the parameters of approximation N, γ, h are determined in an appropriate way. Naturally, this practical procedure is based on the assumptions of convergence of the corresponding methods of approximation of original equation (8.14) and serves as the proof of these assumptions. This procedure is conveniently implemented on the basis of the continuation method with respect to parameters using already calculated solutions for refining subsequent solutions during iterations.

8.2.3 CANM in Spectral Problems

Let us consider the modified Newton evolution process

$$\varphi'(\tilde{z}(t))\frac{dz(t)}{dt} = -\varphi(z(t)), \quad z(0) = z_0,$$

where $\tilde{z}$ is some fixed element in the vicinity of the sought solution z^*. This process yields iterative schemes of the type (8.3), in which the operator $\varphi'(\tilde{z}(t))$ should be inverted just once. In spectral problems, in which the unknown z consists of two components λ and Ψ (eigenvalue and eigenelement), either one or two of these components can be fixed, depending on how well-known the corresponding approximation to the sought solution is.

For the classical spectral problem $(H - \lambda I)\Psi = 0$ with respect to the pair $z = \{\lambda, \Psi\} \in R \times Y$, nonlinear equation (8.2) can be represented in the form

$$\varphi(\lambda, \Psi) = \begin{pmatrix} (H - \lambda I)\Psi \\ F(\lambda, \Psi) \end{pmatrix} = 0. \tag{8.20}$$

Here, H is the operator in the Hilbert space, which in some cases can be represented in the form

$$H \equiv H(g) = H_0 + gH_1, \tag{8.21}$$

where g is the formal parameter (the coupling constant), and $F(\lambda, \Psi)$ is the additional functional, for example,

(a) $(\Psi, \Psi) - 1 = 0$ is the normalization condition,

(b) $(\Psi, (H - \lambda I)\Psi) = 0$ is the orthogonality condition.   (8.22)

For solution of spectral problems (8.20), iterative scheme (8.3) can be applied, which, for a fixed value of the parameter vector $\mathbf{a}$, at each step includes the following system with respect to the residual $\Delta z_k = \{\Delta \lambda_k, \Delta \Psi_k\}$:

$$(H - \lambda_k I)\Delta \Psi_k = -(H - \lambda_k I)\Psi_k + \Delta \lambda_k \Psi_k,$$
$$F'_\lambda(\lambda_k, \Psi_k)\Delta \lambda_k + F'_\psi(\lambda_k, \Psi_k)\Delta \Psi_k = -F(\lambda_k, \Psi_k).$$

The two-component structure of the function φ and the possibility of modifying the form of the functional F in iterations allow one to obtain a wide set of iterative processes with controlled properties.

Depending on the method of solution of this system and the choice of the form of the functional F, different known iterative schemes of the solution of spectral problems can be obtained.

If $\Delta \Psi_k$ is represented in the form

$$\Delta \Psi_k = -\Psi_k + \Delta \lambda_k U_k,$$

where U_k is the solution to the problem

$$(H - \lambda_k I)U_k = \Psi_k,$$

we obtain the following expression for $\Delta \lambda_k$:

$$\Delta \lambda_k = \frac{1 + (\Psi_k, \Psi_k)}{2(\Psi_k, U_k)}.$$

If $\tau_k = 1$ we obtain the following expression for new approximations:

$$\Psi_{k+1} = \Delta \lambda_k (H - \lambda_k I)^{-1}\Psi_k,$$
$$\lambda_{k+1} = \lambda_k + \frac{1 + (\Psi_k, \Psi_k)}{2(\Psi_k, (H - \lambda_k I)^{-1}\Psi_k)}.$$

A known inverse iterative scheme is seen to be obtained. If the functional F in the form (8.22) is used, we obtain the following system with respect to iterative corrections:

$$(H - \lambda_k I)\Delta\Psi_k - \Delta\lambda_k\Psi_k = -(H - \lambda_k I)\Psi_k,$$
$$(\Delta\Psi_k, (H - \lambda_k I)\Psi_k) + (\Psi_k, (H - \lambda_k I)\Delta\Psi_k) - (\Psi_k, \Delta\lambda_k\Psi_k)$$
$$= -(\Psi_k, (H - \lambda_k I)\Psi_k).$$

Using the first equation of this system, we obtain from the second equation

$$(\Delta\Psi_k, (H - \lambda_k I)\Psi_k) = 0.$$

If H is self-conjugate,

$$(\Psi_k, (H - \lambda_k I)\Delta\Psi_k) = 0. \tag{8.23}$$

By substituting the expression for $\Delta\Psi_k$

$$\Delta\Psi_k = -\Psi_k + \Delta\lambda_k(H - \lambda_k I)^{-1}\Psi_k$$

into relation (8.23), we obtain the expression

$$-(\Psi_k, (H - \lambda_k I)\Psi_k) + \Delta\lambda_k(\Psi_k, \Psi_k) = 0.$$

This yields for $\tau_k = 1$

$$\Delta\lambda_k(\Psi_k, \Psi_k) = (\Psi_k, H\Psi_k) - \lambda_k(\Psi_k, \Psi_k)$$

or

$$\Psi_{k+1} = \Delta\lambda_k(H - \lambda_k I)^{-1}\Psi_k, \quad \lambda_{k+1} = \frac{(\Psi_k, H\Psi_k)}{(\Psi_k, \Psi_k)}.$$

This formula results in a known inverse iterative scheme with a Rayleigh shift.

In particular, for the classical spectral problem with the fixed value of $\lambda_k = \tilde\lambda$ and $\tau_k = 1$ the known inverse iterative scheme with a fixed shift providing the convergence to the eigenvalue λ^*, closest to $\tilde\lambda$ is obtained. It is reasonable to use the modified scheme with a fixed shift in successive calculations of elements of the bound part of spectrum of the operator H in combination with additional orthogonalization of the approximation Ψ_{k+1} found at the kth iteration with respect to all already-calculated eigenelements Ψ_k^*, where k is the number of the eigenelement and the shift between the already calculated eigenvalue and the next eigenvalue after the termination of iterations.

8.2.4 Algorithms of Choice of the Iteration Parameter τ_k

In this subsection, we present five algorithms of calculation of the parameters τ_k ($0 < \tau_0 \leq \tau_k \leq 1$), minimizing the residual, these algorithms have proved to be efficient in the solution of a number of problems.

1. $\tau_k \equiv \tau_0$. This algorithm for sufficiently small τ_0 ($\sim 0.1; 0.05; 0.01$) is usually applied for bad initial approximations in order to verify the possibility of convergence from these approximations. In this case, the convergence is very slow. For $\tau_0 \equiv 1$, the classical Newton scheme is obtained.
2. $\tau_k = \min(1, 2\tau_{k-1})$, if $\delta_k < \delta_{k-1}$; $\tau_k = \max(\tau_0, \tau_{k-1}/2)$, if $\delta_k \geq \delta_{k-1}$, where δ_k is defined by formula (8.18) in the grid analogue of the norm in C. This algorithm is similar to the widespread way of choosing the integration step in standard programs of solution of the Cauchy problem and calculation of integrals. It is recommended to apply this algorithm for good initial approximations. It provides fast convergence but is not always stable in the case of bad approximations.
3. $\tau_k = \min(1, \tau_{k-1}\delta_{k-1}/\delta_k)$, if $\delta_k < \delta_{k-1}$; $\tau_k = \max(\tau_0, \tau_{k-1}\delta_{k-1}/\delta_k)$, if $\delta_k \geq \delta_{k-1}$, where δ_k is also calculated by formula (8.18) in the grid analogue of the norm in C. This algorithm, minimizing the transition function for two subsequent residuals [8], is more stable and provides convergence in a sufficiently wide region of initial approximations. However, both the region of initial approximations and the convergence rate depend on the value of τ_0. The smaller τ_0, the wider the region of convergence, and the slower the convergence far away from the solution.
4. $\tau_k = \delta_{k-1}/(\delta_{k-1} + \delta_k(1))$, where $\delta_k(1)$ is the residual at the kth iteration for $\tau_k = 1$. The value of δ_k is calculated by formula (8.18) in the grid analogue of the norm in L_2. This is the algorithm of optimal choice of τ_k proposed in [9]. It is based on the quadratic approximation of δ as a function of τ. It should provide the minimum of the residual at each iteration.
5. The sequence of residuals δ^i is calculated by formula (8.19) on the uniform grid ω_τ of the interval $[0, 1]$ with the step $\Delta\tau$, and the value of τ_k corresponding to the minimum residual is chosen. This algorithm is more general than (8.4), but it requires a larger amount of calculations. The accuracy of finding the optimal step τ_k which provides the minimum residual at each step depends on the choice of the grid ω_τ. This grid can be chosen in such a way that the accuracy of finding τ_k and the processing speed of the algorithm are optimally combined.

8.3 Method of Investigation of Scattering Problem on Basis of Combination of CANM and Variational Approach

8.3.1 Multiparameter Newton Iterative Scheme

The main idea of the construction of the generalized iterative scheme formulated in [10] is to make use of the dependence on physical parameters **a** of the original problem (8.14). The required value of the component $a = a^*$ for which it is necessary to find the sought z^* is fixed in the weak sense by the additional asymptotic condition

$$F(a^*, z) = 0. \tag{8.24}$$

Thus reformulated original problem (8.14)

$$\Phi(a, z) = \{\varphi(a, z), F(a^*, z)\} = 0$$

is solved using the multiparameter Newton iterative scheme

$$\Phi'_a \Delta a + \Phi'_z \Delta z_k = -\Phi(a_k, z_k), \quad a_{k+1} = a_k + \tau_k \Delta a_k, \quad z_{k+1} = z_k + \tau_k \Delta z_k, \tag{8.25}$$

in which $a \to a^*$ is provided by adding asymptotic component (8.24) to (8.14). This scheme, unlike standard scheme (8.4), allows one to find, along with the unknown z, its derivative $\partial z/\partial a|_{a=a^*}$. This circumstance will be used further in calculation of the element z of the trajectory $z(a)$ at the point $a = a^* + \Delta a$, where we will apply a good initial approximation

$$z_0(a) = z^*(a)|_{a=a^*} + \Delta a \left. \frac{\partial z}{\partial a}\right|_{a=a^*}. \tag{8.26}$$

This reduces the number of iterations in process (8.25). If the Newton component $-\Phi(a_k, z_k)$ is excluded from iterative scheme (8.25), we obtain the discrete analogue of the evolution method with respect to the coupling constant [11] or the Davidenko method [4]. For continuum problem (8.20)–(8.22), we can choose, for example, the parameter q, which is the value of momentum in the channel at some value of the spectral parameter λ (or energy $2E = q^2$), as a. Then condition (8.24) takes the form

$$F(q^*, \psi) = (\psi, (H - q^{*2})\psi)) = 0,$$

similar to (8.22). In this case, iterations with respect to the parameter q in the vicinity of $q = q^*$ serve for determination of derivatives with respect to this parameter of interest.

For illustration, we restrict ourselves to consideration of the problem of elastic scattering for models of quantum mechanical systems described by the radial Schrödinger equation on the half-axis $\rho \in (0, \infty)$ with the short-range spherically symmetric potential $V(\rho) \equiv V(\rho, g)$ $(V(\rho) \equiv V(\rho, g = 0) \equiv 0)$ in the n-dimensional space for the given coupling constant $g \geq 0$, momentum $q \geq 0$, and angular momentum l (see [12]),

$$\left(\frac{1}{\rho^{k-1}} \frac{d}{d\rho} \rho^{k-1} \frac{d}{d\rho} - \frac{l(l + k - 2)}{\rho^2} + q^2\right) \Psi_l(\rho) = V(\rho)\Psi_l(\rho). \qquad (8.27)$$

The corresponding boundary conditions for Eq. (8.27) are obtained by the transfer of asymptotic conditions for wave functions from the singular domain $[0, \infty)$

$$\Psi_l(\rho) \underset{\rho \to 0}{\longrightarrow} \rho^l, \quad \Psi_l(\rho) \underset{\rho \to \infty}{\longrightarrow} C\rho^{-\nu}\sin\left(q\rho - \pi\frac{l + \nu - 1}{2} + \delta_l\right), \qquad (8.28)$$

to the finite region of integration $\rho \in [\rho_{\min}, \rho_{\max}]$, where $\nu = (k - 1)/2$, δ_l is the sought phase shift, and C is the normalization coefficient. Problem (8.27) is considered on the whole axis $(-\infty, \infty)$ in the one-dimensional space $(n = 1)$. Then, for the potentials with the asymptotic $V(\rho) \underset{\rho \to \pm\infty}{\longrightarrow} \exp(\pm\rho)$ it is convenient to use the following conditions instead of (8.28)

$$\Psi_l(\rho) \underset{\rho \to -\infty}{\longrightarrow} 0; \quad \Psi_l(\rho) \underset{\rho \to \infty}{\longrightarrow} C\sin(q\rho + \delta). \qquad (8.29)$$

Problem (8.27)–(8.29) is reduced to the finite interval $[0, \rho_m]$ using the homogeneous boundary conditions

$$\phi^{(2)} = \lim_{\rho \to 0} \left(b_{11}\frac{\partial}{\partial\rho} + b_{12}\right) \Psi_l(\rho) = 0,$$

$$\phi^{(3)} = \lim_{\rho \to \infty} \left(b_{21}\frac{\partial}{\partial\rho} + b_{22}\right) \Psi_l(\rho) = 0, \qquad (8.30)$$

where the functions b_{ij}, $i, j = 1, 2$, are defined by asymptotic conditions (8.28) or (8.29). For the Morse potential $(n = 1)$ with two bound states $(\nu = \delta(0)/\pi = 2)$, phase shift δ and its derivative $\partial\delta/\partial q$ as functions of squared momentum are shown in Fig. 8.1a. The accuracy of calculations and the quality of the functions ψ can be checked with the help of the derivative $\partial\delta/\partial q$, using the virial theorem [13]

$$C^2 q^2 \frac{\partial\delta}{\partial q} = \left(\psi, \left(2V + \rho\frac{\partial V}{\partial\rho}\right)\right). \qquad (8.31)$$

Another possibility studied in detail in [10] is the choice of the coupling constant g in (8.21) as a in (8.24). Then, condition (8.24) has the form

$$F(g^*, \psi) = (\psi, (H(g^*) - q^2)\psi)) = 0$$

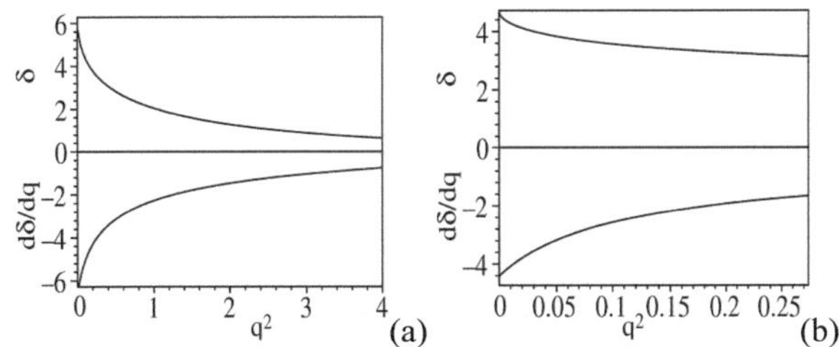

Fig. 8.1 Phase shift δ and its derivative $\partial\delta/\partial q$ as functions of squared momentum: **a** for the potential with two bound states; **b** for the potential with one bound and one semibound state

and, according to the Hellmann-Feynman theorem, it allows one to control the quantity

$$\frac{\partial K}{\partial g} = -\left(\psi, \frac{\partial V(g)}{\partial g}\psi\right),$$

and under certain constraints on the potential V, to obtain one-sided estimates of elements of the K matrix [14]. Iterative schemes (8.24), (8.25) can be applied for more precise determination of different variational calculations in the scattering problem. Indeed, the scattering problem for the Schrödinger equation with the above additional conditions can be reduced to calculation of the functional in the framework of different Hulthen, Kohn, or Schwinger variational principles. Thus, for example, for solution of the quantum problem of a few particles with short-range pair potentials, the Schwinger variational functional is used [15], and different iterative schemes have been developed on its basis. At first sight, the integral formulation of the problem is much simpler than the differential formulation, since it does not require detailed analysis of asymptotic behavior of the sought solution at $g \neq 0$ for calculation of the functions b_{ij}, $i, j = 1, 2$ in (8.30), but uses only known regular and irregular solutions for $g = 0$. However, such schemes for the multichannel scattering problem do not provide stable calculation of the necessary physical parameters in a number of cases. Therefore, the development of stable variational-iterative schemes on the basis of the combination of projection methods, variational principles, and Newton iterative schemes is a topical problem of numerical modeling of quantum mechanical systems.

8.3.2   Multiparameter Newton Iterative Scheme With Schwinger Functional

Boundary value problems (8.27), (8.28) are reduced to the spectral problem for the Fredholm equation [16]:

$$A_l(\rho, \rho')\Psi_l(\rho') = \lambda_l B_l(\rho, \rho')\Psi_l(\rho'), \tag{8.32}$$

$$A_l(\rho, \rho')\Psi_l(\rho') = \Psi_l(\rho) - \int_0^\infty G_l(\rho, \rho')V(\rho')\Psi_l(\rho')\rho'^{n-1}d\rho',$$

$$B_l(\rho, \rho')\Psi_l(\rho') = y_l(\rho)\int_0^\infty y_l(\rho')V(\rho')\Psi_l(\rho')\rho'^{n-1}d\rho', \tag{8.33}$$

where $\lambda_l = -\pi \cot(\delta_l)/2$ is the sought spectral parameter, and the dependence on the normalization coefficient C in the asymptotic of the unknown wave function $\Psi_l(\rho)$ is eliminated. The function $y_l(\rho)$ and the free Green's function $G_l(\rho, \rho')$ are determined in terms of regular and irregular at the point $\rho = 0$ solutions to Eq. (8.27) for $V(\rho) \equiv 0..$ The function $y_l(\rho)$ has the form

$$y_l(\rho) \equiv \rho^{-\mu} J_{l+\mu}(q\rho), \quad \mu = n/2 - 1, \quad n > 1,$$

where J_i is the Bessel function of the first kind. The additional condition of the type (8.24) is used for solution of the problem (8.32), (8.33):

$$(V(\rho)\Psi_l(\rho), (A_l(\rho, \rho') - \lambda_l B_l(\rho, \rho'))\Psi_l(\rho')) = 0, \tag{8.34}$$

which implies the Schwinger variational functional

$$\lambda_l = \frac{(V(\rho)\Psi_l(\rho), A_l(\rho, \rho')\Psi_l(\rho'))}{(V(\rho)\Psi_l(\rho), B_l(\rho, \rho')\Psi_l(\rho'))}, \tag{8.35}$$

where the brackets $(\cdot, \cdot)$ denote a scalar product, i.e., $(f, g) = \int_0^\infty f^* g\, \rho^{n-1}d\rho$. As a result, integral equation (8.32) corresponds to the functional stable with respect to first order variations in Ψ_l, and scattering problem (8.27), (8.28) is formulated as the eigenvalue problem with respect to the pair of unknown functions $z = (\lambda_l, \Psi_l)$, the function of the phase shift λ_l and the wave function Ψ_l. Discretization of problem (8.32), (8.34) on a grid of nodes $\Omega_h \in [\rho_{\min}, \rho_{\max}]$ (using known Bode's quadrature formulas) results in the algebraic generalized eigenvalue problem

$$\varphi(z) = \begin{pmatrix} (A - \lambda B)\Psi, \\ (V\Psi, (A - \lambda B)\Psi) \end{pmatrix} = \mathbf{0}.$$

Then, the iterative scheme for finding the approximations λ_{k+1}, Ψ_{k+1} using the corrections v_k, u_k and μ_k is constructed:

$$\begin{cases} v_k = -\Psi_k, \\ (A - \lambda_k B)u_k = B\Psi_k, \\ \mu_k = \frac{(\Psi_k V, A\Psi_k)}{(\Psi_k V, B\Psi_k)} - \lambda_k, \\ \Psi_{k+1} = \Psi_k + \tau_k(v_k + u_k\mu_k), \\ \lambda_{k+1} = \lambda_k + \tau_k\mu_k, \end{cases} \tag{8.36}$$

where $\{\lambda_0, \Psi_0\}$ is the initial approximation in the vicinity of the sought solution, and the condition of minimization of residual [9] is used for the choice of the iteration step $\tau_k, k = 0, 1, 2, \ldots$. The expression for μ_k coincides with Schwinger variational functional (8.35). The generalization of iterative scheme (8.36) for the multichannel scattering problem is given in [16].

In [17], the convergence of the proposed iterative scheme was demonstrated for elastic scattering problem (8.27)–(8.29) with the Morse potential ($n = 1$), Woods-Saxon potential, and the potential of the spherically symmetric rectangular well ($n = 3$). However, the scheme has the second order of accuracy with respect to the step h of the uniform grid Ω_h, since the first derivative with respect to the argument ρ of the Green's function $G_l(\rho, \rho')$ has a singularity at $\rho = \rho'$. For the potentials considered, the phase shift δ was calculated to an accuracy within six decimal places.

It was also shown in [17] that the application of the asymptotic $\Psi(\rho)$ in the vicinity of the point $\rho_{\min}$ for approximation of solutions allows the construction of schemes of a higher order of accuracy for calculation of the phase shift δ. The efficiency of the implementation of such sixth-order accuracy with respect to the step h scheme on the uniform grid Ω_h was demonstrated by calculation of the phase shift δ to an accuracy of twelve decimal places for the one-dimensional scattering problem with the Morse potential.

8.4 Scattering Problem of Three Bosons on Straight Line

In nuclear physics, methods of bipolar and hyperspherical harmonics and single-parameter surface functions are widely used for correct solution of the problem of few particles with short-range pair interactions. In these approaches, the original problem is reduced by the Galerkin or Kantorovich method to spectral problems for systems of integro-differential or ordinary differential equations with the hyper-radius as an independent variable. Of special interest are problems with singular interactions, for example, centrifugal interactions, interactions of zero radius in the form of δ functions, and problems with boundary conditions of the third kind. The issue is that the differential formulation of such problems causes difficulties. In order to avoid them, the problem is formulated as a system of integral equations, and in this formulation all difficulties are overcome by the choice of the appropriate parameter basis and approximation by the algebraic problem with completely filled matrices. Therefore, it is important to develop stable iterative schemes and algorithms of solution of spectral problems for systems of Fredholm integral equations.

In [18], the developed algorithms are analyzed and tested on integrable models of three bosons on a straight line with zero radius pair interactions, since for these models the energy eigenvalues E of the bound and semi-bound states, and the S matrix of the processes of elastic scattering are known [19]. The phase shift and its derivative as functions of the squared momentum for the process of elastic scattering below the three-particle threshold $E \leq 0$ is shown in Fig. 8.1b. For this model, the Schrödinger equation in polar coordinates ρ and θ for the partial wave function $\Psi_i(\rho, \theta))$ has the form [19]

$$\left(\frac{1}{\rho}\frac{\partial}{\partial\rho}\rho\frac{\partial}{\partial\rho} - h_\rho + 2E\right)\Psi_i(\rho,\theta) = 0, \quad \rho \in \mathcal{R}^1_+, \quad \theta \in \Omega. \tag{8.37}$$

Here, Ψ_i is the sought wave function, E is the energy in the center-of-mass system (in units $\hbar = m = 1$, where m is the boson mass), and h_ρ is the parametric Hamiltonian for each fixed value of ρ:

$$h_\rho = -\frac{1}{\rho^2}\frac{\partial^2}{\partial\theta^2} + \frac{2g}{\rho}\sum_{n=0}^{5}\delta(\theta - \theta_k), \quad \theta_k = \frac{n\pi}{3} + \frac{\pi}{6}, \tag{8.38}$$

where $g = 2c\bar{\kappa}$ is the coupling constant, $\bar{\kappa} = \pi/6$, $c = -1$ corresponds to the attraction of two particles and $c = 1$ to repulsion. The total wave function $\hat{\Psi}$ is sought in the form of Kantorovich expansion in the orthogonal set of surface one-parameter functions $B_j(\rho,\theta)$ and $B_j^{as}(\theta) = B_j(\rho \to \infty, \theta)$ with unknown coefficients $\chi_{ji}(\rho)$:

$$\hat{\Psi} = \sum_{i=0}^{N-1}|\Psi_i\rangle\langle B_i^{as}|, \quad |\Psi_i\rangle = \sum_{j=0}^{N-1}|B_j\rangle\langle B_j|\Psi_i\rangle = \sum_{j=0}^{N-1}B_j(\rho,\theta)\chi_{ji}(\rho). \tag{8.39}$$

The eigenfunctions $B_j(\rho,\theta) \in \mathcal{W}^1_2(\Omega)$ and the eigenvalues $\epsilon_j(\rho)$ of Hamiltonian (8.38) are found for each fixed value of $\rho \in \mathcal{R}^1_+$ from the boundary value problem

$$-\frac{1}{\rho^2}\frac{\partial^2}{\partial\theta^2}B_j(\rho,\theta) = \epsilon_j(\rho)B_j(\rho,\theta),$$

$$\frac{1}{\rho}\frac{\partial}{\partial\theta}B_j(\rho,\theta)\Big|_{\theta=\theta_k^\pm} = \mp c\bar{\kappa}B_j(\rho,\theta_k^\pm), \quad \theta_k^\pm = \pm\frac{\pi}{6} + \frac{\pi n}{3}.$$

By averaging Eq. (8.37) with respect to the basis $B_j(\rho,\theta)$, we obtain the system of N ordinary differential equations on the half-axis $\rho \in \mathcal{R}^1_+$ of the type (8.27) for $n = 2$. The asymptotic boundary conditions with respect to the radial variable depend on the type of physical processes. For example, the asymptotic expression for the radial functions $\chi_{ji}(\rho)$ for $\rho \to \infty$ above the three-particle threshold ($E > 0$) has the form

$$\begin{cases} \chi_{0i}^{as}(\rho) \to -Y_{1/2}(q\rho)\delta_{0i} + J_{1/2}(q\rho)W_{0i}, \\ \chi_{ji}^{as}(\rho) \to J_{6j-3}(k\rho)\delta_{ji} + Y_{6j-3}(k\rho)W_{ji}, \end{cases} \quad \text{for} \quad c = -1,$$

$$\chi_{ji}^{as}(\rho) \to J_{6j+3}(k\rho)\delta_{ji} + Y_{6j+3}(k\rho)W_{ji}, \quad \text{for} \quad c = 1,$$

where $q = \sqrt{\pi^2/36 + k^2}$, $2E = k^2$, and J_i and Y_i are the Bessel and Neumann functions of the first kind, $W_{ji} = K_{ji}^{-1}$ are the elements of the inverse matrix of the reaction. The corresponding two-dimensional scattering problem for Eq. (8.37) in the representation of one-parameter surface functions (8.39) is formulated as multichannel spectral problem (8.2) with respect to the pair of unknown variables $z = (K, \Psi)$ for system of one-dimensional integral equations (8.32 with matrix operators (8.33)

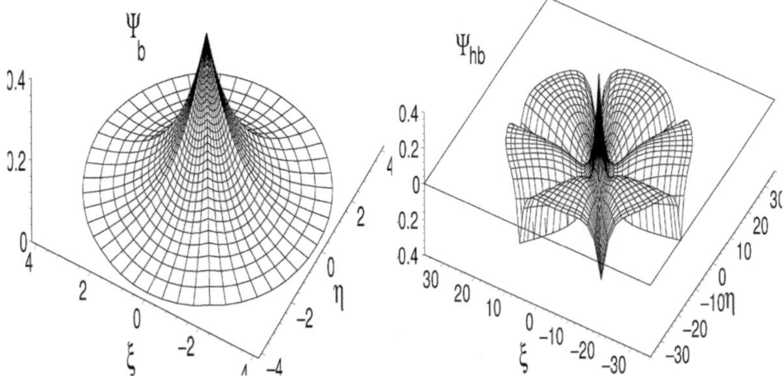

Fig. 8.2 Wave functions Ψ_b of the ground state and Ψ_{hb} of the semibound state for $2E_b \approx 2E^{exact} = -\pi^2/9$ and $2E_{hb} \approx 2E^{hb,exact} = -(\pi/6)^2$, $(q = 0)$

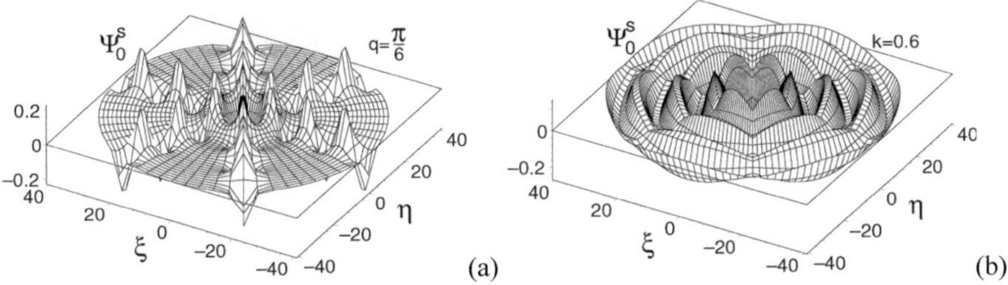

Fig. 8.3 Scattering wave functions Ψ_0^s for $q = \pi/6$ (for the three-particle threshold energy $2E = -k^2 = 0$) for attractive pair potentials with $c = -1$ (**a**), and for $k = 0.6$ for repulsive pair potentials with $c = 1$ (**b**)

and the additional Schwinger variational functional (8.34). The convergence of expansion (8.39) in the Kantorovich method and the efficiency of proposed iterative schemes (8.36) were demonstrated for the model considered below the three-particle threshold ($E < 0$) and in the region above the three-particle threshold ($E > 0$). It is seen from Figs. 8.2 and 8.3 that the wave functions have maximums for attraction ($c = -1$) and minimums for repulsion ($c = 1$) at the boundaries of six sections of the circle Ω, where the first and second derivatives of the solutions have a discontinuity. The elements of the K matrix below the three-particle threshold and in the region above the three-particle threshold were calculated to four and three decimal places, respectively, in the case $N = 6$. In [20], comparison with analytical results was made.

Table 8.1 shows the comparison of the convergence rate of Kantorovich (8.39) and Galerkin

$$\Psi_i(\rho, \theta) = \frac{1}{\sqrt{2\pi}}\bar{\chi}_0(\rho) + \frac{1}{\sqrt{\pi}} \sum_{j=1}^{N-1} \bar{\chi}_j(\rho) \cos(6j\theta) \qquad (8.40)$$

Table 8.1 Comparison of convergence rates of Kantorovich (K) (8.39) and Galerkin (G) (8.40) expansions for calculation of energy $E(N)$ of a bound three-particle state. The first row shows the number of equations N, the second and third rows show the difference $\Delta E(N) = E(N) - E^{exact}$ of calculated $E(N)$ and exact E^{exact} values of energy

N	1	2	3	4	5	6
ΔE^K	1.801(−4)	2.762(−6)	2.697(−7)	5.413(−8)	1.594(−8)	5.949(−9)
ΔE^Γ	9.662(−2)	4.116(−2)	2.573(−2)	1.866(−2)	1.462(−2)	1.201(−2)

expansions, as exemplified by the calculation of the energy $E(N)$ of the bound three-particle state for this model ($c = -1$).

Slow convergence of $\Delta E^\Gamma(N)$ with increasing number N of basis functions (8.40) is explained by the fact that the sufficient conditions of convergence of the classical Galerkin method cannot be weakened here. Moreover, thus reduced problem for any finite N has a false Coulomb spectrum with quantum defect and a qualitatively different threshold behavior of the phase shift [21] on the half-axis. This circumstance should be taken into account if the boundary value problem for the system of three particles with pair interactions allowing bound states is reduced by the Galerkin method.

8.5 High-Precision Calculation of Ground State Energy of Helium Atom

Progress in the development of experimental methods of electron pulsed spectroscopy (EPS) (when at least two fast electrons measured in coincidence are produced in the reaction) achieved in recent years [22], in particular, the development of cold target recoil-ion momentum spectroscopy (COLTRIMS), allows precise kinematically complete investigations of complex atomic collisions [23].

Unique experiments on single and double ionization of atoms and molecules by fast electrons, performed by Japanese researchers in the EPS geometry, showed the necessity of both revision of the theory of the dynamical mechanisms of these processes, and the creation of a new family of variational wave functions as close as possible to the exact solution of the multi-particle Schrödinger equation for an atom or a molecule [24].

For this purpose, instead of the Lippmann-Schwinger equation with a noncompact singular kernel, resolvent type integral equations are formulated and investigated for the system of few particles with Coulomb interaction which have a connected kernel and in which all necessary singularities-two-particle (cluster) singularities and three-particle Coulomb singularities are expressed explicitly. On the basis of such an approach, series of the perturbation theory are constructed in which higher Born terms describing interaction mechanisms are expressed by converging integrals and can be calculated numerically. A new method of construction of sample single-center wave functions of an atom with correlation functions explicitly depending on the distance between the electrons which reproduce the Kato energy of the bound state with high accuracy and satisfy additional conditions of Kato type [25, 26] has been developed. This method is also applicable to molecules in the framework of the method of linear combination of atomic orbitals. The known two-center Coulomb wave functions [27] are used in calculations for the ion of the hydrogen molecule. The symmetry of electron correlations of the molecular hydrogen ion and two atomic systems with two active electrons in ionization processes with excitation by fast electron impact was studied in the Born approximation using the quasi-exact solutions in the basis of two-center Coulomb functions [28–37], which is necessary for adequate description of second order effects in experiment [23]. It is planned to test the above approaches and methods on the helium atom and hydrogen molecule and to use it further for the water and carbon oxide molecules that will be investigated by Japanese experimentalists.

For the investigation of electron correlations, a new single-parameter basis of factorized correlated variational functions necessary for the calculation of energies and wave functions of bound states of helium-like atoms with a predetermined accuracy was constructed and examined in [38]. This problem, unlike the Coulomb two-body problem, does not have an exact analytical solution, but is the basic three-body system for precision calculations and experiments in atomic physics.

The wave function $\Psi(\mathbf{r}_1, \mathbf{r}_2)$ and energy E of bound states of helium-like atoms satisfy the Schrödinger equation

$$(\overline{H} - E)\Psi(\mathbf{r}_1, \mathbf{r}_2) = 0. \tag{8.41}$$

The two-electron non-relativistic Hamiltonian $\overline{H}$ in case of infinity nuclear mass has the form:

$$\overline{H} = -\frac{1}{2}\left(\nabla_{r_1}^2 + \nabla_{r_2}^2\right) - \frac{Z}{r_1} - \frac{Z}{r_2} + \frac{1}{r_{12}}, \tag{8.42}$$

where Z is the nuclear charge, r_i is the distance between the nucleus and the i-th electron, and r_{12} is the distance between the electrons. In spherical coordinates for two-electron S states the Hamiltonian takes the form

$$H(r_1, r_2, r_{12}) = -\frac{1}{2}\left(\frac{1}{r_1^2}\frac{\partial}{\partial r_1}r_1^2\frac{\partial}{\partial r_1} + \frac{1}{r_2^2}\frac{\partial}{\partial r_2}r_2^2\frac{\partial}{\partial r_2}\right) - \frac{1}{r_{12}^2}\frac{\partial}{\partial r_{12}}r_{12}^2\frac{\partial}{\partial r_{12}}$$
$$-\frac{r_1^2 - r_2^2 + r_{12}^2}{2r_1 r_{12}}\frac{\partial^2}{\partial r_1 \partial r_{12}} - \frac{-r_1^2 + r_2^2 + r_{12}^2}{2r_2 r_{12}}\frac{\partial^2}{\partial r_2 \partial r_{12}} - \frac{Z}{r_1} - \frac{Z}{r_2} + \frac{1}{r_{12}}. \tag{8.43}$$

The radial part of the integration volume element can be rewritten in the following way:

$$\int \mathbf{dr_1} \int \mathbf{dr_2} F(r_1, r_2, r_{12})$$

$$= 8\pi^2 \int_0^\infty r_1 dr_1 \int_0^\infty r_2 dr_2 \int_{|r_1-r_2|}^{r_1+r_2} r_{12} dr_{12} F(r_1, r_2, r_{12}). \qquad (8.44)$$

For calculation of the energy of bound states, the variational Rayleigh-Ritz principle and an appropriate set of parametric trial functions in the coordinate representation are used. The upper estimate of the energy of helium atom ($Z = 2$) in the ground state $E = -2.90372\ 43770\ 66$ au in the non-relativistic approximation for the infinite mass of the nucleus was obtained already in 1966 [39]. Recently, better estimates were obtained for the energy of this state, $E = -2.90372\ 43770\ 34119\ 5938$ au [40] and $E = -2.90372\ 43770\ 34119\ 59813$ au [41].

For explicit inclusion of the correlation of two atomic electrons at the distance r_{12} from each other, and at the distances r_1 and r_2 from the atomic nucleus, the perimetric coordinates r_1, r_2, r_{12} are used in variational calculations in [39]. In these coordinates, the radial part of the element of the integration volume is not reduced to a simple product of one-dimensional integrals, therefore, for transition to factorized correlated representation, it is necessary to use special projective coordinates.

In [38], an alternative version of optimization of variational calculations was formulated, which provides a better stability and high accuracy of calculation of energy values. In projective coordinates,

$$s = r_1 + r_2, \quad v = \frac{r_{12}}{r_1 + r_2}, \quad w = \frac{r_1 - r_2}{r_{12}}, \qquad (8.45)$$

the three-dimensional integrals necessary for reduction of the variational problem to the algebraic problem are represented in the form of the product of one-dimensional integrals (see Eq. (8.44)):

$$I = \int_0^\infty s^5 ds \int_0^1 v^2 dv \int_{-1}^1 F(s, v, w) (1 - v^2 w^2) dw. \qquad (8.46)$$

The main advantage of such a formulation of the problem is achieved by the choice of the correlated representation of the variational functions factorized with respect to all three arguments (8.45),

$$\Psi(s, v, w) = \sum_{n=0}^N C_n \psi_n, \quad \psi_n \equiv \psi_{ij,2k} = U_i(s) V_j(v) W_{2k}(w), \qquad (8.47)$$

where $W_{2k}(w)$ are even functions of w on the interval $[-1, 1]$ for the ground S state of helium-like atoms, in which all matrix elements of the Hamiltonian of the original problem with volume element (8.46) are calculated analytically. This problem, after variation of the Rayleigh-Ritz functional, is reduced to the generalized eigenvalue problem. The wave function Ψ should be normalized by the formula (see Eq. (8.44))

$$\langle \Psi(s, v, w) | \Psi(s, v, w) \rangle = 1$$

$$= \pi^2 \int_0^\infty s^5 ds \int_0^1 v^2 dv \int_{-1}^1 (1 - v^2 w^2) \Psi^2(s, v, w) dw. \tag{8.48}$$

For energy calculation, the Newton iterative scheme constructed on the basis of the variational Rayleigh- Ritz functional with the functions (8.47) in the form

$$U_i(s) = N_i e^{-\alpha_i s} L_i^5(2\alpha_i s), \quad V_j(v) = \overline{N}_j P_j^{(0,2)}(2v - 1),$$

$$W_{2k}(w) = \hat{N}_{2k} P_{2k}^{(1,1)}(w), \tag{8.49}$$

where N_i, $\overline{N}_j$ and $\hat{N}_{2k}$ are the normalization constants, L_i^5 are generalized Laguerre polynomials, $P_j^{(q,t)}$ are the Jacoby polynomials, and α_i are the variational parameters, was used. The orthonormal basis U_i with the unique parameter $\alpha \equiv \alpha_i$, whose value was determined from the condition $\partial E(\alpha)/\partial \alpha = 0$, was used in calculations. In this approach, new upper estimates of the ground state of the helium atom were obtained to twenty two decimal places, $E = -2.90372\ 43770\ 34119\ 59829\ 7$ au. Note that calculations with multiparameter correlated exponential variational functions [42] for the ground state of the helium atom $E = -2.90372\ 43770\ 34119\ 59831\ 11594$ au confirmed our results. In [43–45] the ground state of the helium atom was calculated with an accuracy of 36 to 46 decimal places.

8.6   Electron Correlation in Processes of Ionization of Helium Atom

In [24, 46–50], models of processes of ionization of the S ground state of the helium atom by fast electrons and protons in the impulse approximation were investigated. Radial and angular electron correlations were studied, using known variational functions presented in [24]: the single-component Hylleraas function (Hy) [51], Hartree-Fock (HF) [52], Hylleraas-Eckart-Chandrasekhar (HEC) [51, 53, 54], the twelve-parameter Bonham and Kohl function (BK) [55], configuration interaction (CI) [56], and the twelve-component ($N = 12$) single-parameter CVP function (8.47) constructed in [47]. Total energies of these Hy, HF, HEC, BK, CI, and BK wave functions are -2.8477, -2.8617, -2.8757, -2.9035, -2.9031, and -2.9030 au, respectively. The CVP function adequately takes into account electron correlation in the atom and agrees with the leading exponential term $\exp(-\alpha s)$ in the asymptotic expansion of the formal solution of the "exact" wave function of the target [57]. Recently, the processes of double electron impact ionization of the helium atom, single ionization with simultaneous excitation, and double ionization were studied for a large value of the transferred momentum using the energy- and momentum-dispersion binary $(e, 2e)$ spectrometer [24]. The experiment was performed for a collision energy of 2080 eV in the symmetric noncoplanar geometry. Thus, large momentum transfer,

9 au, i.e., a value that has never been achieved before in investigations of double ionization of the helium atom, was achieved. The measured cross sections of $(e, 2e)$ and $(e, 3 - 1e)$ for transitions to the excited $(n = 2)$ He$^+$ state and to doubly ionized He^{2+} state were normalized to the cross sections of transitions to the ground $(n = 1)$ state of He$^+$.

Within the first order plane-wave impulse approximation (PWIA) the triple differential cross section for an $(e, 2e)$ transition of He to an ionic state with a principal quantum number n, which is constituted of energetically unresolved sublevels having different orbital angular momentum quantum numbers ls, can be written as

$$\frac{d^3\sigma_n}{dE_1 d\Omega_1 d\Omega_2} = \frac{p_1 p_2}{p_0} \sigma_{Mott} \sum_{l=0}^{n-1} (2l + 1)|F_{nl}(q)|^2. \tag{8.50}$$

Here σ_{Mott} is the half-off-shell Mott scattering cross section and in the symmetric noncoplanar geometry it is given by

$$\sigma_{Mott} = \frac{1}{4\pi^4} \frac{2\pi\eta}{\exp(2\pi\eta) - 1} \frac{1}{K^4}, \tag{8.51}$$

where $\eta = |\mathbf{p}_1 - \mathbf{p}_2|^{-1}$ and K is the momentum transfer $|\mathbf{p}_0 - \mathbf{p}_1|$. $F_{nl}(q)$ is

$$F_{nl}(q) = (4\pi)^2 \int_0^\infty \int_0^\infty r_1^2 r_2^2 j_l(qr_1)\varphi_{nl}(r_2)\Phi_l(r_1, r_2)dr_1 dr_2, \tag{8.52}$$

where j_l and φ_{nl} are the spherical Bessel function and the radial function of a nl orbital for the He$^+$ ion and $\Phi_l(r_1, r_2)$ is a component of the following expansion of the He ground state wave function $\Phi(\mathbf{r}_1, \mathbf{r}_2)$:

$$\Phi(\mathbf{r}_1, \mathbf{r}_2) = \sum_{l=0}^\infty (2l + 1)\Phi_l(r_1, r_2) P_l(\cos\theta_{12}), \tag{8.53}$$

with P_l being the Legendre polynomial. For $(e, 3 - 1e)$ reactions of He, the corresponding four-fold differential cross section is given by [58]

$$\frac{d^4\sigma}{dE_1 dE_3 d\Omega_1 d\Omega_2} = \frac{2}{\pi} \frac{p_1 p_2 p_3}{p_0} \sigma_{Mott} \sum_{l=0}^\infty (2l + 1)|F_l(q, E_3)|^2, \tag{8.54}$$

where

$$F_l(q, E_3) = (4\pi)^2 \int_0^\infty \int_0^\infty r_1^2 r_2^2 j_l(qr_1)\varphi_l(E_3, r_2)\Phi_l(r_1, r_2)dr_1 dr_2. \tag{8.55}$$

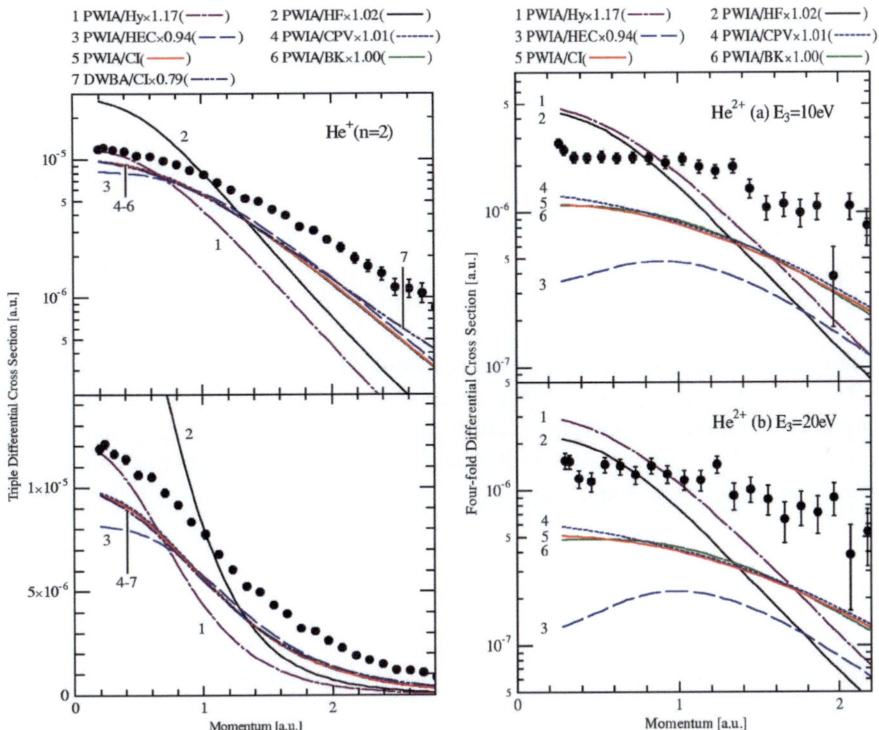

Fig. 8.4   Comparison of experimental three- and four-fold differential cross sections of the processes $(e, 2e)$ (**a**) and $(e, 3 - 1e)$ **b** as functions of transferred momentum with theoretical calculations in PWIA and DWBA for different variational functions: (1) PWIA/Hy; (2) PWIA/HF; (3) PWIA/HEC; (4) PWIA/CPV; (5) PWIA/CI; (6) PWIA/BK; (7) DWBA/CI

Here φ_l is the partial one-electron state with energy E_3 in the Coulomb field of the He^{2+} ion. In the above formulas the $\mathbf{p}_j$'s and E_j's ($j = 0, 1, 2, 3$) are momenta and kinetic energies of the incident and outgoing electrons, respectively.

Figure 8.4a, b show the comparison of calculations of the $(e, 2e)$ and $(e, 3 - 1e)$ reaction cross sections with the variational functions Hy, HF, HEC, CPV, CI, BK. The corresponding numerical results for the normalized cross sections were obtained in the PWIA and in the distorted-wave Born approximation (DWBA) using various wave functions of the ground state of the helium atom (with the purpose of investigation of models of the process dynamics and the structure of the quantum correlation of two-electron states). It is seen that the shape of the dependence of experimental cross sections of $(e, 2e)$ and $(e, 3 - 1e)$ on the transferred momentum is well-reproduced by PWIA calculations only when strongly correlated electron wave functions of the helium atom are used.

In Refs. [59–62] there was a discussion about the influence of cusp conditions on the differential cross sections of $(e, 3e)$ reactions. In this model the final state of helium is treated using a 3C function [63], which contains a product of three two-particle Coulomb waves, and the ground state is given by the Pluvinage (PL) wave function [64, 65]

$$\Psi_P = N_P e^{-Z(r_1+r_2)} e^{-\iota k r_{12}} {}_1F_1\left(1 - \iota\frac{1}{2k}, 2, 2\iota k r_{12}\right),\qquad(8.56)$$

where ${}_1F_1$ is the confluent hypergeometric function. This function satisfies cusp conditions of Kato [26] (see Eq. (8.59)), however, from the spectroscopic point of view, it gives a poor value for the ground state energy, $E_{PL} = -2.8788$ au. At the same time, the proposed model surprisingly well describes the most of experimental data by Lahmam-Bennani et al. [66]. In Ref. [59] two other functions also giving poor energy values while satisfying Kato's cusp conditions have been proposed, which have described the experiment similarly well. In this connection, the employment of the "poor" functions is mainly conditioned by hardships in constructing a variational wave function that gives an accurate energy value while satisfies the cusp conditions.

To put things right in the developing discussion, we have constructed two relatively simple wave functions satisfying Kato's cusp conditions and yielding an accurate energy value. The variational wave function has been chosen in the form which was intensively exploited in Refs. [42, 67]

$$\Psi(r_1, r_2, r_{12}) = \sum_{j=1}^{N} D_j \left(e^{-\alpha_j r_1 - \beta_j r_2} + e^{-\alpha_j r_2 - \beta_j r_1}\right) e^{-\gamma_j r_{12}},\qquad(8.57)$$

where the real parameters α_j, β_j, and γ_j generated in a quasirandom manner:

$$\alpha_j = \left\{\frac{1}{2}j(j+1)\sqrt{p_\alpha}\right\}(A_2 - A_1) + A_1, \qquad 0 \le A_1 < A_2,$$

$$\beta_j = \left\{\frac{1}{2}j(j+1)\sqrt{p_\beta}\right\}(B_2 - B_1) + B_1, \qquad 0 \le B_1 < B_2,\qquad(8.58)$$

$$\gamma_j = \left\{\frac{1}{2}j(j+1)\sqrt{p_\gamma}\right\}(C_2 - C_1) + C_1, \qquad 0 \le C_1 < C_2.$$

Here $\{x\}$ is a fractional part of x. $[A_1, A_2]$, $[B_1, B_2]$ and $[C_1, C_2]$ are variational intervals that are to be optimized. p_α, p_β and p_γ are some prime numbers. We use the values $p_\alpha = 2$, $p_\beta = 3$ and $p_\gamma = 5$. It should be noted that the final results for α_j, β_j, and γ_j are practically insensitive to the variation of the above prime numbers.

In principle, one can supplement the variational procedure with Kato's cusp conditions [26]:

$$\frac{\left.\dfrac{\partial \Psi(r_1, r_2, r_{12})}{\partial r_i}\right|_{r_i \to 0}}{\Psi(r_1, r_2, r_{12})|_{r_i \to 0}} = -2, \quad i = 1, 2,$$

$$\frac{\left.\dfrac{\partial \Psi(r_1, r_2, r_{12})}{\partial r_{12}}\right|_{r_{12} \to 0}}{\Psi(r_1, r_2, r_{12})|_{r_{12} \to 0}} = \frac{1}{2}.\qquad(8.59)$$

However, this is extremely difficult to realize in practice, since in that case the wave function (8.57) should satisfy the two-particle cusp ratios locally, i.e., in every two-particle coalescence point of configuration space. Therefore we utilize in the variational procedure the averaged two-particle cusp ratios, which follow from Eq. (8.59) and traditionally occur in variational calculations (see, for instance, Ref. [67]):

$$
\nu_1 = \nu_2 = \frac{\left\langle \Psi(r_1, r_2, r_{12}) \left| \delta(\mathbf{r}_1) \frac{\partial}{\partial r_1} \right| \Psi(r_1, r_2, r_{12}) \right\rangle}{\langle \Psi(r_1, r_2, r_{12}) | \delta(\mathbf{r}_1) | \Psi(r_1, r_2, r_{12}) \rangle}
$$

$$
= - \frac{\sum\limits_{i,j=1}^{N} D_i D_j \left(\frac{\alpha_i}{(\beta_i+\gamma_i+\beta_j+\gamma_j)^3} + \frac{\beta_i}{(\alpha_i+\gamma_i+\beta_j+\gamma_j)^3} + \frac{\alpha_i}{(\beta_i+\gamma_i+\alpha_j+\gamma_j)^3} + \frac{\beta_i}{(\alpha_i+\gamma_i+\alpha_j+\gamma_j)^3} \right)}{\sum\limits_{i,j=1}^{N} D_i D_j \left(\frac{1}{(\beta_i+\gamma_i+\beta_j+\gamma_j)^3} + \frac{1}{(\alpha_i+\gamma_i+\beta_j+\gamma_j)^3} + \frac{1}{(\beta_i+\gamma_i+\alpha_j+\gamma_j)^3} + \frac{1}{(\alpha_i+\gamma_i+\alpha_j+\gamma_j)^3} \right)},
$$

$$
\nu_{12} = \frac{\left\langle \Psi(r_1, r_2, r_{12}) \left| \delta(\mathbf{r}_{12}) \frac{\partial}{\partial r_{12}} \right| \Psi(r_1, r_2, r_{12}) \right\rangle}{\langle \Psi(r_1, r_2, r_{12}) | \delta(\mathbf{r}_{12}) | \Psi(r_1, r_2, r_{12}) \rangle} \qquad (8.60)
$$

$$
= - \frac{\sum\limits_{i,j=1}^{N} D_i D_j \frac{\gamma_i}{(\alpha_i+\beta_i+\alpha_j+\beta_j)^3}}{\sum\limits_{i,j=1}^{N} D_i D_j \frac{1}{(\alpha_i+\beta_i+\alpha_j+\beta_j)^3}}.
$$

These two-particle cusp conditions can be regarded as "weak", since they are less stringent than those given by Eq. (8.59). The wave function satisfying the two-particle cusp conditions given by Eq. (8.60) will be referred to as 2PC. One can also derive from Kato's cusp conditions (8.59) the cusp ratios in the three-particle coalescence point. These are given by [57, 68]:

$$
\mu_1 = \mu_2 = \frac{\left. \dfrac{\partial \Psi(r_1, r_2, r_{12})}{\partial r_1} \right|_{r_1,r_2,r_{12}\to 0}}{\Psi(r_1, r_2, r_{12})|_{r_1,r_2,r_{12}\to 0}} = - \frac{\sum\limits_{j=1}^{N} D_j \left(\alpha_j + \beta_j \right)}{2 \sum\limits_{j=1}^{N} D_j},
$$

$$
\mu_{12} = \frac{\left. \dfrac{\partial \Psi(r_1, r_2, r_{12})}{\partial r_{12}} \right|_{r_1,r_2,r_{12}\to 0}}{\Psi(r_1, r_2, r_{12})|_{r_1,r_2,r_{12}\to 0}} = - \frac{\sum\limits_{j=1}^{N} D_j \gamma_j}{\sum\limits_{j=1}^{N} D_j}. \qquad (8.61)
$$

We will refer to the wave function satisfying these three particle cusp conditions as 3PC.

For finding variational parameters A_1, A_2, B_1, B_2, C_1 and C_2 we used Newton-type minimization method [69]. Table 8.2 presents the values of the parameters A_1, A_2, B_1, B_2, C_1 and C_2 optimized in the case of two-particle cusp conditions (8.60)

Table 8.2 The optimized values of the parameters A_1, A_2, B_1, B_2, C_1 and C_2 as functions of N in the case of the 2PC function

N	A_1	A_2	B_1	B_2	C_1	C_2	$\nu_1 = \nu_2$	ν_{12}	$\mu_1 = \mu_2$	μ_{12}	E
10	1.4692	1.8233	1.5396	2.1185	0.0001	0.6290	-1.9999	0.4999	-1.9319	0.4425	-2.9036300
20	1.3372	1.3701	1.2095	1.9006	0.1259	1.1792	-2.0000	0.5000	-1.9356	0.4112	-2.9037049
30	1.4066	1.9011	1.2911	2.0469	0.0546	1.1871	-2.0000	0.5000	-1.9594	0.4238	-2.9037200
40	0.7485	3.8705	1.0572	2.8256	0.0000	1.1305	-1.9999	0.4999	-1.9538	0.4557	-2.9037208
50	1.5364	2.4587	0.6015	1.9085	0.2798	1.3094	-2.0000	0.5000	-1.9513	0.4476	-2.9037231
60	0.9475	2.3578	1.0382	2.1392	0.0948	1.9572	-1.9999	0.4999	-1.9550	0.4625	-2.9037242

Table 8.3 The same as in Table 8.2, but in the case of the 3PC function

N	A_1	A_2	B_1	B_2	C_1	C_2	$\mu_1 = \mu_2$	μ_{12}	$\nu_1 = \nu_2$	ν_{12}	E
10	0.7663	2.2086	1.8551	2.0062	0.0000	1.3536	-1.9999	0.4999	-1.9925	0.6125	-2.9033159
20	0.4192	4.5346	1.2151	1.7010	0.0000	1.8615	-2.0000	0.4999	-2.0034	0.5288	-2.9036494
30	0.2561	4.6882	1.1742	1.7592	0.0000	1.7725	-2.0000	0.4999	-2.0106	0.4803	-2.9036911
40	0.5470	3.1393	0.3584	3.2475	0.1225	1.7708	-1.9999	0.5000	-1.9991	0.5135	-2.9036961
50	0.5866	3.1204	0.6441	3.2376	0.2488	1.7700	-1.9999	0.4999	-2.0062	0.5345	-2.9037083
60	0.3370	4.5752	0.2431	2.6201	0.0582	2.3883	-2.0000	0.4999	-2.0020	0.5019	-2.9037211

for different values of N in Eq. (8.57). Here E is a minimized energy value (in au) for the ground state of helium. The calculated values of $\nu_1 = \nu_2$, ν_{12} and $\mu_1 = \mu_2$, μ_{12} indicate a quality of both cusp ratios. Table 8.3 presents the analogous results in the case of three-particle cusp conditions (8.61). As can be seen, in both cases the calculated cusp ratios for $N = 60$ are in a good agreement with exact ones, $\nu_1 = \nu_2 = \mu_1 = \mu_2 = -2$ and $\nu_{12} = \mu_{12} = 0.5$, and yield the energy values $E = -2.903724$ au and $E = -2.903721$ au, respectively. Functions satisfying the two-particle and three-particle cusp ratios, (8.60) and (8.61), are referred to as 2PC and 3PC, respectively.

In calculations of the $(e, 3e)$ experiments by Lahmam-Bennani et al. [66] a 3C model [63] has been employed for the final, doubly ionized state of helium

$$\Psi_{3C}^{-}(\mathbf{p}_1, \mathbf{p}_2; \mathbf{r}_1, \mathbf{r}_2) = e^{-i\mathbf{p}_{12}\mathbf{r}_{12}}\varphi_1^{-}(\mathbf{p}_1, \mathbf{r}_1)\varphi_2^{-}(\mathbf{p}_2, \mathbf{r}_2)\varphi_{12}^{-}(\mathbf{p}_{12}, \mathbf{r}_{12}), \qquad (8.62)$$

where $\mathbf{p}_1$ and $\mathbf{p}_2$ are the asymptotic electron momenta, $\mathbf{p}_{12} = (\mathbf{p}_1 - \mathbf{p}_2)/2$, and φ^{-} is an outgoing Coulomb wave. The five-fold differential cross section evaluates as

$$\frac{d^5\sigma}{dE_1 dE_2 d\Omega_1 d\Omega_2 d\Omega_s} = \frac{2p_s p_1 p_2}{(2\pi)^8 p_0} |T|^2, \qquad (8.63)$$

where p_0 and p_s are the momenta of incident and scattered electrons, respectively. The amplitude T is given by

$$T(\mathbf{Q}; \mathbf{p}_1, \mathbf{p}_2) = -\frac{4\pi}{Q^2}\sum_j D_j \frac{\partial^3}{\partial a_j \partial b_j \partial \gamma_j} \int \frac{d\mathbf{p}}{(2\pi)^3} I_{12}(\gamma_j; \mathbf{p}_{12}, \mathbf{p} + \mathbf{p}_{12})$$

$$\times [I_1(a_j; \mathbf{p}_1, \mathbf{Q} - \mathbf{p})I_2(b_j; \mathbf{p}_2, \mathbf{p}) + I_1(a_j; \mathbf{p}_1, -\mathbf{p})I_2(b_j; \mathbf{p}_2, \mathbf{Q} + \mathbf{p})$$

$$- 2I_1(a_j; \mathbf{p}_1, -\mathbf{p})I_2(b_j; \mathbf{p}_2, \mathbf{p})], \qquad (8.64)$$

where

$$I_n(\lambda; \mathbf{p}_n, \mathbf{k}) = \int \frac{d\mathbf{r}}{r}\varphi^{-*}(\mathbf{p}_n, \mathbf{r})e^{-\lambda r + i\mathbf{k}\mathbf{r}}$$

$$= 4\pi N^*(\xi_n)\frac{[(\lambda - ip_n)^2 + k^2]^{i\xi_n}}{[(\mathbf{k} - \mathbf{p}_n)^2 + \lambda^2]^{(1+i\xi_n)}} \qquad (n = 1, 2, 12). \qquad (8.65)$$

Here ξ_n is a corresponding Coulomb number and $N(\xi) = e^{-\pi\xi/2}\Gamma(1 - i\xi)$. For the first time such method for calculating with a 3C function was proposed in Ref. [70], where it was applied to the double photoionization using a BK function [55] for the ground state of helium.

Figure 8.5 displays numerical results for the $(e, 3e)$ experiments by Lahmam-Bennani et al. [66]. Both functions 2PC and 3PC give almost identical cross sections that are visually indistinguishable. Our results are very close to those using a BK function [59]. Also are shown the results using a helium function by Le Sech [68]. Such a choice is due to the fact that the BK function gives an accurate energy value

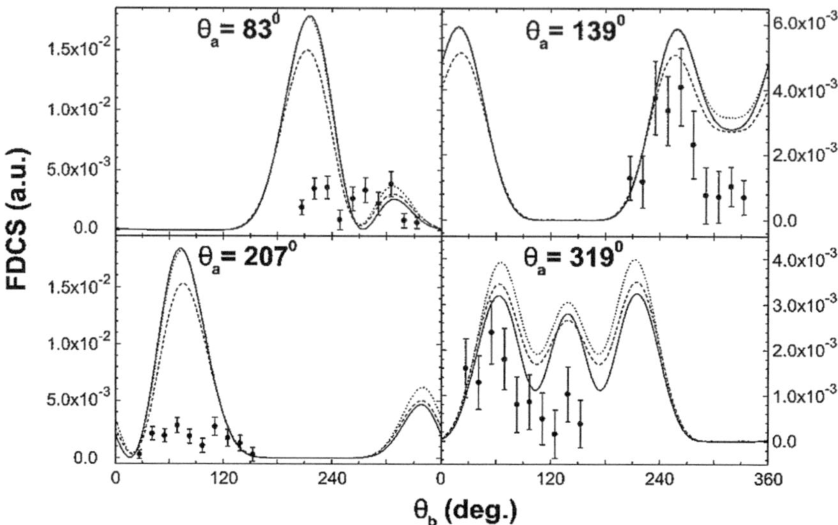

Fig. 8.5 The five-fold differential cross sections for the experiment [66]. Theoretical curves are the results of 3C calculations using the following wave functions of helium: 2PC (solid line), BK (dashed line), Le Sech (dotted line)

($E = -2.9035$ au), but not exact cusp condition in the triple coalescence point, while the Le Sech function giving not so accurate energy value ($E = -2.9020$ au) exactly satisfies the cusp condition in the triple coalescence point. Note that in the cases $\theta_a = 83°$, $207°$ the results using the PL function, which are presented in Refs. [59, 62], exhibit similar departures from the experiment.

Figure 8.6 shows the numerical results for the recent symmetric noncoplanar $(e, 3 - 1e)$ and $(e, 2e)$ experiments by Watanabe et al. [24]. In these experiments two fast outgoing electrons, the scattered and ejected ones, having equal energies $E_s = E_a = 1000$ eV and polar angles $\theta_s = \theta_a = 45°$ are detected. In the $(e, 3 - 1e)$ case, the slow ejected electron, which is not detected, has energy $E_b = 10$ eV. The $(e, 2e)$ and $(e, 3 - 1e)$ cross sections are studied as functions of the momentum $\mathbf{q} = \mathbf{p}_s + \mathbf{p}_a - \mathbf{p}_0$. The calculations have been performed in PWIA, and these results have been folded with the experimental momentum resolution. In the $(e, 3 - 1e)$ case the results using the 2PC and 3PC are very close to each other and also to those using the BK function. Interestingly, the PL function better reproduces the $(e, 3 - 1e)$ experiment on an absolute scale, but it clearly fails in the case of $(e, 2e)$ transition to the $n = 2$ excited state of He$^+$. The marked discrepancies between theory and experiment in Fig. 8.6 show that the PWIA model is insufficient for a description of ionization-excitation and double ionization in the considered kinematical regime. These observations call for new experiments at higher incident energies in order to test the validity of the PWIA.

Fig. 8.6 The triple (upper panel) and four-fold (lower panel) differential cross sections for the $(e, 2e)$ and $(e, 3 - 1e)$ experiments [24], respectively. The curves are the results of PWIA calculations, convoluted with the experimental resolution function, using the following wave functions of helium: 2PC [25] (solid line), BK (dashed line), and PL (dotted line)

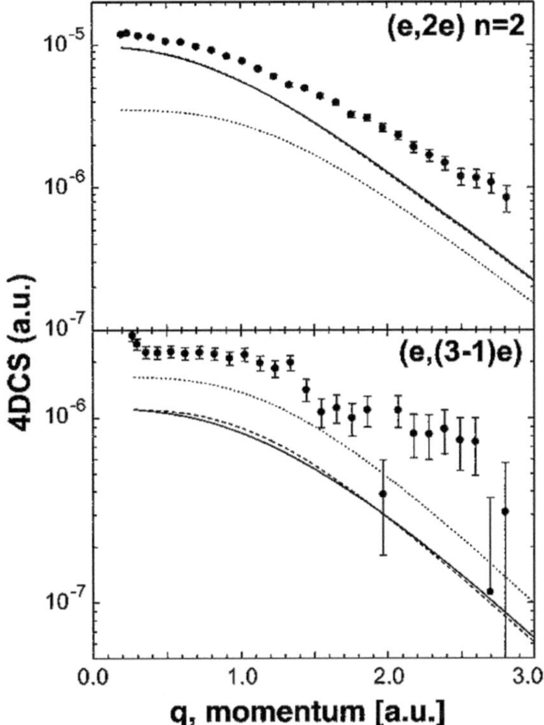

From the results of the present study we can conclude that in the angular and energy regions reachable by the nowadays $(e, 3e)$ experiments the Kato's cusp conditions do not give any advantage. In addition, the functions giving an accurate energy value for the ground state of helium give similar results for cross sections in the framework of the 3C and PWIA models.

Note that in [49], different models of the capture reaction were used, with ionization $p + \mathrm{He} \rightarrow \mathrm{H} + \mathrm{He}^{++} + e$ at very small hydrogen scattering angles $\theta_p = 0.1 - 0.5$ mrad and the proton energy $E_p = 0.15 - 1.4$ MeV. It was proposed in [71] to use them in order to obtain new information on the structure of the wave function of the target in the impulse approximation. An assumption was made in [72] that upon construction of the theoretical model, the requirement of resonance capture can be omitted, and only the pole mechanism can be considered. However, in calculations with different variational functions and calculations with exact functions for the charge transfer reaction between the hydrogen atom and fast protons, second order effects should be taken into account for quantitative agreement with experimental data [73].

8.7   Spectrum of Beryllium Dimer in Ground $X^1\Sigma_g^+$ state

In quantum chemical calculations, effective potentials of interatomic interaction are presented in the form of numerical tables calculated with a limited accuracy and defined on a nonuniform mesh of nodes in a finite range of interatomic distances. It is important that the proposed finite element method with the interpolation Hermite polynomials ensures smooth matching of the tabulated potential with its analytical asymptotic expression, as well as high-quality smooth approximation of eigenfuntions [74]. Consider, e.g., the Schrödinger equation for a diatomic beryllium molecule in the adiabatic approximation, commonly referred to as Born-Oppenheimer approximation

$$\left(-s_2 \frac{1}{r^2} \frac{d}{dr} r^2 \frac{d}{dr} + V_J(r) - E \right) \Phi_J(r) = 0,$$

$$V_J(r) = V(r) + s_2 \frac{J(J+1)}{r^2}, \quad s_2 = \frac{\hbar^2}{2m} \frac{1}{\text{Å}^2}. \tag{8.66}$$

Here J is the total angular momentum quantum number, $m = M/2 = 4.506$ Da is the reduced mass of beryllium molecule expressed in 1 Da = 931.494061 MeV/c^2 atomic mass unit (u) [75], 1 eV = 8065.54429 cm^{-1}, $\hbar c = 1973.269718$ eV Å, and $\hbar^2/(2m) = 3.741151852 \cdot 10^{-8}$ Å. The factor $1/\text{Å}^2$ in s_2 means that the distance r between atoms is expressed in Å, and $s_2 = 3.741151852$ cm^{-1}. Also E is the energy in cm^{-1} and $V(r)$ is potential energy curve (PEC) in cm^{-1}.

In Ref. [74] the potential $V(r)$ (in cm^{-1}) is given by the modified expanded Morse oscillator (MEMO) potential function which is an approximation of the MEMO tabular values $\{V(r_i)\}_{i=1}^{76}$ in interval $r \in [r_1 = 1.5, r_{76} = 48]$. Here and below, the value r is given in units of Å unless otherwise is indicated. These tabular values were chosen to provide better approximation of the potential $V(r)$ by the fifth-order Lagrange interpolation polynomials (LIPs) of the variable r in subintervals $r \in [r_{5k-4}, r_{5k+1}], k = 1, \ldots, 15$. Note that the tabular values give a smooth approximation till $r_{49} = 12$, where the approximate PEC is matched with the asymptotic potential $V_{as}^{BO}(r) = s_2 U_{as}^{BO}(r)$ given analytically by the expansions [76]

$$U_{as}^{BO}(r) = s_1 \tilde{V}_{as}^{BO}(r), \quad \tilde{V}_{as}^{BO}(r) = -\left(\frac{214}{Z^6} + \frac{10230}{Z^8} + \frac{504300}{Z^{10}} \right), \tag{8.67}$$

where $s_1 = \text{aue}/s_2 = 58664.99239$ or $s_1 s_2 = \text{aue} = 219474.6314$ cm^{-1}, $Z = r/s_3$ and $s_3 = 0.52917$ is the Bohr radius in Å. This fact allows considering the interval $r \in [r_{\text{match}} \geq 12, \infty)$ as possible for using the asymptotic potential $V_{as}^{BO}(r)$ at large r and executing conventional calculations based on tabular values of $V(r)$ in the finite interval $r \in [r_1, r = 12]$ (see also [77]).

In Ref. [77], the potential $V(r)$ (in cm^{-1}) is given by the BO potential function plus relativistic potential function marked as STO with tabular values $\{V(Z_i)\}_{i=1}^{28}$ in the interval $Z \in [Z_1 = 3.75, Z_{28} = 25]$ au which corresponds to $r \in [r_1 = 1.9843,$

Table 8.4 Comparison of the vibrational spectra $E_{vJ=0} - E_{v=0J=0}$ (in cm^{-1}) for the $X^1\Sigma_g^+$ state of the beryllium dimer: eigenenergies $E_{vJ=0}$ of vibrational-rotational bound states (in cm^{-1}) calculated by means STO and MEMO PECs and experimental (Exp) results. Here D_e is the well depth and $D_0 = -E_{v=0J=0}$ is the dissociation energy in cm^{-1}, r_e is the equilibrium internuclear distance in Å, rms is root-mean-square discrepancy between the theoretical and experimental data

v	STO	MEMO	Exp
r_e	2.447	2.4534	2.4536
D_e	934.4	929.804	929.7±2
D_0	807.7	806.07	807.4
1	223.5	222.50	222.6
2	398.2	397.34	397.1
3	519.3	517.71	518.1
4	595.7	594.89	594.8
5	652.2	651.91	651.5
6	699.3	698.92	698.8
7	738.1	737.72	737.7
8	768.6	768.27	768.2
9	790.1	789.74	789.9
10	802.6	801.66	802.6
11	807.2	805.74	
rms	0.7	0.4	

$r_{28} = 13.229$]. One can see that these tabular values were chosen to provide the best approximation of the potential $V(r)$ by the fourth-order LIPs of the variable r in subintervals $r \in [r_{4k-3}, r_{4k+1}]$, $k = 1, \ldots, 6$. On interval $Z \in [Z_{25}, Z_{\text{match}} = 27.5]$ au we consider the approximation of the potential $V(r)$ by the fifth-order HIP using the values of the potential $V(Z)$ at the points $Z = \{Z_{25} = 17.5, Z_{26} = 20.0, Z_{27} = 22.5, Z_{28} = 25.0\}$ au and the values of the asymptotic potential $V_{\text{as}}(r)$ and its derivative $dV_{\text{as}}(Z)/dZ = s_3 dV_{\text{as}}(r)/dr$ at the point $Z = Z_{\text{match}} = 27.5$ au In the interval $r \in [r_{\text{match}} = 14.552, \infty)$ the potential $V_{\text{as}}(r) = s_2 U_{\text{as}}(r)$ is approximated by the asymptotic expansion

$$U_{\text{as}}(r) = s_1 \tilde{V}_{\text{as}}(r), \quad \tilde{V}_{\text{as}}(r) = \tilde{V}_{\text{as}}^{\text{BO}}(r) + \tilde{V}_{\text{as}}^{\text{rel}}(r),$$

$$\tilde{V}_{\text{as}}^{\text{rel}}(r) = -\left(\frac{1.839 \cdot 10^{-4}}{Z^4} + \frac{0.11944}{Z^6} + \frac{19.582}{Z^8} - \frac{1323.5}{Z^{10}} \right),$$

where $\tilde{V}_{\text{as}}^{\text{BO}}(r)$ and $\tilde{V}_{\text{as}}^{\text{rel}}(r)$ are given by Eq. (8.67) and Ref. [77], respectively.

The results were obtained using KANTBP 4M [78] and KANTBP 3.0 [79] codes.

The potential functions MEMO and STO $V_J(r)$ from $J = 0$ to $L = 36$ support 252 and 253 vibrational-rotational energy levels E_{vL}, respectively. The calculated $v = 1 - 10$ vibrational spectra $E_{vJ=0} - E_{v=0J=0}$ (in cm^{-1}) are in good agreement with the experimental results [80] (see the Table 8.4). Note that there is no energy

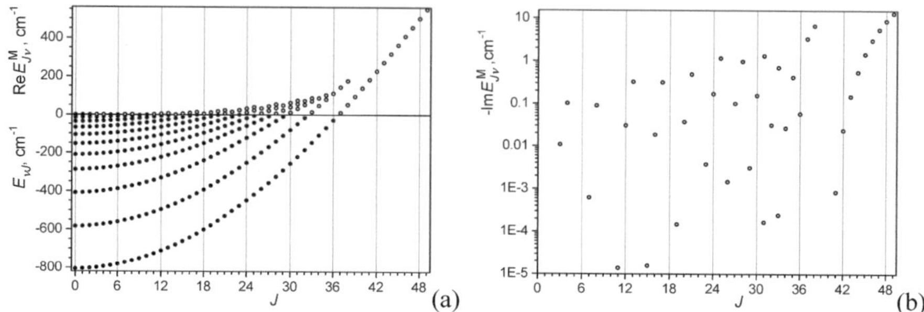

Fig. 8.7 **a** Eigenenergies E_{vJ} of vibrational-rotational bound states (black dots), the real part $\Re E_{Jv}^{M}$ (empty circles) and **b** minus the imaginary part $-\Im E_{Jv}^{M}$ of complex eigenenergies $E_{Jv}^{M} = \Re E_{Jv}^{M} + \imath \Im E_{Jv}^{M}$ of rotational-vibrational metastable states in cm^{-1}

level $E_{v=8,J=14}$ for MEMO PEC while $E_{v=8,J=14} = -1.4$ (in cm^{-1}) for STO PEC. Figure 8.7a (black dots) shows also the vibrational-rotational spectrum E_{vJ} (in cm^{-1}) of Be$_2$ vs J for MEMO PEC.

The potential functions MEMO and STO $V_J(r)$ at $J = 3, 4, 7, 8, 9, 11, \ldots, 49$ support 62 and 64 rotational-vibrational metastable energy levels $E_{Jv}^{M} = \Re E_{Jv}^{M} + \imath \Im E_{Jv}^{M}$, respectively. The rotational-vibrational metastable E_{Jv}^{M} (in cm^{-1}) vs L for MEMO PEC are shown in Fig. 8.7a (real parts $\Re E_{Jv}^{M}$ are empty circles) and Fig. 8.7b (minus the imaginary part $-\Im E_{Jv}^{M}$).

8.8　Accurate Calculations for the Dirac Electron in the Field of Two-Center Coulomb Field

The one-electron wave function $\Psi(\mathbf{r})$ is a solution of the stationary Dirac equation

$$H_D \Psi(\mathbf{r}) = E \Psi(\mathbf{r}), \tag{8.68}$$

where $E = E_R$ is the relativistic energy, and H_D is the two-center Dirac hamiltonian in atomic units ($e = m_e = \hbar = 1$):

$$H_D = c\boldsymbol{\alpha}\mathbf{p} + c^2\beta + U(\mathbf{r})\mathbf{I}. \tag{8.69}$$

Here c is the speed of light, $\boldsymbol{\alpha}$ and β are Dirac matrices:

$$\alpha = \begin{pmatrix} 0 & \sigma \\ \sigma & 0 \end{pmatrix}, \quad \sigma = (\sigma_x, \sigma_y, \sigma_z), \quad \beta = \begin{pmatrix} I & 0 \\ 0 & -I \end{pmatrix},$$

$$\sigma_x = \begin{pmatrix} 0 & 1 \\ 1 & 0 \end{pmatrix}, \quad \sigma_y = \begin{pmatrix} 0 & -\imath \\ \imath & 0 \end{pmatrix}, \quad \sigma_z = \begin{pmatrix} 1 & 0 \\ 0 & -1 \end{pmatrix}, \tag{8.70}$$

$$\mathbf{I} = \begin{pmatrix} I & 0 \\ 0 & I \end{pmatrix}, \quad I = \begin{pmatrix} 1 & 0 \\ 0 & 1 \end{pmatrix}.$$

The two-center attractive Coulomb potential $U(\mathbf{r})$ is defined by

$$U(\mathbf{r}) = V(r_1) + V(r_2), \quad V(r) = -\frac{Z}{r},$$

where Z is the point charge of the centers 1 and 2, $\mathbf{r}_1 = \mathbf{r} + \rho/2$, $\mathbf{r}_2 = \mathbf{r} - \rho/2$, $\rho = (0, 0, \rho)$, and ρ is the internuclear distance.

The Slater-type spinor orbitals centered on the nuclei are given as in [81] by

$$\psi_{njlm}(\mathbf{r}) \equiv \psi_{n\kappa m}(\mathbf{r}) = \begin{cases} \imath P_{n\kappa}(r)\Omega_{+\kappa m}(\theta, \varphi) \\ Q_{n\kappa}(r)\Omega_{-\kappa m}(\theta, \varphi). \end{cases} \tag{8.71}$$

Here the spinor spherical harmonics

$$\Omega_{jlm}(\theta, \varphi) \equiv \Omega_{\kappa m}(\theta, \varphi)$$

$$= \begin{cases} \begin{cases} +\sqrt{\frac{l+m+1/2}{2l+1}}\, Y_{lm-1/2}(\theta, \varphi), \\ +\sqrt{\frac{l-m+1/2}{2l+1}}\, Y_{lm+1/2}(\theta, \varphi), \end{cases} & \kappa < 0, \quad l = -\kappa - 1, \\[2mm] \begin{cases} -\sqrt{\frac{l-m+1/2}{2l+1}}\, Y_{lm-1/2}(\theta, \varphi), \\ +\sqrt{\frac{l+m+1/2}{2l+1}}\, Y_{lm+1/2}(\theta, \varphi), \end{cases} & \kappa > 0, \quad l = \kappa, \end{cases} \quad j = |\kappa| - 1/2, \tag{8.72}$$

are eigenfunctions of the operator $\sigma \mathbf{L}$:

$$(-\sigma \mathbf{L} - I)\Omega_{\kappa m}(\theta, \varphi) = \kappa \Omega_{\kappa m}(\theta, \varphi), \tag{8.73}$$

where $\mathbf{L}$ is the orbital angular momentum, and $Y_{lm}(\theta, \varphi)$ is the spherical harmonic, according to the Condon and Shortley phase convention [82]. The spinor spherical harmonics satisfy the following equations:

$$\sigma \hat{r}\, \Omega_{+m}(\theta, \varphi) = -\Omega_{-\kappa m}(\theta, \varphi), \quad \hat{r} = \mathbf{r}/r, \tag{8.74}$$

$$\sigma \mathbf{p}\, \Omega_{+\kappa m}(\theta, \varphi) = \imath \left(\frac{\partial}{\partial r} + \frac{1+\kappa}{r} \right) \Omega_{-\kappa m}(\theta, \varphi). \tag{8.75}$$

When the potential $U(\mathbf{r}) \equiv V(r)$ is spherically symmetrical, the large $P_{n\kappa}(r)$ and small $Q_{n\kappa}(r)$ radial components satisfy the following coupled equations

$$+c\left[\frac{\partial}{\partial r} + \frac{1-\kappa}{r}\right] Q_{n\kappa}(r) + \left(V(r) + c^2 - E_{n\kappa}\right) P_{n\kappa}(r) = 0,$$

$$-c\left[\frac{\partial}{\partial r} + \frac{1+\kappa}{r}\right] P_{n\kappa}(r) + \left(V(r) - c^2 - E_{n\kappa}\right) Q_{n\kappa}(r) = 0. \tag{8.76}$$

The particularity of our approach is that, for the two-center problem considered, the one center basis large $P_{n\kappa}(r)$ and small $Q_{n\kappa}(r)$ radial components located on each center are chosen in the form

$$P_{n\kappa}(r) = (-1)^l \sum_{n_p=1}^{n_{p\,\mathrm{max}}} c_{n_p\kappa}\, p_{n_p\kappa}(r),$$

$$Q_{n\kappa}(r) = (-1)^l \sum_{n_q=1}^{n_{q\,\mathrm{max}}} d_{n_q\kappa}\, q_{n_q\kappa}(r), \tag{8.77}$$

Here $(-1)^l$ insures the the inversion symmetry for the Dirac wave function (see [83] for a detailed explanation of the relativistic molecular symmetry spinors for diatomics). $p_{n_p\kappa}(r)$ and $q_{n_q\kappa}(r)$ are the normalized Slater type orbitals with non-integer principal quantum numbers (NISTOs) [84]:

$$p_{n_p\kappa}(r) = \frac{(2\lambda_{p\kappa})^{\gamma_\kappa+n_p-1/2}}{\sqrt{\Gamma(2\gamma_\kappa + 2n_p - 1)}} r^{\gamma_\kappa+n_p-2} \exp(-\lambda_{p\kappa} r),$$

$$q_{n_q\kappa}(r) = \frac{(2\lambda_{q\kappa})^{\gamma_\kappa+n_q-1/2}}{\sqrt{\Gamma(2\gamma_\kappa + 2n_q - 1)}} r^{\gamma_\kappa+n_q-2} \exp(-\lambda_{q\kappa} r), \tag{8.78}$$

$\lambda_{p\kappa} > 0$ and $\lambda_{q\kappa} > 0$ are variational parameters. These functions satisfy the following uncoupled system of equations

$$\left[\frac{\partial}{\partial r} + \frac{1+\kappa}{r}\right] p_{n_p\kappa}(r) = \left[-\lambda_{p\kappa} + \frac{\gamma_\kappa + n_p + \kappa - 1}{r}\right] p_{n_p\kappa}(r),$$

$$\left[\frac{\partial}{\partial r} + \frac{1-\kappa}{r}\right] q_{n_q\kappa}(r) = \left[-\lambda_{q\kappa} + \frac{\gamma_\kappa + n_q - \kappa - 1}{r}\right] q_{n_q\kappa}(r). \tag{8.79}$$

The non-integer parameter γ_κ is chosen from the asymptotic behavior of the large $P_{n\kappa}(r)$ and small $Q_{n\kappa}(r)$ components near the corresponding center

$$\gamma_\kappa = \gamma_{-\kappa} = \sqrt{\kappa^2 - \frac{Z^2}{c^2}}. \tag{8.80}$$

The total Slater-type spinor function centered on a nucleus has the following compact form

Table 8.5 Total numbers of NISTO depending on the maximal principal quantum number $N_{\max}$

κ	States	$N_{\max} = 2$		$N_{\max} = 3$		$N_{\max} = 4$		$N_{\max} = 5$	
		$n_{p\,\max}$	$n_{q\,\max}$	$n_{p\,\max}$	$n_{q\,\max}$	$n_{p\,\max}$	$n_{q\,\max}$	$n_{p\,\max}$	$n_{q\,\max}$
-1	$s_{1/2}$	2	2	3	3	4	4	5	5
1	$p_{1/2}$	2	3	3	4	4	5	5	6
-2	$p_{3/2}$	1	1	2	2	3	3	4	4
2	$d_{3/2}$			2	3	3	4	4	5
-3	$d_{5/2}$			1	1	2	2	3	3
3	$f_{5/2}$					2	3	3	4
-4	$f_{7/2}$					1	1	2	2
4	$g_{7/2}$							2	3
-5	$g_{9/2}$							1	1
Total number		11		24		41		62	

$$
\psi_{N_{\max}}(\mathbf{r}) = \begin{cases} \iota \sum_{n=1}^{N_{\max}} \sum_{\kappa=-n,\,\kappa\neq0}^{n-1} P_{n\kappa}(r)\,\Omega_{+\kappa m}(\theta, \varphi), \\ \sum_{n=1}^{N_{\max}} \sum_{\kappa=-n,\,\kappa\neq0}^{n-1} Q_{n\kappa}(r)\,\Omega_{-\kappa m}(\theta, \varphi), \end{cases} \tag{8.81}
$$

where $N_{\max}$ is a maximal principal quantum number. To avoid spurious solutions the highest power $n_{p\,\max}$, $n_{q\,\max}$ of r of Eq. (8.77) are chosen as [85]

$$
n_{p\,\max} = n - |\kappa| + 1, \quad n_{q\,\max} = \begin{cases} n_{p\,\max}, & \kappa < 0, \\ n_{p\,\max} + 1, & \kappa > 0. \end{cases} \tag{8.82}
$$

The total numbers of NISTOs depending on the $N_{\max} = 2 - 5$ are presented in Table 8.5, where we can see that the total number of nonlinear parameters $\lambda_{p\kappa}$ and $\lambda_{q\kappa}$ is $2N_{\max} - 1$, i.e. the one which brings the total number of nonlinear parameters to $4N_{\max} - 2$.

The one-electron wave function $\Psi(\mathbf{r})$ for Eq. (8.68) has the form

$$
\Psi(\mathbf{r}) = \psi_{N_{\max}}(\mathbf{r}_1) + \psi_{N_{\max}}(\mathbf{r}_2), \tag{8.83}
$$

for the $\Psi(\mathbf{r}) = \Psi(-\mathbf{r})$ gerade case with the condition $\langle \Psi(\mathbf{r}) | \Psi(\mathbf{r}) \rangle = 1$. Finally, the minimax formulation [86] of the Dirac equation (8.68) is

$$
-c^2 \leq E_R = \min_{P_{n\kappa}(r)\neq0} \max_{Q_{n\kappa}(r)} \langle H_D \rangle = \min_{\lambda_{p\kappa}} \max_{\lambda_{q\kappa}} \langle H_D \rangle \leq c^2. \tag{8.84}
$$

Making the Rayleigh quotient (8.84) stationary with respect to the variations in $c_{n_p\kappa}$ and $d_{n_q\kappa}$, leads to the generalized eigenvalue problem

$$\mathbf{A}\begin{pmatrix}\mathbf{c}\\\mathbf{d}\end{pmatrix} = E_R\mathbf{B}\begin{pmatrix}\mathbf{c}\\\mathbf{d}\end{pmatrix}. \tag{8.85}$$

The determination of the matrix elements of $\langle\Psi(\mathbf{r})|\Psi(\mathbf{r})\rangle$ and $\langle\Psi(\mathbf{r})|H_D|\Psi(\mathbf{r})\rangle$ needs the determination of two center integrals such as

$$Q_{l_1 l_2}^{m_1 m_2}(a, v, \rho) = \int d\mathbf{r}\, r_1^{-1} r_2^{v-1} e^{-ar_2} Y_{l_1 m_1}^*(\mathbf{r}_2) Y_{l_2 m_2}(\mathbf{r}_2)$$

$$= (-1)^{l_1-l_2} \int d\mathbf{r}\, r_2^{-1} r_1^{v-1} e^{-ar_1} Y_{l_1 m_1}^*(\mathbf{r}_1) Y_{l_2 m_2}(\mathbf{r}_1), \tag{8.86}$$

and

$$F_{l_1 l_2}^{m_1 m_2}(a_1, v_1, a_2, v_2, \rho) = \int d\mathbf{r}\, r_1^{v_1-1} e^{-a_1 r_1} Y_{l_1 m_1}^*(\mathbf{r}_1) r_2^{v_2-1} e^{-a_2 r_2} Y_{l_2 m_2}(\mathbf{r}_2)$$

$$= (-1)^{l_1-l_2} \int d\mathbf{r}\, r_2^{v_1-1} e^{-a_1 r_2} Y_{l_1 m_1}^*(\mathbf{r}_2) r_1^{v_2-1} e^{-a_2 r_1} Y_{l_2 m_2}(\mathbf{r}_1). \tag{8.87}$$

Here a, a_1, a_2 are positive numbers; v, v_1, v_2 are non-integer numbers; $v > -1$, $\max(v_1, v_2) > 0$, $\min(v_1, v_2) > -1$. Using the Fourier transform [84]

$$Q_{l_1 l_2}^{m_1 m_2}(a, v, \rho) = 8 \sum_{l=|l_1-l_2|,2}^{l_1+l_2} (-1)^l Y_{l0}(\boldsymbol{\rho})\, \Upsilon_{ll_1 l_2}^{0m_1 m_2}$$

$$\times \int_0^\infty dp\, j_l(p\rho) g_l(a, v, p), \tag{8.88}$$

and

$$F_{l_1 l_2}^{m_1 m_2}(a_1, v_1, a_2, v_2, \rho) = 8 \sum_{l=|l_1-l_2|,2}^{l_1+l_2} (-1)^{\frac{3l+l_1-l_2}{2}} Y_{l0}(\boldsymbol{\rho})\, \Upsilon_{ll_1 l_2}^{0m_1 m_2}$$

$$\times \int_0^\infty dp\, p^2 j_l(\rho p) g_{l_1}(a_1, v_1, p) g_{l_2}(a_2, v_2, p), \tag{8.89}$$

where

$$\Upsilon_{ll_1 l_2}^{mm_1 m_2} = \int d\Omega_p Y_{lm}(\mathbf{p}) Y_{l_1 m_1}^*(\mathbf{p}) Y_{l_2 m_2}(\mathbf{p}), \tag{8.90}$$

which has a non-zero value for

$$m = m_2 - m_1, \quad l = |l_1 - l_2|, |l_1 - l_2| + 2, \ldots, l_1 + l_2. \tag{8.91}$$

Here the sign $*$ denotes the complex conjugate, and $j_l(p)$ is a spherical Bessel function [87]. The integral (8.90) is calculated analytically using the Clebsch-Gordan coefficients. It can also be calculated directly using MAPLE without loss of precision. The function $g_l(a, v, p)$ has the form

$$g_l(a, v, p) = \int_0^\infty dr\, r^{v+1} \exp(-ar) j_l(pr), \quad v > -2. \tag{8.92}$$

Using the following relation for the integral of the product of two spherical Bessel functions [88]

$$\int_0^\infty dp\, j_l(p\rho) j_l(pr) = \frac{\pi}{2(2l+1)} \begin{cases} \rho^{-l-1} r^l, & \rho \geq r, \\ \rho^l r^{-l-1}, & \rho < r, \end{cases} \tag{8.93}$$

we obtain

$$\int_0^\infty dp\, j_l(p\rho) g_l(a, v, p) = \int_0^\infty dr\, r^{v+1} \exp(-ar) \int_0^\infty dp\, j_l(p\rho) j_l(pr)$$

$$= \frac{\pi}{2(2l+1)} \left(\frac{1}{\rho^{l+1}} \int_0^\rho dr\, r^{v+l+1} \exp(-ar) + \rho^l \int_\rho^\infty dr\, r^{v-l} \exp(-ar) \right)$$

$$= \frac{\pi}{2(2l+1)a^{v+1}} \left(\frac{1}{(a\rho)^{l+1}} \gamma(v+l+2, a\rho) + (a\rho)^l \Gamma(v-l+1, a\rho) \right), \tag{8.94}$$

where $\Gamma(a, x)$ and $\gamma(a, x)$ are respectively the upper and lower incomplete gamma functions [87]. This permits the analytical calculation of the integral in (8.88).

Now let us consider the integral (8.89), which will be calculated numerically. Using the following recurrence relations of the spherical Bessel function [87]

$$j_l(pr) = \frac{(2l-1)}{pr} j_{l-1}(pr) - j_{l-2}(pr) = \frac{l-1}{pr} j_{l-1}(pr) - \frac{1}{p}\frac{d}{dr} j_{l-1}(pr), \tag{8.95}$$

we obtain the following recurrence relations for the integral (8.92)

$$g_l(a, v, p) = \frac{a}{p}\frac{2l-1}{v-l+1} g_{l-1}(a, v, p) - \frac{l+v}{v-l+1} g_{l-2}(a, v, p) \tag{8.96}$$

with

$$g_0(a, v, p) = \frac{\Gamma(v+1)}{p\sqrt{(a^2+p^2)^{v+1}}} \sin\left((v+1) \arcsin\left(\frac{p}{\sqrt{a^2+p^2}} \right) \right),$$

$$g_1(a, v, p) = \frac{1+v}{p} g_0(a, v-1, p) - \frac{a}{p} g_0(a, v, p). \tag{8.97}$$

For finding variational parameters $\lambda_{p\kappa}$ and $\lambda_{q\kappa}$ we used minimax optimization method [89] based on Newton-type minimization method [69] in combination with negative curvature direction methods for non-convex-non-concave, convex-non-

Table 8.6 The relativistic $1\sigma_g$ state energy $E_e = E_R - c^2$ at $\rho = 2/Z$

Z	Ion	$N_{\max}$	Energy	Energy [91]
1	H_2^+	2	$-1.102\,248\,990$	
		3	$-1.102\,624\,606$	
		4	$-1.102\,640\,853$	
		5	$-1.102\,641\,574$	$-1.102\,641\,581$
2	He_2^{3+}	2	$-4.409\,083\,664$	
		3	$-4.410\,586\,724$	
		4	$-4.410\,651\,814$	
		5	$-4.410\,654\,700$	$-4.410\,654\,728$
10	Ne_2^{19+}	2	$-110.297\,363\,139$	
		3	$-110.335\,418\,709$	
		4	$-110.337\,130\,064$	
		5	$-110.337\,203\,677$	$-110.337\,204\,410$
20	Ca_2^{39+}	2	$-442.073\,489\,140$	
		3	$-442.231\,839\,925$	
		4	$-442.239\,683\,071$	
		5	$-442.239\,984\,905$	$-442.239\,997\,265$
30	Zn_2^{59+}	2	$-998.024\,713\,682$	
		3	$-998.404\,961\,752$	
		4	$-998.426\,014\,019$	
		5	$-998.426\,773\,718$	$-998.426\,763\,032$
40	Zr_2^{79+}	2	$-1782.802\,217\,068$	
		3	$-1783.539\,606\,894$	
		4	$-1783.585\,615\,512$	
		5	$-1783.587\,315\,610$	$-1783.587\,355\,445$
50	Sn_2^{99+}	2	$-2803.286\,356\,845$	
		3	$-2804.566\,122\,743$	
		4	$-2804.656\,452\,590$	
		5	$-2804.659\,770\,918$	$-2804.659\,807\,931$
60	Nd_2^{119+}	2	$-4069.063\,950\,398$	
		3	$-4071.139\,360\,212$	
		4	$-4071.304\,123\,628$	
		5	$-4071.309\,804\,433$	$-4071.309\,830\,161$

concave and non-convex-concave cases [69, 90]. In all our calculations the mass of the electron $m_e = 1$, and the speed of light $c = 137.0359895$ both in atomic units.

In Tables 8.6 and 8.7 are presented, the calculated relativistic $1\sigma_g$ state energy $E_e = E_R - c^2$ for different ions at $\rho = 2/Z$ versus the number $N_{\max}$ and compared to results given in [91]. From the results presented in Tables 8.6 and 8.7, we see that our calculated energies give the upper bounds of the exact energies with monotonic convergence with increasing $N_{\max}$. For $N_{\max} = 5$ our results are comparable to those [91] (which uses more than 600 basis sets) with a relative error of $10^{-7} - 10^{-8}$.

Table 8.7 Continuation of Table 8.6

Z	Ion	N_{max}	Energy	Energy [91]
70	Yb_2^{139+}	2	−5593.251 717 146	
		3	−5596.462 436 297	
		4	−5596.746 534 092	
		5	−5596.754 858 416	−5596.754 864 752
80	Hg_2^{159+}	2	−7393.964 617 394	
		3	−7398.751 717 018	
		4	−7399.218 916 979	
		5	−7399.228 761 754	−7399.228 805 892
90	Th_2^{179+}	2	−9497.103 757 814	
		3	−9504.013 069 800	
		4	−9504.748 342 655	
		5	−9504.756 578 763	−9504.756 746 927
92	U_2^{183+}	2	−9957.149 867 657	
		3	−9964.557 040 570	
		4	−9965.357 892 685	
		5	−9965.365 278 947	−9965.365 468 058
100	Fm_2^{199+}	2	−11942.178 005 611	
		3	−11951.832 987 584	
		4	−11952.939 381 324	
		5	−11952.941 727 610	−11952.941 940 110
110	Ds_2^{219+}	2	−14796.324 879 456	
		3	−14809.322 202 267	
		4	−14810.852 901 248	
		5	−14810.898 911 675	
118	Og_2^{235+}	2	−17461.232 762 069	
			−17477.122 912 713	
		4	−17479.073 413 320	
		5	−17479.125 249 624	
121	$_2^{241+}$	2	−18577.427 339 685	
		3	−18594.349 381 894	
		4	−18596.454 717 052	
		5	−18596.509 996 585	

References

1. I.V. Puzynin, T.L. Boyadzhiev, S.I. Vinitskii, E.V. Zemlyanaya, T.P. Puzynina, O. Chuluunbaatar, Methods of computational physics for investigation of models of complex physical systems. Phys. Part. Nucl. **38**, 70–116 (2007)
2. L.V. Kantorovich, V.I. Krylow, *Approximate Methods of Higher Analysis, Moscow, Fizmatgiz, 1962* (Wiley, New York, 1964)
3. I.V. Puzynin, I.V. Amirkhanov, E.V. Zemlyanaya, V.N. Pervushin, T.P. Puzynina, T.A. Strizh, V.D. Lakhno, The generalized continuous analog of Newton's method for the numerical study of some nonlinear quantum-field models. Phys. Part. Nucl. **30**, 87–110 (1999)
4. D.F. Davidenko, On application of method of variation of parameter to the theory of nonlinear functional equations. Ukr. Mat. Zh. **7**, 18–28 (1955)

5. S.I. Vinitskii, I.V. Puzynin, Yu.S. Smirnov, High precision calculations of the multichannel scattering problem for processes involving mesic atoms. Phys. At. Nucl. **55**, 1830–1838 (1992)
6. S. Ul'm, Iterative methods with sequential approximation of inverse operator. Izv. Akad. Nauk Est. SSR **16**, 403–411 (1967)
7. J.M. Ortega, *Introduction to parallel and vector solution of linear systems* (Mir, Moscow, 1991). ((in Russian))
8. T. Zhanlav, I.V. Puzynin, The convergence of iteration based on a continuous analogue of Newton's method. Comput. Math. Math. Phys. **32**, 729–737 (1992)
9. V.V. Ermakov, N.N. Kalitkin, The optimal step and regularization for Newton's method. USSR Comput. Math. Math. Phys. **21**, 235–242 (1981)
10. S.I. Vinitskii, I.V. Puzynin, T.P. Puzynina, L.I. Ponomarev, *Newton's Process in Perturbation Theory with Continuous Inclusion of Interaction, Preprint OIYaI R4–10942* (Joint Institute for Nuclear Research, Dubna, 1977)
11. D.A. Kirzhnits, N.G. Takibaev, New approach in problem of three and more bodies. Sov. J. Nucl. Phys. **25**, 370–376 (1977)
12. S.I. Vinitskii, I.V. Puzynin, Yu.S. Smirnov, Solving of scattering on the base of multiparametric Newton's schemes. Sov. J. Nucl. Phys. **52**, 746–754 (1990)
13. Yu.N. Demkov, *Variational Principles in the Theory of Collisions, Moscow, Fizmatgiz, 1958* (Macmillan, New York, 1963)
14. M. Gailitis, Extremal properties of approximate methods of collision theory in the presence of non-elastic processes. Sov. Phys. J. Exp. Theor. Phys. **20**, 107–111 (1965)
15. A.L. Zubarev, *Schwinger Variational Principle in Quantum Mechanics* (Energoatomizdat, Moscow, 1981). ((in Russian))
16. O. Chuluunbaatar, I.V. Puzynin, S.I. Vinitsky, Newtonian iteration scheme with the Schwinger variational functional for solving a scattering problem. J. Comput. Methods Sci. Eng. **2**, 37–49 (2002)
17. O. Chuluunbaatar, Newtonian variation-iteration schemes for computational investigation of three-particle quantum systems, Candidate's dissertation in mathematical physics (Joint Institute for Nuclear Research, Dubna, 2002), 11–2002–209
18. O. Chuluunbaatar, I.V. Puzynin, D.V. Pavlov, A.A. Gusev, S.Y. Larsen, S.I. Vinitsky, Newtonian iteration schemes for solving the three-body scattering problem on a line. Proc. SPIE **4706**, 155–165 (2002)
19. A. Amaya-Tapia, S.Y. Larsen, J.J. Popiel, Three body phase shift in one-dimensional 2 + 1 scattering. Few-Body Syst. **23**, 87–109 (1997)
20. O. Chuluunbaatar, A.A. Gusev, S.Y. Larsen, S.I. Vinitsky, Three identical particles on a line: comparison of some exact and approximate calculations. J. Phys. A **35**, L513–L525 (2002)
21. W.G. Gibson, S.Y. Larsen, J.J. Popiel, Hyperspherical harmonics in one dimension: adiabatic effective potentials for three particles with δ-function interactions. Phys. Rev. A **15**, 4919–4929 (1987)
22. V.G. Neudachin, Yu.V. Popov, Yu.F. Smirnov, Electron momentum spectroscopy of atoms, molecules, and thin films. Phys. Usp. **42**, 1017–1044 (1999)
23. A. Lahmam-Bennani, A. Duguet, S. Roussin, Observation of non-first-order effects in an $(e, 3--1e)$ investigation of the double ionization of helium and molecular hydrogen. J. Phys. B **35**, L59–L63 (2002)
24. N. Watanabe, Y. Khajuria, M. Takahashi, Y. Udagawa, P.S. Vinitsky, Yu.V. Popov, O. Chuluunbaatar, K.A. Kouzakov, $(e, 2e)$ and $(e, 3--1e)$ studies on double processes of He at large momentum transfer. Phys. Rev. A **72**, 032705-1–9 (2005)
25. O. Chuluunbaatar, I.V. Puzynin, P.S. Vinitsky, Yu.V. Popov, K.A. Kouzakov, C. Dal Cappello, Role of the cusp conditions in electron-atom double ionization. Phys. Rev. A **74**, 014703-1–4 (2006)
26. T. Kato, On the eigenfunctions of many-particle systems in quantum mechanics. Commun. Pure Appl. Math. **10**, 151–177 (1957)
27. D.V. Pavlov, I.V. Puzynin, B.B. Joulakian, S.I. Vinitsky, Wave functions of continuous spectrum of the coulomb two-center problem. J. Comput. Methods Sci. Eng. **2**, 261–269 (2002)

28. O. Chuluunbaatar, B.B. Joulakian, Kh. Tsookhuu, S.I. Vinitsky, Two center electron continua: application to the dissociative ionization of H_2^+ by fast electron. J. Phys. B **37**, 2607–2616 (2004)
29. O. Chuluunbaatar, B.B. Joulakian, I.V. Puzynin, Kh. Tsookhuu, S.I. Vinitsky, Modified two-center continuum wave function: application to the dissociative double ionization of H_2 by electron impact. J. Phys. B **41**, 015204-1–6 (2008)
30. V.V. Serov, B.B. Joulakian, D.V. Pavlov, I.V. Puzynin, S.I. Vinitsky, $(e, 2e)$ ionization of H_2^+ by fast electron impact: Application of the exact nonrelativistic two-center continuum wave. Phys. Rev. A **65**, 062708-1–7 (2002)
31. V.V. Serov, B.B. Joulakian, V.L. Derbov, S.I. Vinitsky, Ionization excitation of diatomic systems having two active electrons by fast electron impact: a probe to electron correlation. J. Phys. B **38**, 2765–2773 (2005)
32. E.M. Staicu-Casagrande, A. Naja, F. Mezdari, A. Lahmam-Bennani, P. Bolognesi, B. Joulakian, O. Chuluunbaatar, O. Al-Hagan, D.H. Madison, D.V. Fursa, I. Bray, $(e, 2e)$ ionization of helium and the hydrogen molecule: signature of two-centre interference effects. J. Phys. B **41**, 025204-1–7 (2008)
33. O. Chuluunbaatar, A.A. Gusev, B. Joulakian, The correlated two-centre double continuum and the double ionization of H_2 and N_2 by fast electron impact. J. Phys. B **45**, 015205-1–6 (2012)
34. O. Chuluunbaatar, A.A. Gusev, B. Joulakian, The double ionization of H_2 by fast electron impact: influence of the final state electron-electron correlation. Phys. At. Nucl. **76**, 121–125 (2013)
35. A. Naja, E.M. Staicu-Casagrande, A. Lahmam-Bennani, M. Nekkab, F. Mezdari, B. Joulakian, O. Chuluunbaatar, D.H. Madison, Triply differential $(e, 2e)$ cross sections for ionisation of the nitrogen molecule at large energy transfer. J. Phys. B **40**, 3775–3783 (2007)
36. A.A. Bulychev, O. Chuluunbaatar, A.A. Gusev, B. Joulakian, $(\gamma, 2e)$ photo-double ionization of N_2 molecule for equal energy sharing. J. Phys. B **46**, 185203-1–9 (2013)
37. P. Bolognesi, B. Joulakian, A.A. Bulychev, O. Chuluunbaatar, L. Avaldi, Photo-double-ionization of the nitrogen molecule. Phys. Rev. A **89**, 053405-1–5 (2014)
38. O. Chuluunbaatar, I.V. Puzynin, S.I. Vinitsky, Uncoupled correlated calculations of helium isoelectronic bound states. J. Phys. B **34**, L425–L432 (2001)
39. K. Frankowski, C.L. Pekeris, Logarithmic terms in the wave functions of the ground state of two-electron atoms. Phys. Rev. **146**, 46–49; 150, 336 (1966)
40. S.P. Goldman, Uncoupling correlated calculations in atomic physics: very high accuracy and ease. Phys. Rev. A **57**, R677–R680 (1998)
41. G.W.F. Drake, High precision theory of atomic helium. Phys. Scr. **83**, 83–92 (1999)
42. V.I. Korobov, Nonrelativistic ionization energy for the helium ground state. Phys. Rev. A **66**, 024501-1–2 (2002)
43. D.T. Aznabaev, A.K. Bekbaev, V.I. Korobov, Nonrelativistic energy levels of helium atoms. Phys. Rev. A **98**, 012510-1–7 (2018)
44. H. Nakashima, H. Nakatsuji, Solving the Schrödinger equation for helium atom and its isoelectronic ions with the free iterative complement interaction (ICI) method. J. Chem. Phys. **127**, 224104-1–14 (2007)
45. C. Schwartz, Experiment and theory in computations of the he atom ground state. Int. J. Mod. Phys. E **15**, 877–888 (2006); Further Computations of the He Atom Ground State (2008). https://arxiv.org/abs/math-ph/0605018
46. O. Chuluunbaatar, K.A. Kouzakov, S.A. Zaytsev, A.S. Zaytsev, V.L. Shablov, Yu.V. Popov, H. Gassert, M. Waitz, H.-K. Kim, T. Bauer, A. Laucke, Ch. Müller, J. Voigtsberger, M. Weller, J. Rist, K. Pahl, M. Honig, M. Pitzer, S. Zeller, T. Jahnke, LPh.H. Schmidt, H. Schmidt-Böcking, R. Dörner, M.S. Schöffler, Single ionization of helium by fast proton impact in different kinematical regimes. Phys. Rev. A **99**, 062711-1–11 (2019)
47. O. Chuluunbaatar, Y.V. Popov, S.I. Vinitskii, Factorized correlated variation function at application to calculation $(e, 2e)$ and $(e, 3e)$ helium atom ionization reactions, Soobshch. OIYaI R4–2002–134 (Joint Institute for Nuclear Research, Dubna, 2002)

48. O. Chuluunbaatar, S.A. Zaytsev, K.A. Kouzakov, A. Galstyan, V.L. Shablov, Yu.V. Popov, Fully differential cross sections for singly ionizing 1-MeV p+He collisions at small momentum transfer: beyond the first Born approximation. Phys. Rev. A **96**, 042716-1–7 (2017)

49. Yu.V. Popov, O. Chuluunbaatar, S.I. Vinitsky, L.U. Ancarani, C. Dal Cappello, P.S. Vinitsky, Theoretical investigation of the p + He $\rightarrow$ H + He$^+$ and p + He $\rightarrow$ H + He^{++} + e reactions at very small scattering angles of hydrogen. J. Exp. Theor. Phys. **95**, 620–624 (2002)

50. M.S. Schöffler, H.-K. Kim, O. Chuluunbaatar, S. Houamer, A.G. Galstyan, J.N. Titze, T. Jahnke, LPh.H. Schmidt, H. Schmidt-Böcking, R. Dörner, Yu.V. Popov, A.A. Bulychev, Transfer excitation reactions in fast proton-helium collisions. Phys. Rev. A **89**, 032707-1–9 (2014)

51. E.A. Hylleraas, Neue berechnung der energie des heliums im grundzustande, sowie des tiefsten terms von ortho-helium. Z. Phys. **54**, 347–366 (1929)

52. E. Clementi, C. Roetti, Roothaan-Hartree-Fock atomic wavefunctions: basis functions and their coefficients for ground and certain excited states of neutral and ionized atoms, $Z \leq 54$. At. Data Nucl. Data Tables **14**, 177–478 (1974)

53. S. Chandrasekhar, Some remarks on the negative hydrogen ion and its absorption coefficient. Astrophys. J. **100**, 176–180 (1944)

54. C. Eckart, The theory and calculation of screening constants. Phys. Rev. **36**, 878–892 (1930)

55. R.A. Bonham, D.A. Kohl, Simple correlated wavefunctions for the ground state of heliumlike atoms. J. Chem. Phys. **45**, 2471–2473 (1966)

56. J. Mitroy, I.E. McCarthy, E. Weigold, A natural orbital analysis of the helium (e, $2e$) spectrum. J. Phys. B **18**, 4149–4157 (1985)

57. Yu.V. Popov, L.U. Ancarani, Rigorous mathematical study of the He bound states. Phys. Rev. A **62**, 042702-1–9 (2000)

58. Yu.V. Popov, C. Dal Cappello, K. Kuzakov, (e,3e) electronic momentum spectroscopy: perspectives and advantages. J. Phys. B **29**, 5901–5908 (1996)

59. L.U. Ancarani, T. Montagnese, C. Dal Cappello, Role of the helium ground state in (e, $3e$) processes. Phys. Rev. A **70**, 012711-1–10 (2004)

60. S. Jones, J.H. Macek, D.H. Madison, Test of the Pluvinage wave function for the helium ground state. Phys. Rev. A **70**, 012712-1–7 (2004)

61. S. Jones, J.H. Macek, D.H. Madison, Three-Coulomb-wave Pluvinage model for Compton double ionization of helium in the region of the cross-section maximum. Phys. Rev. A **72**, 012718-1–3 (2005)

62. S. Jones, D.H. Madison, Role of the ground state in electron-atom double ionization. Phys. Rev. Lett. **91**, 073201-1–4 (2003)

63. M. Brauner, J.S. Briggs, H. Klar, Triply-differential cross sections for ionisation of hydrogen atoms by electrons and positrons. J. Phys. B **22**, 2265–2287 (1989)

64. P. Pluvinage, Fonction d'onde approchée à un paramètre pour l'état fondamental des atomes à deux électrons. Ann. Phys. Paris **5**, 145–152 (1950)

65. P. Pluvinage, Nouvelle famille de solutions approchées pour certaines équations de Schrödinger non séparables. Applications à l'état fondamental de l'hélium. J. Phys. Radium **12**, 789–792 (1951)

66. A. Lahmam-Bennani, I. Taouil, A. Duguet, M. Lecas, L. Avaldi, J. Berakdar, Origin of dips and peaks in the absolute fully resolved cross sections for the electron-impact double ionization of He. Phys. Rev. A **59**, 3548–3555 (1999)

67. A.M. Frolov, Two-stage strategy for high-precision variational calculations. Phys. Rev. A **57**, 2436–2439 (1998)

68. C. Le Sech, Accurate analytic wavefunctions for two-electron atoms. J. Phys. B **30**, L47–L50 (1997)

69. P.E. Gill, W. Murray, Newton-type methods for unconstrained and linearly constrained optimization. Math. Programm. **7**, 311–350 (1974)

70. M.A. Kornberg, J.E. Miraglia, Double photoionization of helium: Use of a correlated two-electron continuum wave function. Phys. Rev. A **48**, 3714–3719 (1993)

71. V. Mergel, R. Dörner, Kh. Khayyat, M. Achler, T. Weber, O. Jagutzki, H.J. Lüdde, C.L. Cocke, H. Schmidt-Böcking, Strong correlations in the He ground state momentum wave function

observed in the fully differential momentum distributions for the p + He transfer ionization process. Phys. Rev. Lett. **86**, 2257–2260 (2001)

72. H.T. Schmidt, A. Fardi, R. Schuch, S.H. Schwartz, H. Zettergren, H. Cederquist, L. Bagge, H. Danared, A. Källberg, J. Jensen, K.-G. Rensfelt, V. Mergel, L. Schmidt, H. Schmidt-Böcking, C.L. Cocke, Double-to-single target ionization ratio for electron capture in fast p—He collisions. Phys. Rev. Lett. **89**, 163201-1–4 (2002)

73. P.S. Vinitsky, Yu.V. Popov, O. Chuluunbaatar, Fast proton-hydrogen charge exchange reaction at small scattering angles. Phys. Rev. A **71**, 012706-1–9 (2005)

74. V.L. Derbov, G. Chuluunbaatar, A.A. Gusev, O. Chuluunbaatar, S.I. Vinitsky, A. Góźdź, P.M. Krassovitskiy, I. Filikhin, A.V. Mitin, Spectrum of beryllium dimer in ground $X^1\Sigma_g^+$ state. J. Quant. Spectrosc. Radiat. Transfer **262**, 107529-1–10 (2021)

75. Atomic Spectroscopy Databases (2009). https://www.nist.gov/pml/atomic-spectroscopy-databases

76. S.G. Porsev, A. Derevianko, High-accuracy calculations of dipole, quadrupole, and octupole electric dynamic polarizabilities and van der Waals coefficients C_6, C_8, and C_{10} for alkaline-earth dimers. J. Exp. Theor. Phys. **102**, 195–205 (2006)

77. M. Lesiuk, M. Przybytek, J.G. Balcerzak, M. Musiał, R. Moszynski, Ab initio potential energy curve for the ground state of beryllium dimer. J. Chem. Theor. Comput. **15**, 2470–2480 (2019)

78. G. Chuluunbaatar, A.A. Gusev, O. Chuluunbaatar, S.I. Vinitsky, L.L. Hai, KANTBP 4M Program for solving the scattering problem for a system of ordinary second-order differential equations. EPJ Web Conf. **226**, 02008-1–4 (2020)

79. A.A. Gusev, O. Chuluunbaatar, S.I. Vinitsky, A.G. Abrashkevich, KANTBP 3.0: new version of a program for computing energy levels, reflection and transmission matrices, and corresponding wave functions in the coupled-channel adiabatic approach. Comput. Phys. Commun. **185**, 3341–3343 (2014)

80. J.M. Merritt, V.E. Bondybey, M.C. Heaven, Beryllium dimer-caught in the act of bonding. Science **324**, 1548–1551 (2009)

81. W.R. Johnson, K.T. Cheng, M.H. Chen, Accurate relativistic calculations including QED contributions for few-electron systems. Theor. Comput. Chem. **14**, 120–187 (2004)

82. E.U. Condon, G.H. Shortley, *The Theory of Atomic Spectra* (Cambridge at the University Press, 1970)

83. J. Oreg, G. Malli, Relativistic molecular symmetry spinors for diatomics. J. Chem. Phys. **61**, 4349–4356 (1974)

84. O. Chuluunbaatar, B.B. Joulakian, G. Chuluunbaatar, J. Buša Jr., G.O. Koshcheev, Accurate calculations for the Dirac electron in the field of two-center Coulomb field: application to heavy ions. Chem. Phys. Lett. **784**, 139099-1–9 (2021)

85. A. Kolakowska, J.D. Talman, K. Aashamar, Minimax variational approach to the relativistic two-electron problem. Phys. Rev. A. **53**, 168–177 (1996)

86. J.D. Talman, Minimax principle for the dirac equation. Phys. Rev. Lett. **57**, 1091–1094 (1986)

87. M. Abramovits, I.A. Stegun, *Handbook of mathematical functions* (Dover, New York, 1972)

88. I.S. Gradshteyn, I.M. Ryzhik, *Table of Integrals, Series, and Products*, 7th ed. (Academic Press is an imprint of Elsevier, 2007)

89. G. Chuluunbaatar, Computational schemes for solving quantum mechanical problems, Ph.D. thesis, Dubna, Russia (2023) (in Russian)

90. L. Adolphs, H. Daneshmand, A. Lucchi, Th. Hofmann, Local saddle point optimization: a curvature exploitation approach. PMLR **89**, 486–495 (2019)

91. I.I. Tupitsyn, D.V. Mironova, Relativistic calculations of ground states of single-electron diatomic molecular ions. Opt. Spectrosc. **117**, 351–357 (2014)

Correction to: New Developments of Newton-Type Iterations for Solving Nonlinear Problems

Correction to:
T. Zhanlav and O. Chuluunbaatar, *New Developments*
of Newton-Type Iterations for Solving Nonlinear Problems,
Mathematical Engineering,
https://doi.org/10.1007/978-3-031-63361-4

The book was inadvertently published without updating the following corrections:

Chapter 06:

Equation 6.3 on Springer Link is displayed incorrectly, which has now been corrected, and Reference 23: In the original version of this paper, reference 23 was available, but it is missing from Springer Link, which has now been corrected.

Chapter 08:

1. Author provided belated correction for chapter 8 to increase the equation font size for 8.94 and deletion of eqn array text.

2. The paragraph indentation differs between the print and Springer Link publications, which has now been updated.

3. The link for Reference 32 was broken due to a spelling error in the book title, which has now been corrected. The correction chapters and the book have been updated with the changes.

The updated versions of these chapters can be found at
https://doi.org/10.1007/978-3-031-63361-4_6
https://doi.org/10.1007/978-3-031-63361-4_8

Index